Nanoscopy and Nanospectroscopy

This book builds a narrative on the near-field optical and spectroscopic studies with an emphasis on plasmonic- and photonic-assisted nano-optics as a tool for superlensing. Deliberations on near-field studies using confined light in various applications are included along with their commercial implications. Single-molecule detection utilizing efficient surface-enhanced Raman scattering phenomenon in the far-field and plasmonic tip-enhanced Raman scattering studies in the near-field measurements for fast analysis up to trace level is discussed.

Features:

- Covers the broad area of nano-optical spectroscopy from the perspective of putting the concepts and innovations in the field to use.
- Discusses entire spectra of near-field optics and spectroscopy using light.
- Explores gas/chemical sensing using surface plasmon resonance (SPR) in the Kretschmann configuration.
- Includes dielectric nano-photonics and optical confinement.
- Studies phonon behaviour using confined light for the analysis of chemical, biological, and other materials.

This book is aimed at graduate students and researchers in material science, analytical chemistry, nanotechnology, and electrical engineering.

Nanoscopy and Nanospectroscopy

Edited by
Sandip Dhara, Deep Jariwala, and Soumen Das

CRC Press
Taylor & Francis Group
Boca Raton London New York

CRC Press is an imprint of the
Taylor & Francis Group, an **informa** business

Designed cover image: © Shutterstock; Sandip Dhara, Deep Jariwala, and Soumen Das

First edition published 2023
by CRC Press
6000 Broken Sound Parkway NW, Suite 300, Boca Raton, FL 33487-2742

and by CRC Press
4 Park Square, Milton Park, Abingdon, Oxon, OX14 4RN

CRC Press is an imprint of Taylor & Francis Group, LLC

ISBN: 9781032163888 (hbk)
ISBN: 9781032163895 (pbk)
ISBN: 9781003248323 (ebk)

DOI: 10.1201/9781003248323

Typeset in Times
by codeMantra

Contents

Editors

Prof. Dr. Sandip Dhara completed his Ph.D. on Magneto-optic Recording Materials from National Physical Laboratory, New Delhi in 1994 and joined Indira Gandhi Centre for Atomic Research (IGCAR), Kalpakkam in 1996. Presently he is the Head of the Surface and Sensors Studies Division, Materials Science Group and a Professor in Homi Bhabha National Institute. He was also a visiting Professor in the Institute of Atomic and Molecular Sciences, Academia Sinica, Taiwan. Dr. Dhara has more than 300 research publications including several review articles, books, book chapters, several conference presentations, etc. He has delivered more than 100 invited and plenary talks and is the recipient of Homi Bhabha Gold Medal for 2019 and Department of Atomic Energy-Science Research Council (DAE-SRC) outstanding researcher award for 2012. He is also in the editorial board of many national and international scientific journals. Dr. Dhara is specialized in studies of light-matter interaction at the nanoscale in the sub-diffraction limit. He is a Fellow of Institute of Physics (FInstP) and a Fellow of the Royal Society of Chemistry (FRSC). He is among World's Top 2% Scientists in 2021.

Deep Jariwala is an Assistant Professor in the Department of Electrical and Systems Engineering at the University of Pennsylvania (Penn). His research interests broadly lie at the intersection of new materials, surface science, and solid-state devices for computing, sensing, opto-electronics, and energy-harvesting applications. Deep completed his undergraduate degree in Metallurgical Engineering from the Indian Institute of Technology, Banaras Hindu University in 2010. He completed his Ph.D. in Materials Science and Engineering at Northwestern University in 2015. At Northwestern, Deep made contributions to the study of charge transport and electronic applications of two-dimensional (2D) semiconductors and pioneering the study of gate-tunable, mixed-dimensional, and Van der Waals' heterostructures. He was Resnick Prize Postdoctoral Fellow at Caltech from 2015 to 2017 working on nanophotonic devices and ultrathin solar cells before joining Penn in 2018. Deep's research has earned him awards from multiple professional societies including the Russell and Sigurd Varian Award and Paul H. Holloway Award of the American Vacuum Society, the Richard L. Greene Dissertation Award of the American Physical Society, Johannes and Julia Weertman Doctoral Fellowship, the Hilliard Award, the Army Research Office Young Investigator Award, Nanomaterials Young Investigator Award, TMS Frontiers in Materials Award, Intel Rising Star Award, IEEE Young Electrical Engineer of the Year Award, IEEE Photonics Society Young Investigator Award, and IUPAP Early Career Scientist Prize in Semiconductors in addition to being named in Forbes Magazine list of 30 scientists under 30 and is an invitee to Frontiers of Engineering conference of the National Academy of Engineering as well as a recipient of the Sloan Fellowship. In addition, he has also received the S. Reid Warren Jr. award. He also serves on the editorial board of the journals *Electronics* and *Micromachines*. He has published over 100 journal papers with more than 14,000 citations and 7 patents.

Dr. Soumen Das is a Scientific Officer at Indira Gandhi Centre for Atomic Research (IGCAR), Kalpakkam, Tamil Nadu, India. His research interests involve semiconductor, ferroelectric and magnetic oxide thin films, high-temperature materials, protective coating, and nanostructures. At present, he is a program leader in the Materials Chemistry and Metal Fuel Chemical Group, IGCAR, looking after materials processing, qualification, and studying LASER-based surface phenomena.

Contributors

Sharad Ambardar
Department of Medical Engineering,
University of South Florida, Tampa, Florida

D. N. Basov
Department of Physics, Columbia University,
New York, New York

Debanjan Bhowmik
Ramanujan Faculty Scientist, Transdisciplinary
Biology, Rajiv Gandhi Centre for
Biotechnology, Thiruvananthapuram, India

Hsun-Jen Chuang
Materials Science and Technology Division,
Naval Research Laboratory, Washington,
District of Columbia and
Nova Research, Inc., Washington, District of
Columbia

Thomas Darlington
Department of Mechanical Engineering,
Columbia University, New York, New York

Sandip Dhara
Surface and Sensors Studies Division,
Materials Science Group, Indira Gandhi
Centre for Atomic Research, A CI of Homi
Bhabha National Institute, Kalpakkam,
India

Sreetosh Goswami
National University of Singapore, Singapore

James C. Hone
Department of Mechanical Engineering,
Columbia University, New York,
New York Nan Jiang
Department of Chemistry, University of
Illinois, Chicago, Illinois

Berend T. Jonker
Materials Science & Technology Division,
Naval Research Laboratory, Washington,
District of Columbia

Naresh Kumar
Department of Chemistry and Applied
Biosciences, ETH Zurich, Zurich,
Switzerland

Andrey Krayev
Horiba Scientific, Novato, California

Kishore K. Madapu
Surface and NanoScience Division,
Materials Science Group, Indira
Gandhi Centre for Atomic Research,
Kalpakkam, India

Kathleen McCreary
Materials Science & Technology Division,
Naval Research Laboratory, Washington,
District of Columbia

Sandip Mondal
Nano-Optics and Mesoscopic Optics
Laboratory, Tata Institute of Fundamental
Research, Mumbai, India

Sushil Mujumdar
Nano-Optics and Mesoscopic Optics
Laboratory, Tata Institute of
Fundamental Research,
Mumbai, India

Chandrabhas Narayana
Rajiv Gandhi Centre for Biotechnology,
Thiruvananthapuram, India. On
Deputation from Professor Chemistry and
Physics of Materials Unit,
School of Advanced Materials,
Jawaharlal Nehru Centre for
Advanced Scientific Research, Bangalore,
India

Avinash Patsha
Department of Materials Science and
Engineering, Tel Aviv University, Tel Aviv,
Israel

T. R. Ravindran
Condensed Matter Physics Division,
 Materials Science Group, Indira
Gandhi Centre for Atomic Research, A
CI of Homi Bhabha National Institute,
Kalpakkam, India

Matthew R. Rosenberger
Department of Aerospace and Mechanical
 Engineering, University of Notre Dame,
 Notre Dame, Indiana

Pratap K. Sahoo
School of Physical Sciences,
 National Institute of Science
 Education and Research,
 An OCC of Homi Bhabha
 National Institute, Bhubaneswar, India

Binaya Kumar Sahu
School of Physical Sciences,
 National Institute of Science
 Education and Research,
 An OCC of Homi Bhabha National Institute,
 Bhubaneswar, India

Rabisankar Samanta
Nano-Optics and Mesoscopic Optics
 Laboratory, Tata Institute of Fundamental
 Research, Mumbai, India

P. James Schuck
Department of Mechanical Engineering,
 Columbia University, New York, New York

Jeremy F. Schultz
Nanoscale Spectroscopy Group, National
 Institute of Standards and Technology,
 Gaithersburg, Maryland

Agnès Tempez
HORIBA France SAS, Palaiseau, France

Dmitri V. Voronine
Department of Medical Engineering,
 University of South Florida, Tampa, Florida
 and
Department of Physics, University of South
 Florida, Tampa, Florida

Emanuil Yanev
Department of Mechanical Engineering,
 Columbia University, New York, New York

Kaiyuan Yao
Department of Mechanical Engineering,
 Columbia University, New York, New York

Shuai Zhang
Department of Physics, Columbia University,
 New York, New York

Preface

The prime attraction of the present book is the elaborate discussion on optical microscopy and Raman spectroscopy in the nanoscale of molecular dimension with an additional emphasis on its applications to decipher the cause-and-effect relations of physical, chemical, and biological samples of interest. The unique perspective of the discussion sets the present book apart from other books available in the market. There are books on plasmonics, photonics, and the basics of nano-optics phenomenon, but they hardly shed light on its applications in contemporary research. Interestingly, many books are available on the surface-enhanced Raman spectroscopy (SERS) for analytical, biomolecular, and medical diagnostics applications, yet there is no book available for tip-enhanced Raman spectroscopy (TERS) and its applications. Most discussion on TERS is available barely as review chapters in journals or sporadic discussion as book chapters. In contrast, the present book is a near-comprehensive collection of relevant topics to comprehend and analyze properties at the nanoscale using TERS.

Fundamentals of phonon properties in material and Raman spectroscopy as a diagnostic tool are deliberated in great detail, covering molecular symmetry and group theoretical analysis. A brief analysis of elastic properties in nanomaterials and its application in finding structural modes are discussed. The size and shape of nanomaterials are determined using confinement of acoustic phonon mode and radial breathing mode is analyzed for the atomically thin wall of a nanotube. In this regard, low-frequency bosonic mode (spin wave) is also discussed in the magnetic lattice. The essence of plasmonics, photonics, and near-filed optics is also included while discussing the latest developments. Similarly, the principle and various models of SERS and TERS are discussed in great detail for the state-of-the-art development of the technique as well as its applications. We envisage that doctoral and graduate students looking for comprehensive guidance, referral, and suggestion may find this book immensely helpful. For example, simple techniques in assigning Raman mode will attract Master's students doing their advanced projects and researchers investigating inter-disciplinary subjects dealing with Raman spectroscopy as a tool. A comprehensive discussion on all techniques, including near-field studies for analyzing properties of nanomaterials and characterization at the nanoscale, will serve as a ready reference for most readers. This book reviews contemporary plasmonics, photonics, and their advanced optical and optical-mode spectroscopic applications which will be very attractive topics for many readers.

The book is a chronicle on the present status and progress related to nanoscopy and nanospectroscopy. The detail of Raman spectroscopy probing different materials, including nanomaterials, are described in Chapter 1. Plasmonic-based research and applications found tremendous impetus in modern science and technology in the last decade. Abbe's diffraction limit was overcome and made technologically acceptable by confining light. The confinement of light using plasmonics and later photonics brought a renaissance in optical nanoscopy that resulted in the development of superlensing, leading to the Nobel Prize in Chemistry in 2014. The subjects are detailed in Chapters 2 and 3. Near-field studies using confined light have made nanoscopy available for in-vitro studies of biological samples and have helped in observing surface plasmon polariton waves, achieving sub-micron lithography at sub-diffraction limit and medical diagnostic applications (Chapter 4). Earlier, confinement of light was achieved by utilizing efficient SERS phenomenon in the far-field measurements to study a single molecule. The technique has proved to be so versatile that nowadays, it is routinely used for compositional analysis of biological samples in the pharmaceutical industry with trace-level detection limits (Chapter 5). In Chapter 6, the techniques have been made further sophisticated by combining surface probe microscopy (SPM), where a plasmonic tip was raster scanned for site-specific analysis in inorganic crystal and semiconducting nanostructures of strategic importance, e.g., strained Si and quantum dots, along with other layered MXenes and Van der Waals' bonded materials in the TERS technique. At low-temperature and ultrahigh-vacuum conditions, TERS study can achieve sub-nanometer resolution for single nanostructure analysis and chemical sensitive imaging for studying various catalytic activities. The use of TERS for both DNA and RNA sequencing, including the study of bacteria and viruses, is in the ambit of the book. The nano-spectroscopic study is further extended to imaging 2D transition metal dichalcogenide stacking as excitons greatly enhance near-field second harmonic generation efficiency. Finally, the chapter concludes with its important application of tip-enhanced photoluminescence measurements for understanding excitonic properties of semiconductors. The final Chapter 7 offers a summary of the expected future developments.

Plasmonics- and photonics-assisted "nanoscopy and nanospectroscopy" are very efficient material characterization techniques for surface, sub-surface, and trace-level detection up to the level of molecular imaging. There is a persistent interest in developing optical-based characterization techniques at the sub-nanometer or atomic scale (superlensing) for quick and non-destructive studies of material. Optical characterization techniques are popular for their ease of use in batch processing, attracting researchers and scholars alike. This book thus will be suitable as a reference for graduate, undergraduate, and doctoral students looking for quick guidance for material characterizations at the nanoscale using optical techniques.

1 Theory of Light Scattering and Applications

T. R. Ravindran and Sandip Dhara

CONTENTS

1.1 RAMAN SPECTROSCOPY

1.1.1 CLASSICAL AND QUANTUM THEORY

Raman spectroscopy is a tool to study the various chemical bonds in molecules and solids by detecting the shift in the frequency of laser light scattered from a sample. Atoms in a molecule are held together by chemical bonds that can be, e.g., covalent or ionic. Depending on the bond strength and the mass of the bonded atoms, they execute vibrational motion that can be approximated as a simple

DOI: 10.1201/9781003248323-1

1

harmonic motion. The frequency of vibration of a diatomic molecule can be derived straightforwardly when Hooke's law is obeyed, i.e., the oscillations are small amplitude. In that case, $F=-kx$, where k is the spring constant or force constant of the bond and x is the displacement. From Newton's second law, $F=m(dx^2/dt^2)$, and hence

$$m\left(dx^2 / dt^2\right) + kx = 0 \tag{1.1}$$

This is a simple harmonic equation, and a harmonic function can be tried as a trial solution for x. Plugging in $x(t)=A\cos(\omega t)$ into equation (1.1), one gets

$$\left(dx^2 / dt^2\right) + w^2 x = 0 \tag{1.2}$$

Comparing equations (1.1) and (1.2), one gets the frequency of the harmonic oscillator as $\omega = \sqrt{k / m}$. For a diatomic molecule, m is defined as the reduced mass, $1/m = (1/m_1) + (1/m_2)$.

Thus, the frequency of vibration is directly proportional to the square root of the force constant or bond strength and inversely proportional to the square root of the reduced mass. Stronger bonds such as double and triple bonds result in a higher frequency compared to single bonds of the same species of atoms, and more massive atoms result in lower frequencies. Since molecules and solids are made up of atoms of different masses, bonded by chemical bonds of varying strengths, they exhibit correspondingly several different vibrational frequencies. Every molecule or crystalline solid could have a distinct set of frequencies, much like the X-ray diffraction (XRD) patterns of crystalline materials that are unique because of the unique set of planes of atoms. It so happens that these vibrational frequencies are in the infrared (IR) frequency region of the electromagnetic spectrum, 10^{12}–10^{14} Hz (wavelengths of 2.5–15 μm: ~4,000–400 cm^{-1}).

When a band of electromagnetic waves in the IR region is incident on a material, the frequencies of such molecular vibrations are resonantly absorbed. In the transmitted waves, we get an IR spectrum characteristic of the material. The phenomenon is similar to atomic absorption spectroscopy in the ultraviolet-visible (UV-Vis) spectral region of the electromagnetic spectrum. These molecular vibrations can be studied by such a direct IR absorption technique or by an indirect scattering method called Raman spectroscopy in honor of the Indian scientist who have had demonstrated this kind of scattering in 1928. We do not discuss IR spectroscopy in detail in this chapter other than making the following observations. However, we discuss in some detail a few of these issues later in the context of a comparison between Raman and IR techniques: (i) IR spectroscopy was first demonstrated as early as the year 1881 by Abney and Festing, who recorded the spectra using photographic plates and related them to the chemical composition of the liquids investigated by them [1]. In 1905, William Coblentz published the IR spectra of many compounds he recorded, thus creating the first IR catalog that was instrumental in motivating further research in this area [2]. (ii) Though commercial IR spectrometers were available only in the late 1950s, IR spectroscopic techniques were more advanced than Raman spectroscopy, and chemists preferred IR over Raman for several reasons such as ease of measurements and lower cost. The atlas of IR spectra were probably more extensive than the Raman atlas, which contributed to the popularity of IR. (iii) Fourier transform IR techniques (based on the Michelson interferometer) evolved quickly with the help of mini computers that became available in the late 1960s and 1970s, and the earlier dispersive IR instruments became history. (iv) Before the advent of lasers, Raman spectroscopy was almost a rarity, used by some physicists for the extra information in the studies of crystals it provided over IR. Even after powerful Ar$^+$ lasers were employed to excite Raman spectra, the technique was too specialized for normal, synthetic chemists to employ it routinely, since the double spectrometers required to suppress the background Rayleigh intensity also reduced the Raman intensities. Only after the integration of Raman spectroscopy and microscopy in the late 1970s, followed by developments in optical filter technologies, it has become a versatile tool to record spectra hassle-free and quickly.

In a nutshell, Raman scattering is the inelastic scattering of light by chemical bonds in a material. Though phenomena such as magnon scattering in magnetic systems exist, we like to limit ourselves to the more common vibrational spectroscopy, including the far IR region of translations and librations (i.e., restricted rotations) of atomic and molecular species.

When an intense laser light of frequency ν_0 is incident on a material, most of the light is scattered elastically without change in energy or frequency, called Rayleigh scattering. About 1 in 10 million (10^7) photons interact with the chemical bonds in the material, resulting in either giving up some of their energy to increase the amplitude of the vibrations and getting scattered as a photon with lower energy, or gaining some energy from the vibrations to result in photons of higher energy. The quanta of energy thus lost or gained represent the vibrational energy levels of the molecules, represented in simple harmonic approximation as

$$E_n = (n + 1/2)\hbar\omega \tag{1.3}$$

More photons get scattered with lower energy since more molecules are in the ground state at room temperature. Fewer molecules are in higher energy states (corresponding to larger amplitudes of vibration), following Boltzmann distribution $p_i \propto \exp(-E_i/k_BT)$ where p_i is the probability of the molecule being in state i, E_i is the energy of that state, k_B is Boltzmann's constant, and T is the temperature. Photons scattered with lower energy, i.e., longer wavelength $(\nu_0 - \Delta\nu_j)$, are called Stokes Raman shifted photons, in analogy with the phenomenon of luminescence where the emitted photons are always of longer wavelength than the wavelength of the incident photons. Photons scattered with higher energy $(\nu_0 + \Delta\nu_j)$ are called anti-Stokes Raman shifted photons (Figure 1.1). Though the terminology Stokes shift is borrowed from luminescence, the analogy ends there since luminescence is of electronic origin, where electrons are excited to a higher energy state by photons of energy typically of a few eV. However, lasers in the visible wavelength range such as 532 nm (2.3 eV) are generally used to excite Raman spectra. In practice, since both Raman scattering and photoluminescence are the emissions of photons from the material analyzed, both types of photons may be present in the scattered/emitted signal. Luminescence can sometimes be significantly stronger than the feeble Raman scattering, and it is a practical problem to eliminate luminescence spectra so that Raman spectra can be recorded. This can be particularly challenging in biomedical, life science, and pharmaceutical research, where the analyte molecules could emit strong photoluminescence that may submerge or flood the weak Raman bands. Since Raman scattering is an almost instantaneous process (time scale 10^{-13} to 10^{-12} s), whereas luminescence involves absorption and re-emission of photons that is a significantly slower process (10^{-10} to 10^{-7} s), time-resolved Raman spectroscopy can be used to eliminate the luminescence signal [3]. Since time-resolved experiments can be complex, other methods such as using a long-wavelength laser (such as 785 nm) to excite the Raman spectra without exciting luminescence bands or UV Raman spectroscopy (e.g., 257 or 325 nm laser excitation) have been traditionally used to circumvent the problem. Under UV excitation, though strong luminescence occurs since the Raman bands (as with any laser excitation) are close to the excitation line (within ~0.5 eV from the laser line) and luminescence bands are significantly longer wavelengths in the visible range, the latter do not interfere with the Raman bands. Though in the schematic in Figure 1.1, the Stokes, anti-Stokes, and the Rayleigh components are shown as distinct for clarity, the scattering is uniform all over the space, and all photons are spatially mixed. It is the function of a Raman spectrometer to suppress the intense Rayleigh component and resolve the different shifted frequencies $(\nu_0 \pm \Delta\nu_j)$ of photons corresponding to the molecular frequencies.

The molecular vibrations mentioned above can be symmetric stretch modes, asymmetric stretch modes, or bending modes in molecules consisting of N number of atoms. A non-linear molecule with N atoms has $3N$ degrees of freedom that are distributed as three translations, three rotations, and $3N-6$ vibrations. In the case of linear molecules, the rotational degree of freedom is 2, and we have $3N-5$ vibrational degrees of freedom. For example, oxygen molecule (O_2) has $(3\times2)-5=1$ vibrational degree of freedom, which is stretching vibration (at 1,580 cm^{-1}). CO_2 molecule has

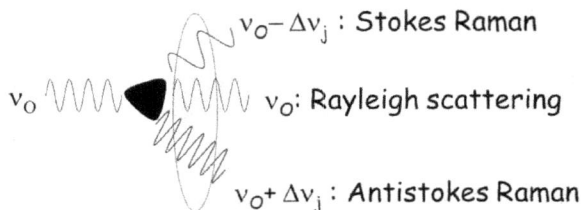

FIGURE 1.1 Inelastic scattering of light from a material. ν_0 is the incident laser frequency, and $\Delta\nu_j$ are the vibrational frequencies of the molecule or the solid. Stokes and anti-Stokes Raman shifted photons are scattered in all directions along with the intense unshifted Rayleigh scattered photons.

$(3\times3)-5=4$ vibrational degrees of freedom. They are C–O symmetric stretching ($1{,}390\ \text{cm}^{-1}$), C–O asymmetric stretching ($2{,}350\ \text{cm}^{-1}$), and a doubly degenerate O–C–O bending mode ($526\ \text{cm}^{-1}$).

Since only ~1 in 10^7 photons get scattered inelastically, the Raman effect is feeble. Historically this was one of the reasons scientists had difficulty detecting it. The equally important problem is detecting this feeble signal in the large background of Rayleigh scattered laser light at the excitation frequency. We need photon detectors that are sensitive enough and a monochromator that can resolve the closely spaced vibrational modes. Vibrational frequencies are stated in units of cm^{-1}. Raman shift is defined as follows:

$$\Delta\nu\left(\text{cm}^{-1}\right) = \frac{1}{\lambda_{\text{incident}}\left(\text{cm}\right)} - \frac{1}{\lambda_{\text{scattered}}\left(\text{cm}\right)} \tag{1.4}$$

where $\lambda_{\text{incident}}(\text{cm})$ is the wavelength of the laser expressed in cm and $\lambda_{\text{scattered}}(\text{cm})$ is the scattered wavelength in cm. The cm^{-1} is named wavenumber and is a convenient scale in vibrational spectroscopy to avoid writing large fractional numbers if wavelength or frequency is used. To get a feel for this wavenumber unit, we express a laser wavelength of Ar$^+$ laser, 514.531 nm in cm^{-1}, which is

$$\left(\frac{1}{514.5\ \text{nm} \times 10^{-7}\text{cm}\,/\,\text{nm}}\right)^{-1} = 19{,}436\ \text{cm}^{-1}$$

When Ar$^+$ laser is used to excite the Raman spectra of a sample, the absolute wavenumber of $19{,}436\ \text{cm}^{-1}$ is taken as the reference or zero point from which the Raman shifted wavenumber of a photon is measured. If the Raman shift of a particular molecular vibration is, e.g., $3{,}000\ \text{cm}^{-1}$, this means the laser wavelength is Stokes shifted from 514.5 nm to $(19{,}436-3{,}000)\ \text{cm}^{-1} = 608.4$ nm. The green wavelength photons lose some of their energy to the molecular vibration and get scattered as photons of red color. C–H vibrational frequencies occur in this region of $3{,}000\ \text{cm}^{-1}$. However, we cannot see the shifted photons with unaided eyes since such Raman shifted photons are too few, and the elastically scattered green photons swamp our eyesight. It is interesting to note that historically, C. V. Raman used the technique of complementary filters to quickly assess visually dozens of organic liquid samples to see if any of them is a promising candidate to show the Raman effect [4]: "A blue-violet filter, when coupled with a yellow-green filter and placed in the incident light, completely extinguished the track of the light through the liquid or vapor. The reappearance of the track when the yellow filter is transferred to a place between it and the observer's eye is proof of the existence of a modified scattered radiation" (Figure 1.2) [5]. It can take some hours for human eyes to fully adapt in a dark laboratory room to detect faint signals from the crossed filters; Raman employed several students for these observations, and under the guidance of Raman, it was K. S. Krishnan's keen eyes that detected the "new type of secondary radiation" that was later named "Raman scattering".

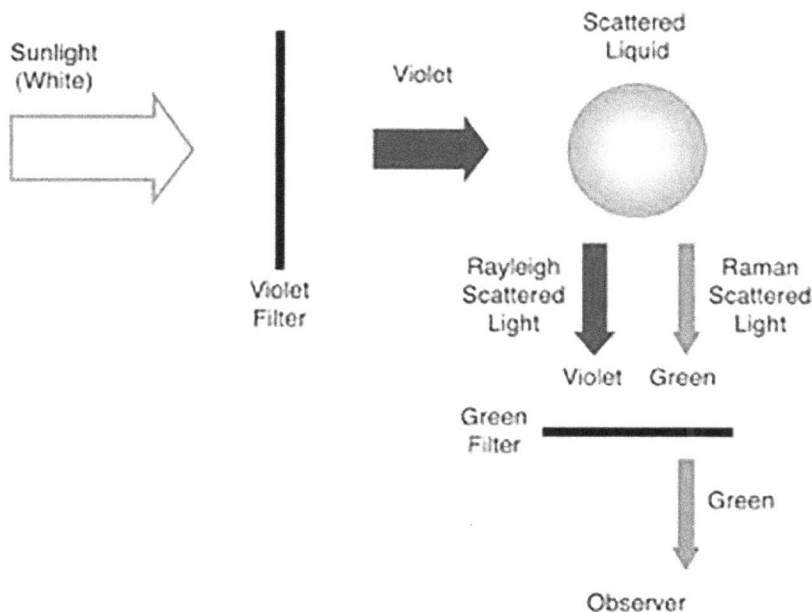

FIGURE 1.2 Method of complementary filters to visually observe Raman shift. (Adapted for the internet from "the Raman effect", produced by the National Historic Chemical Landmarks program of the American Chemical Society in 1998 [4].)

1.1.1.1   Selection Rules for IR Absorption and Raman Scattering

1.1.1.1.1   IR Absorption

Absorption of radiation at the vibrational frequencies of the molecule/solid occurs only if the vibrational mode is accompanied by a change in the dipole moment of the molecule. This is called a "Selection rule": Dipole moment $P = q \times d$, where d is the displacement vector. Because of the vectorial nature of d, symmetric stretch vibrations produce zero dipole moment and hence no IR absorption. In other words, symmetric stretch vibrations are IR inactive. For example, in the symmetric stretch vibration of the CO_2 molecule, the dipole moment (P) produced by the oxygen atom stretching on the right-hand side annuls the P of the oxygen atom stretching on the left side: $O=C=O \rightarrow O \leftarrow C \rightarrow O$. On the other hand, asymmetric stretch and bending modes of CO_2 molecules do produce a net dipole moment, and hence they are IR active. Another selection rule that is often cited is, for a harmonic oscillator, transitions are allowed only when the quantum number n changes by one, i.e., $\Delta n = \pm 1$ in equation (1.3). This means that higher harmonics or overtones of the fundamental vibrational frequencies are not permitted in the harmonic approximation. The fact that experimentally we have overtones indicates that the interaction potentials are generally anharmonic.

1.1.1.1.2   Raman Scattering

A mode is likely to be be Raman active if it induces a change in the electronic polarizability (α) of the molecule. The induced dipole moment $P = \alpha E$. α is a second-order tensor, and at least one of the tensor elements should change under a vibrational motion for Raman activity to occur. The magnitude of α can be directly related to the volume of the electron cloud around the particular bond. Change in polarizability $\Rightarrow$ Change in the volume of the electron cloud. Since electronic motion follows the nuclei of the atoms, symmetric stretch modes produce the largest change in the volume of the electron cloud, and hence they are intensely Raman active.

1.1.1.2 Mutual Exclusion Principle

It is found that the symmetric stretch mode of the CO_2 molecule is Raman active, and it is not IR active as discussed earlier. The other two IR active modes of CO_2 are found to be not Raman active. This mutual exclusion of Raman and IR activities is generally true in centrosymmetric molecules and crystal structures with an inversion center. Other obvious examples of this rule are symmetric molecules such as benzene and ethylene. The mutual exclusion rule can sometimes help resolve ambiguities in crystal structures, as demonstrated some years ago in the case of $Zn(CN)_2$ [6].

The mutual exclusion makes IR absorption and Raman spectroscopic techniques complementary to each other. In the absence of centrosymmetry of molecules or inversion symmetry of crystals, Raman modes will generally also be IR active and vice versa. Induced dipole moment can be written as $\mathbf{P} = \alpha \mathbf{E} + q\mathbf{d}$.

The magnitude of change in the dipole moment or the polarizability of a particular vibrational mode may be low, in which case we may not be able to experimentally detect the expected IR or Raman activity. Such modes are called optically silent. In the solid state, the irreducible representation for a particular structure can be calculated from the Character Table of its point group. The Character Tables of all 32 point groups are available in textbooks [7–9]. In this context, the phonon irreducible representations (IRR) are the number of Raman and IR modes of different symmetries. Symmetries of the phonon modes represent whether that mode vibrates symmetrically or anti-symmetrically. Some of the modes would be translations and some restricted rotations called librations. This chapter does not go into the basics of symmetry and group theory. However, we present two explanatory examples of the calculation of IRR.

1.1.1.3 Classical Theory of Raman Scattering

An oscillating electromagnetic field $E = E_0 \cos(\omega t)$ induces a dipole moment P in a molecule

$$P = \alpha E = \alpha E_0 \cos(\omega t) \tag{1.5}$$

If there is a change in the polarizability α of the molecule due to a vibrational motion,

$$P = (\alpha_0 + \beta \cos(\omega_{vib} t)) \times E_0 \cos(\omega t) \tag{1.6}$$

where $\beta = \partial \alpha / \partial q$, q = nuclear displacement.

Using $\cos(A)\cos(B) = \frac{1}{2}\{\cos(A - B) + \cos(A + B)\}$,

$$P = \alpha_0 E_0 \cos(\omega t) + \tfrac{1}{2}\beta E_0 \{\cos(\omega - \omega_{vib})t + \cos(\omega + \omega_{vib})t\} \tag{1.7}$$

where the first term is the Rayleigh component, the second is Stokes Raman, and the third is anti-Stokes Raman shifted photons.

If $\beta = 0$, there is no change in polarizability of the molecule during the particular vibration motion: that mode will not be Raman active.

Though such a classical picture can explain Raman shift, other aspects of Raman scattering, such as the intensities and molecular rotations, can be explained only by a quantum mechanical theory.

1.1.2 Raman Spectroscopy Instrumentation

We now discuss in some details of the experimental set-ups that are used to record Raman spectra, with particular emphasis on a modern micro-Raman instrument. We first briefly discuss high-resolution Raman spectrometers that use a double monochromator. A monochromator is the dispersive component of a spectrometer. When a laser excitation source and a detector are integrated with a monochromator, we get a spectrometer. A few decades ago, with intense laser excitation of spectra, the monochromator of choice for physicists was a double monochromator. In addition to high resolution, due to their high Rayleigh rejection, these spectrometers were capable of recording low wavenumber Raman spectra, down to as low as 10 cm^{-1} from suitable (~crystalline) samples with low stray (Rayleigh) light scattering. This functionality was not offered by single grating spectrometers till about 10 years ago. Since the far IR region (10 to ~200 cm^{-1}) offered an insight into 'external modes' such as translations and rotations of molecular species in solids, this range is important in the study of phase transformations in solids. Apart from a laser (typically in the visible range, 532 nm) to excite the spectra, such a high-resolution spectrometer consists of: (i) a 'fore-monochromator' with a suitable grating to disperse the scattered signal collected from the sample using a lens with a short focal length and high numerical aperture (large diameter, typically a camera lens) and focused on the entrance slit of the double monochromator. The first stage of the monochromator served to remove the Rayleigh signal to a large extent, and (ii) a second monochromator with another grating that would disperse the signal further to collect spectra, one data point at a time at the exit slit with (iii) a photomultiplier tube (PMT) detector. Integration times were typically of the order of 5–10 s per point, depending upon the signal intensity. The monochromator was programmed to move the gratings in suitable steps so that spectra could be recorded, e.g., every half a wavenumber or one wavenumber, and the spectra were collected for 30 min to a few hours depending upon the wavenumber range desired, and the signal intensity. Spectrometer resolution, among other things, is determined by the length of the spectrometer, the gratings that are used, and the slits used for recording the spectra. For example, the high-resolution spectrometer marketed by Horiba Jobin Yvon, model U1000 (focal length 1,000 mm), with 1,800 L/mm gratings and 514.5 nm laser excitation, has a spectral bandpass of ~9 cm^{-1}/mm at the exit slit. For a typical exit slit opening of 300 μm, thus a ~3 cm^{-1} band is integrated into a single data point in the spectrum by the PMT detector. If a better resolution is desired, one can use gratings with more lines, such as 2,400 l/mm, or more often, reduce the slit width, with a corresponding reduction in intensity. One advantage of double grating instruments is that several different laser excitations can be interchangeably used, whereas single monochromators rely upon optical filters to suppress the Rayleigh line from the scattered signal before it is sent to the grating, and each laser line requires its Rayleigh filter (Figure 1.3).

Micro-Raman spectrometers are single grating instruments that are used in conjunction with a microscope to focus a laser beam to a diameter of about a micron on the sample and a CCD detector at the other end. We will look at the example of a Renishaw micro-Raman spectrometer for discussion here (Figure 1.4). The excitation laser (not shown) beam enters into the spectrometer box through a set of four neutral density filters of different optical densities with which one can change the laser intensity from 100% (without any filters in the beam path) to 5×10^{-8}%. The laser beam is expanded and directed out of the spectrometer box into the microscope via a set of beam directors and a beam expander. The microscope (not shown) focuses the laser beam on the sample into a spot size of ~1 μm (depending upon the objective used), and the scattered light. The microscope objective with a high numerical aperture also collects the scattered signal and sends it into the monochromator. For normal measurements at ambient conditions, a 50× objective is used. Soon after entering the monochromator, the signal beam encounters a set of Rayleigh rejection (edge) filters. Most of the job of a fore-monochromator in a double monochromator is done by a set of two edge filters. This filtered signal is then focused on an entrance slit, normally set to an optimal opening of ~65 μm. Then this signal is collimated into a parallel beam and directed to the diffraction grating (typically 1,800 l/mm for visible laser excitations). The diffracted signal spatially ordered into a spectrum is detected by a CCD detector displayed by a PC that controls the spectrometer via proprietary software.

The advent of micro-Raman spectrometers has revolutionized the field of Raman spectroscopy and almost consigned the older type of spectrometers to history. The biggest advantage of micro-Raman spectrometers over those without a matching microscope attachment is the high optical throughput or optical conductance [10] of micro-Raman spectrometers. The Raman signal in a micro-Raman instrument can be a few orders of magnitude higher than the old Raman instruments.

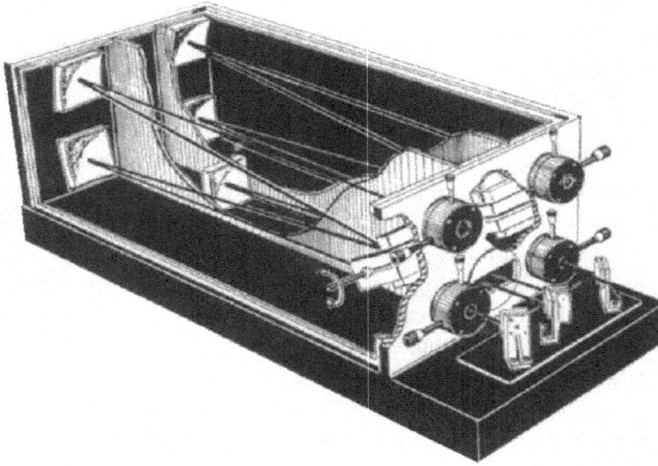

FIGURE 1.3 Horiba Jobin Yvon U1000 high-resolution monochromator. It consists of two holographic gratings that are fixed on the same horizontal rotation axis next to the slits and a set of mirrors that focus the collected signal on the gratings. Here the top left slit is the entrance slit, top right one is the exit slit on which a PMT is fitted, and the bottom slits are intermediate slits used to further reduce stray light entering into the PMT. Typical slit settings are 300–400 to 400–300 µm, where the intermediate slits are set to 400 µm. U1000 has 2×1 m focal length with a high-precision drive mechanism that drives the gratings. (Reprinted with permission from Horiba Jobin Yvon Catalog.)

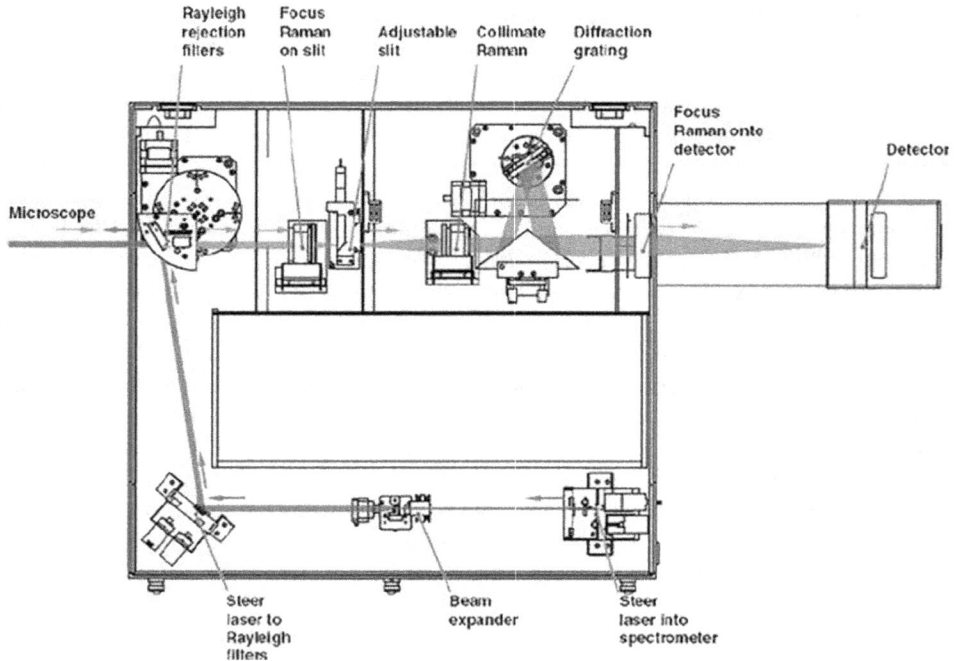

FIGURE 1.4 Schematic diagram of a Renishaw Raman microscope. (Diagram of Renishaw's inVia confocal Raman microscope. © Copyright Renishaw Plc. All rights reserved. Image reproduced with the permission of Renishaw.)

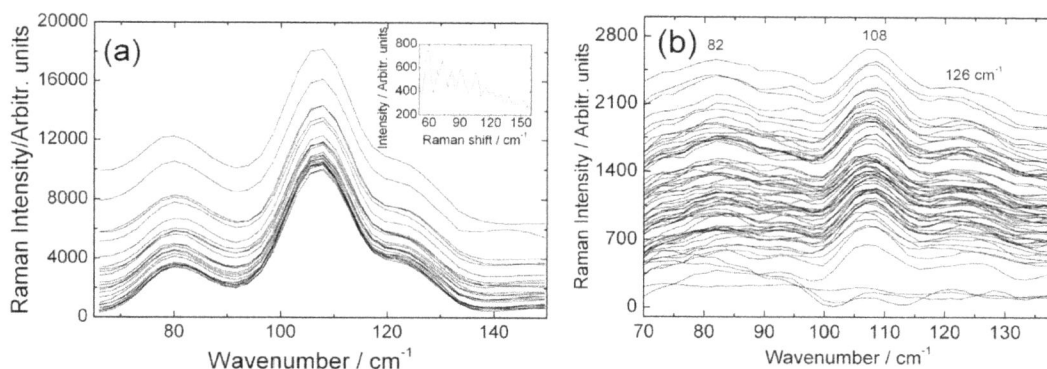

FIGURE 1.5 Raman spectra (background subtracted) of uranium coated with 5 nm gold film with (a) 785 nm and (b) 514 nm laser excitation at different spots on the sample. The higher intensity of spectra with 785 nm excitation is due to the larger sampling volume and laser intensity. Three spectra from smooth surfaces are shown at the bottom of (b). Inset (a) Raman spectrum of uncoated uranium showing rotational Raman bands of N_2 and O_2. (Reprinted with permission from [11]. Copyright © 2011 John Wiley & Sons, Ltd.)

High throughput enables not only significantly faster recording of spectra but also obtaining spectra from a variety of samples that were difficult or impossible earlier. For example, the first Raman spectra of metallic U were recorded using a Renishaw Raman microscope [11] through a surface-enhanced Raman spectroscopy (SERS) technique. The main advantage of double or triple spectrometers over micro-Raman instruments was the formers' ability to record low wavenumber spectra, down to 10 cm^{-1} in some samples, whereas in a micro-Raman instrument till a couple of decades ago, the closest to the Rayleigh line that one could go was ~200 cm^{-1}. Advances in filter technology, including the advent of Volume Bragg Gratings, have enabled routine measurements in a micro-Raman instrument down to 10 cm^{-1}, with significantly less intensity loss than older instruments (Figure 1.5).

1.1.3 RAMAN AND IR TECHNIQUES: A COMPARISON

Raman and IR spectroscopic techniques have their own set of advantages and disadvantages.
    We will first look at the advantages of Raman spectroscopy over IR:

1. The presence of moisture in a sample does not pose any problem to record its Raman spectra since the spectra of water are weak; however, IR radiation is absorbed heavily in the range, and the IR absorption spectra of a moist sample is more likely to show no sample peaks.
2. Sample preparation for Raman spectroscopy is minimal for most samples. Spectra can be recorded from solid pieces, powder samples, liquid samples in cuvettes, or even from a drop of a liquid on a substrate. Sometimes polished surfaces of solid samples are preferred to avoid high Rayleigh background. Raman spectra can be recorded from, e.g., specific spots on a large sample such as a painting to obtain information on the color pigments used. This is unlike for IR spectra, where, e.g., a small quantity of the sample (say, 2 mg or less) is mixed with a large quantity (~200 mg) of an IR-transparent base alkali halide such as KBr or CsI, ground into a fine powder, and then pressed into a thin pellet of 10–13 mm diameter using a suitable press. The Attenuated Total Reflection (ATR) method is a specialized IR technique that can be used under certain circumstances without much sample preparation. In recent times, this method has been widely used and even preferred over the traditional absorption method.

3. It is fairly simple to carry out Raman spectroscopic measurements as a function of pressure and/or temperature. For measurements at high pressure in a diamond anvil cell, a sample of ~100 μm lateral dimensions can be loaded into a hole of a diameter ~250 μm along with a pressure marker such as specks of ruby and a pressure transmitting media such as a mixture of methanol and ethanol. A laser can be focused to a ~1 μm by a 50× objective in a regular micro-Raman spectrometer, and the sample spectra can be measured at different pressures that can be measured by ruby luminescence spectra that can be excited by the same laser (namely, a green laser of wavelength 514 or 532 nm). In contrast, IR waves are of several μm wavelength, and they cannot be focused into spots of diameter less than the wavelength due to diffraction limit, reducing the spatial resolution of IR spectroscopy.

4. High-temperature Raman spectroscopy is straightforward: depending on the temperature range desired, a sample can be heated on an enclosed heater with a suitable window in an inert atmosphere, if necessary, and subsequent to excitation, spectra can be collected by a micro-Raman spectrometer using a long focal length objective. At high temperatures, however, the blackbody IR radiation emission background of a solid sample could interfere with/overshadow the IR absorption bands of the sample, leading to difficulty in recording the spectra.

5. To record an FTIR spectrum from 10 to 4,000 cm^{-1}, at least two different beam splitters have to be used in the different ranges. The commonly used KBr beam splitter for the mid-IR range is sensitive to moisture and needs a dry nitrogen atmosphere to be always maintained. Obtaining far IR spectra (e.g., using a mylar beam splitter) is non-trivial. In contrast, Raman spectra in this range can be recorded without such experimental demands.

6. Polarized Raman spectroscopy of crystals enables one to obtain the symmetry properties of normal vibrations and their assignments. In comparison, polarized IR spectroscopy, though possible, is not widely practiced due to technical difficulties.

There are some advantages of IR spectroscopy over Raman spectroscopy too:

1. As mentioned earlier, fluorescence or photoluminescence is a major problem for Raman spectroscopy on samples that contain fluorescing species or impurities. Fluorescence intensities could be orders of magnitude more than Raman scattering intensities. For example, in the exploratory synthesis of harder-than-diamond material by laser heating a suitable precursor at high pressure, the retrieved sample gave out a huge fluorescent background and no Raman bands under visible laser excitation. There is no such problem in IR spectroscopy since IR energies are lower than the electronic energies needed to excite the luminescent spectra. In the current example, this problem is circumvented by using deep UV laser (257 nm) excitation, and useful Raman spectra can be obtained [12]. Another obvious way to avoid unwanted fluorescence that is more routinely used is to excite the Raman spectra with a longer wavelength laser such as 785 or 1,064 nm. However, recording an IR spectrum is simpler in the case of samples with luminescent impurities unless the sample contains moisture, like in biomedical samples.

2. Mid-IR absorption is an intense effect, which is why less than 1% of the sample is mixed in a non-absorbing medium such as KBr to record the IR spectra. Mid-IR absorption is a highly sensitive method for detecting different chemical bonds and functional groups. In contrast, the Raman effect is feeble. Enormous efforts have been made, and a host of techniques such as resonance Raman spectroscopy, SERS, and coherent anti-Stokes Raman spectroscopy have been mainly developed to address this singular lack of Raman intensity.

3. Raman spectrometers may be more expensive compared to IR spectrometers with equivalent capabilities.

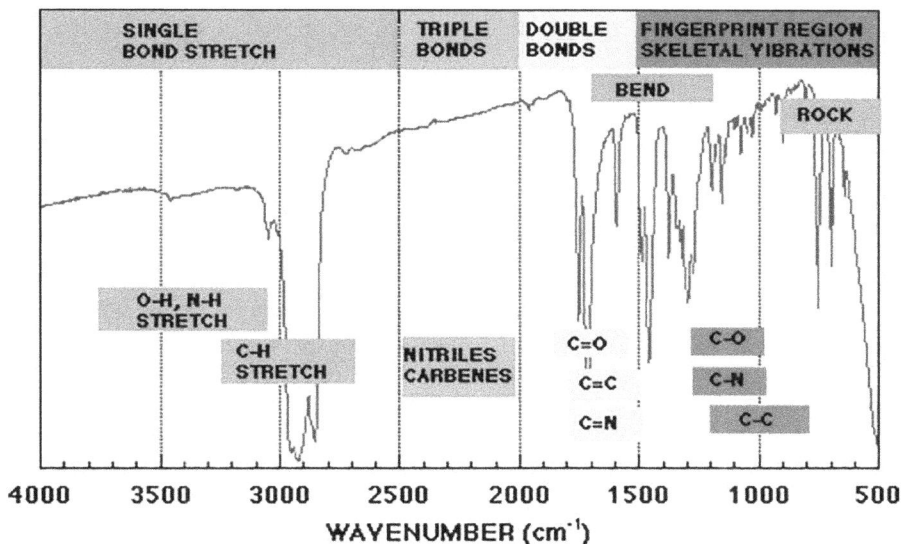

FIGURE 1.6 Group frequency regions of the various important species in a molecule. (From ref. [13].)

1.1.3.1 Characteristic or Fingerprint Spectra

We note that each molecule or solid has its characteristic IR and Raman spectra (collectively known as vibrational spectra), much like each material has a unique set of reflections in an XRD pattern. Characteristic vibrational spectra arise due to each molecule having a unique set of atomic species with different masses and the nature of bonds such as a single bond, double bond, and triple bond. In contrast, the uniqueness of XRD patterns is due to the distinct sets of planes of atoms in a crystal. The characteristic IR or Raman spectrum (or a specific section such as 600–1,600 cm^{-1}) of a molecule is sometimes called a fingerprint spectrum to emphasize its uniqueness. A pictorial representation showing the various species and their wavenumber regions is given in Figure 1.6 [13]. It is possible to identify a molecule or a functional group by comparing its IR or Raman spectrum to a data bank or a catalog of spectra, which is routinely done. However, XRD databases (such as the International Centre for Diffraction Data (ICDD) and the Cambridge Crystallographic Data Centre (CCDC)) are highly evolved, voluminous, more universal, and widely used. Nevertheless, it has been observed that Raman spectroscopy is an excellent tool for routine characterization, with a success rate comparable to XRD [14]. Also, for suitable samples, recording a Raman spectrum is 10–100 times quicker than XRD and more convenient, establishing Raman spectroscopy as an important tool to study structural phase transformations of materials as a function of temperature and pressure, among other things.

1.1.4 ELEMENTARY LATTICE DYNAMICS IN THE CONTEXT OF VIBRATIONAL SPECTROSCOPY

Metallic systems are an important category of materials where vibrational spectroscopy is either not applicable or challenging to undertake. From elementary lattice dynamics, it is known that for a linear monoatomic chain, the dispersion relation between the vibrational frequency ω and the wave vector k ($=2\pi/\lambda$, 'λ' being the wavelength of lattice vibrations) is

$$\omega = \sqrt{\frac{4K}{M}} \left| \sin\frac{ka}{2} \right| = \omega_m \left| \sin\frac{ka}{2} \right|$$

where K is the force constant of the bonds that connect the atoms, M is the mass of the atoms, a is the interatomic distance, and ω_m is the maximum frequency. For a linear chain of N atoms, there are

N–1 solutions (modes) to this equation (Figure 1.7). When N becomes a large number like in real solids, the dispersion curves are essentially continuous.

For a linear diatomic chain with atoms of two different masses but with the same spring constant between the nearest neighbors, there are two branches in the dispersion line; in addition to the one for the monoatomic case, we have an optical branch at higher frequencies (Figure 1.8):

$$\omega^2 = K\left(\frac{1}{M_1} + \frac{1}{M_2}\right) \pm K\sqrt{\left(\frac{1}{M_1} + \frac{1}{M_2}\right)^2 - \frac{4\sin^2(ka)}{M_1 M_2}}$$

Whereas the model linear chain above has a single optical branch and one acoustic branch, a real three-dimensional (3D) solid with N atoms per unit cell has three acoustic modes and $3N$–3 optical modes. For example, titanium trisulfide (TsS$_3$) crystallizes in a monoclinic structure (space group P2$_1$/m) with two formula units in the unit cell. Its 24 degrees of freedom are distributed at the Brillouin zone center as follows: $\Gamma = 8A_g + 4B_g + 4A_u + 8B_u$, where the A_g modes are Raman active, $A_u + 2B_u$ are the three acoustic modes, and the rest of the 'u' modes are IR active. All modes are expected to be non-degenerate. In 1-D, phonon density of states (DOS) $G(\omega) = (Na/2\pi)/(d\omega/dq)$; a = lattice constant. $G(\omega) \rightarrow$ infinity when $d\omega/dq = 0$, i.e., at the Brillouin zone center and the edges, where the dispersion curves are flat. This singularity in phonon DOS is called the Van Hove singularity. In reality, this means that there are a large number of phonons with these ω

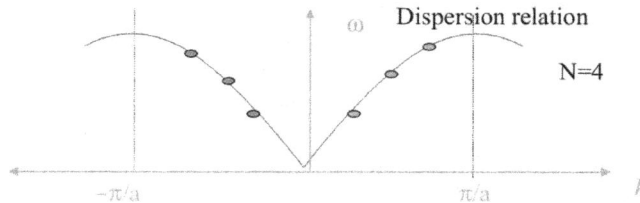

FIGURE 1.7 Dispersion curves for a linear chain of four atoms. The region between $-\pi/a$ and $+\pi/a$ is called the first Brillouin zone. Beyond this region, the curves repeat themselves, and there is no new information. (From ref. [15].)

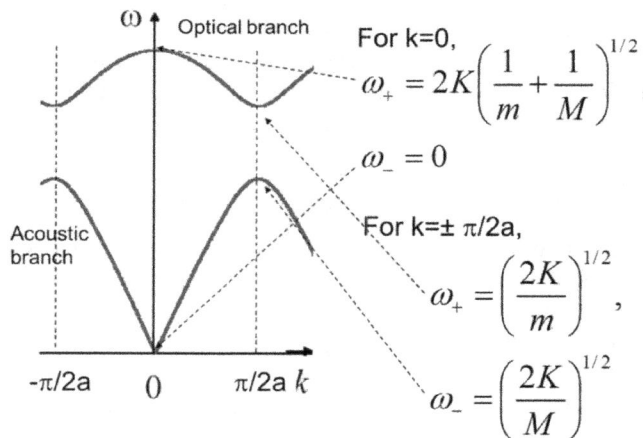

For k=0,
$$\omega_+ = 2K\left(\frac{1}{m} + \frac{1}{M}\right)^{1/2},$$
$$\omega_- = 0$$

For k=± π/2a,
$$\omega_+ = \left(\frac{2K}{m}\right)^{1/2},$$
$$\omega_- = \left(\frac{2K}{M}\right)^{1/2}$$

FIGURE 1.8 Dispersion curves for a linear diatomic chain with atoms of different masses but with the same spring constant between the nearest neighbors. Optical techniques can probe the optical branch.

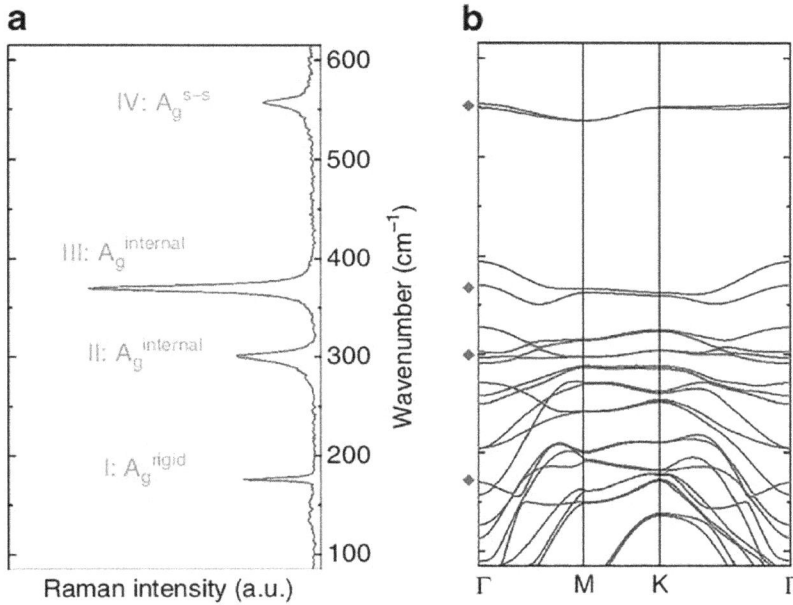

FIGURE 1.9 (a) The Van Hove singularities at the critical points in the Brillouin zone center appear as peaks in the Raman spectra of TiS_3 whiskers. (b) Phonon dispersion spectrum calculated using DFT of TiS_3. The Raman active modes are highlighted by blue diamond symbols. (Adapted from ref. [16]. Licensed under a Creative Commons Attribution 4.0 International License.)

values. They show up as peaks in experimental measurements of Raman and IR spectroscopies. Out of the 12 Raman active modes in TiS_3, only four are observed experimentally (Figure 1.9a) [16]. However, DFT calculations reveal that some A_g modes are 'accidentally degenerate', i.e., there are two or three closely spaced Raman active modes around 300 cm^{-1}, leading to a larger width of this Raman band. The band at 556 cm^{-1} also exhibits a large FWHM due to a similar reason [16]. Other (6 or 7) expected Raman bands are not observed experimentally since they could be weak in intensity.

Extending the linear monoatomic chain model to 3D, we have a solid with one atom per primitive cell: there are no optical branches, i.e., no Raman or IR modes. As much as half the metals in the periodic table are bcc (Ac, Al, Ca, Ce, Cu, Au, Ir, Pb, Ni, Pd, Pt, Rh, Ag, Sr, Th, Yb) or fcc (Ba, Cs, Cr, Eu, Fr, Fe, Li, Mn, Mo, Nb, K, Ra, Rb, Na, Ta, W, V): they have one atom per primitive cell, and hence no optical modes are observed. Non-cubic elemental metal with more than one atom in the primitive cell does have optical branches, and they are amenable to Raman spectroscopic studies. In fact, in addition to the example of U above, Raman spectra of several metallic elements such as Be, Ga, Ti, and Os have been reported in the literature. In metals, however, the Raman signal (if present) is weak: sampling volume is small due to the low penetration depth of excitation laser due to dynamic screening by electrons.

Due to their low momentum, the photons (in Raman and IR measurements) cannot probe the entire Brillouin zone but can probe only the phonons at or near the Brillouin zone center (Figure 1.10):

- Wavevector of light $k_0 = 2\pi/\lambda \sim 10^5$ cm^{-1}
- Law of conservation of momentum demands $k_i + q = k_s$
- This implies that the maximum change in momentum of this photon that can occur is 2×10^5 cm^{-1} when a phonon with this energy back-scatters the photon
- Brillouin zone boundary $= \pi/2a$ (a = lattice parameter) $\sim 10^8$ cm^{-1}

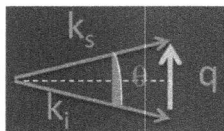

FIGURE 1.10 Momentum conservation of phonon wavevectors: $k_i \approx k_s = k_0$; $\sin(\theta/2) = (q/2)/k_0$; $q = 2k_0 \sin(\theta/2)$; maximum possible q is when $\theta = 180° =$ back scattering of photon; then $k_s = -2k_i$; $q = 2k_0$.

1.1.5 AN INTRODUCTION TO FACTOR GROUP ANALYSIS TO DETERMINE THE PHONON IRREDUCIBLE REPRESENTATIONS OF CRYSTALLINE SOLIDS

We intend to describe the Bhagavantam and Venkatarayudu (BV) method [17,18] which is the earliest procedure to calculate the molecular vibrational modes in a molecule and phonon IRR in a crystalline solid. Other methods, such as the site symmetry method or correlation method [7] of Hornig and Halford [19–21], may be practically less complex since no visualization of symmetry operations is needed, but the notation can be challenging. It has also been noted that there are cases where the use of the site symmetry method can give results that conflict with experiments [22], but given the success of this method, we expect such cases to be uncommon. However, the BV method has been automated by Adams and Newton [22], and the contribution of each atom in a unit cell to a representation of the factor group has been determined for all 230 space groups using a computer program that was written for this purpose. The inputs needed to obtain IRR are only the Wyckoff positions of all atoms in the structure, which simplifies the entire procedure. We obtain the IRR for a few crystals using their published Tables [23]. The Bilbao crystallographic website (https://www. cryst.ehu.es/rep/sam.html) may also be helpful in obtaining the IRR in some cases quickly. It is noteworthy that density functional theory-based (first principles) computational methods can be used to obtain IRRs, including mode assignments. In general, mode assignments could be difficult to determine experimentally since they involve growing single crystals, determining their orientation, and carrying out polarized Raman spectroscopic studies. However, computational work can itself be specialized, and they, of course, are based on the availability of suitable high-end computational facilities and expertise. In the absence of such infrastructure, 'hand calculations' of IRRs are still useful. We will not discuss the first principles computational methods in this chapter.

1.1.5.1 Example 1: Carbon Tetrachloride

We will first obtain the IRRs of the tetrahedral molecule CCl_4 to demonstrate the procedure. Its point group is T_d.

Step 1: We write the character Table 1.1 of the point group T_d from a standard textbook such as [7]:

Here A_1, A_2, E, F_1, and F_2 are called the symmetry species (or Mulliken symbols) of the point group. They represent how the positions of the atoms in the molecule transform under the various symmetry operations such as C_3, S_4, etc. A and B denote one-dimensional (1D) representations, meaning

TABLE 1.1

Character Table of the Point Group T_d

T_d	E	$8C_3$	$3C_2$	$6S_4$	$6\sigma_d$	Activity
A_1	+1	+1	+1	+1	+1	Raman
A_2	+1	+1	+1	−1	−1	Nil
E	+2	−1	+2	0	0	Raman
F_1	+3	0	−1	+1	−1	Nil
F_2	+3	0	−1	−1	+1	Raman, IR

the corresponding vibrational mode is non-degenerate; E is two-dimensional (2D), i.e., the vibrational mode is doubly degenerate, and F modes are triply degenerate. The modes those are symmetric with respect to the rotation operation C_n (with character+1 in Table 1.1) are denoted by 'A', and 'B' modes are antisymmetric w.r.t. C_n (with character −1). The subscripts 1 and 2 on A and B denote whether the mode is symmetric or antisymmetric with respect to a rotation around a two-fold axis (C_2) normal to C_n.

Step 2: The next task is to obtain m_j, the number of atoms that remain fixed or are transformed into translationally equivalent atoms under the various symmetry operations, by looking at a diagram of the molecule (Figure 1.11):

Under the identity operation E, all five atoms are unshifted; under the C_3 operations (the axis passing through C and one Cl atom), two atoms are unshifted. Under the C_2 operations, only the C atom is unshifted. Under the S_4 operation (C_4 rotation followed by reflection), only the C atom is unmoved, and under the σ_d operation, three atoms are unshifted. So we write (Table 1.1.1)

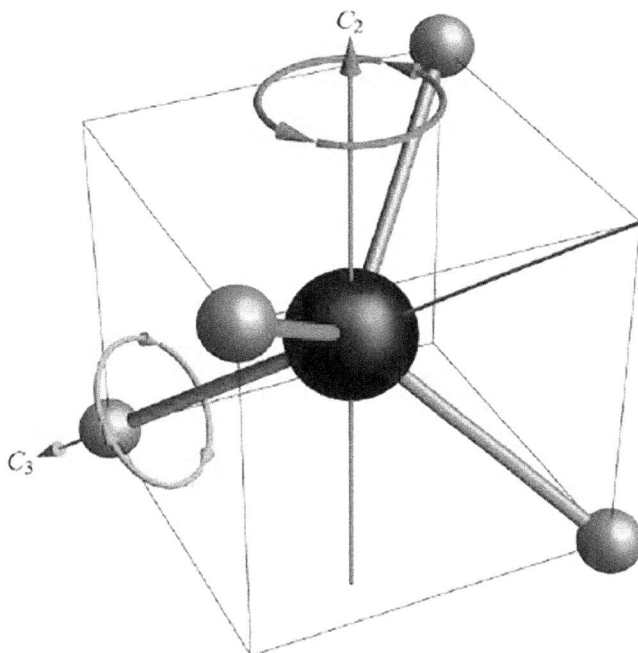

FIGURE 1.11 Rotational symmetry axes of CCl_4 in tetrahedral symmetry (T_d point group) showing one of the three equivalent C_2 axes of symmetry with the red arrow and one of the four equivalent C_3 axes with the green arrow. (Republished with permission from Royal Society of Chemistry [24].)

TABLE 1.1.1
Obtaining m_j Values for CCl_4 Molecule

T_d	E	$8C_3$	$3C_2$	$6S_4$	$6\sigma_d$	Activity
A_1	+1	+1	+1	+1	+1	Raman
A_2	+1	+1	+1	−1	−1	Nil
E	+2	−1	+2	0	0	Raman
F_1	+3	0	−1	+1	−1	Nil
F_2	+3	0	−1	−1	+1	Raman, IR
m_j	5	2	1	1	3	

TABLE 1.1.2

Universal Characters for Translations of the Various Symmetry Operations

	$\chi_j(\text{trans})$
E	3
C_2	-1
C_3	0
C_4	1
C_6	2
$S_1 = \sigma_h$	1
$S_2 = i$	-3
S_3	-2
S_4	-1
S_6	0
σ_v, σ_d	1

Step 3: From the knowledge of m_j, we can obtain the characters of the total irreducible representation by using the formula

$$\chi_j(\text{tot}) = m_j(2\cos\varphi_j \pm 1) \tag{1.8}$$

where φ_j is the angle of the proper $(+)$ or improper $(-)$ rotation. The factor $(2\cos\varphi_j \pm 1)$ represents the characters for translation, $\chi_j(\text{trans})$. The various symmetry operations are listed in Table 1.1.2.

We can then obtain the characters of the rotational transformations under the various operations given by

$$\chi_j(\text{rot}) = \pm 2\cos\varphi_j + 1 \quad (\text{for non-linear molecules}) \tag{1.9}$$

The characters of the vibrational motion can then be determined from

$$\chi_j(\text{vib}) = \chi_j(\text{tot}) - \chi_j(\text{trans}) - \chi_j(\text{rot}) \tag{1.10}$$

We can now write up these various characters in the character table (Table 1.1.3):

Step 4: The final step is to use the 'magic formula' derived from the 'great orthogonality theorem' and obtain the IRR:

$$n_\alpha = \frac{1}{g}\sum_{j=1}^{k} g_j \chi_j^\alpha \chi_j \tag{1.11}$$

where n_α are the coefficients of the symmetry species (i.e., the number of modes of symmetry A, B, E, etc.), g is the total number of symmetry operations in the point group (in the present example it is $1+8+3+6+6=24$, see the top row of the character table), g_j is the coefficient of the symmetry operation in the jth column on the top row of the character table. j is the character of the irreducible representation of the symmetry operation in the jth column of the character table, and j is the character in the jth column of the reducible representations such as $\chi_j(\text{tot})$ and $\chi_j(\text{trans})$ that we obtained above. We thus obtain the total irreducible representation $\Gamma(\text{tot})$ as follows: for A_1, n_α $(A_1) = 1/24(1\times1\times15+8\times1\times0+3\times1\times(-1)+6\times1\times(-1)+6\times1\times3) = 1/24(15-3-6+18) = 1$. Thus we have $1A_1$. Similarly, n_α $(A_2) = 1/24(15-3+6-18) = 0A_2$; n_α $(E) = 1/24(30-6) = 1E$, n_α $(F_1) = 1/24(45+3-6-18) = 1F_1$; n_α $(F_2) = 1/24(45+3+6+18) = 3F_2$. Thus we have

TABLE 1.1.3

Obtaining the Characters of Vibrational Motion for the Various Symmetry Operations

T_d		E	$8C_3$	$3C_2$	$6S_4$	$6\sigma_d$	Activity
A_1		$+1$	$+1$	$+1$	$+1$	$+1$	Raman
A_2		$+1$	$+1$	$+1$	-1	-1	Nil
E		$+2$	-1	$+2$	0	0	Raman
F_1		$+3$	0	-1	$+1$	-1	Nil
F_2		$+3$	0	-1	-1	$+1$	Raman, IR
m_j		5	2	1	1	3	
$\chi_j(\text{trans}) = (2\cos\varphi_j \pm 1)$		3	0	-1	-1	1	
$\chi_j(\text{tot}) = m_j(2\cos\varphi_j \pm 1)$		15	0	-1	-1	3	
$\chi_j(\text{rot}) = \pm 2\cos\varphi_j + 1$		3	0	-1	1	-1	
$\chi_j(\text{vib}) =_j(\text{tot}) -_j(\text{trans}) -_j(\text{rot})$		9	0	1	-1	3	

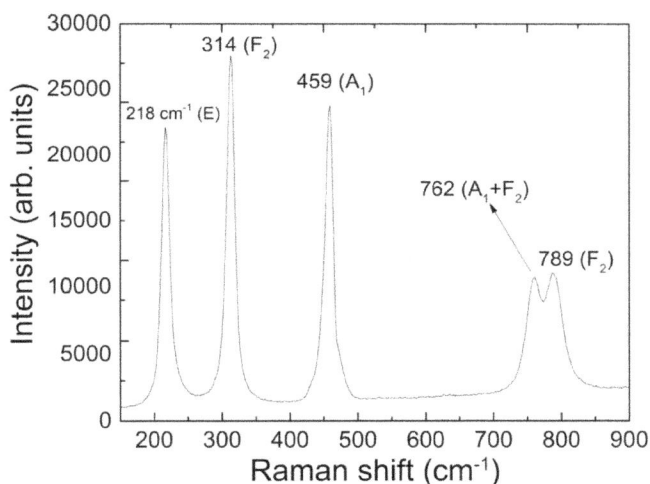

FIGURE 1.12 Stokes Raman spectrum of CCl_4 liquid, recorded in low resolution. Anti-Stokes spectra, on the negative side of the Raman shift, are usually not recorded in a modern micro-Raman instrument. (InVia, Renishaw UK, inVia.)

$$\Gamma(\text{tot}) = A_1 + E + F_1 + 3F_2$$

We can calculate in a similar manner the IRR for the rotational modes $\Gamma(\text{rot}) = F_1$, and the translational IRR $\Gamma(\text{trans}) = F_2$. $\Gamma(\text{vib}) = \Gamma(\text{tot}) - \Gamma(\text{rot}) - \Gamma(\text{trans}) = A_1 + E + 2F_2$.

According to the character table (Table 1.1.1) for the T_d point group, A_1, E, and the F_2 modes are Raman active, and the F_2 mode is also IR active. So we expect to obtain in the Raman spectrum four Raman bands, viz., one non-degenerate A_1 mode, one doubly degenerate E mode, and two triply degenerate F_2 modes.

Experimentally we obtain the above spectrum (Figure 1.12), where we have simply stated the various mode assignments without proof. We have a combination mode at 762 cm^{-1} and a F_2 band at 789 cm^{-1}. Mode assignments are done by polarized Raman spectroscopy in which the orientation of

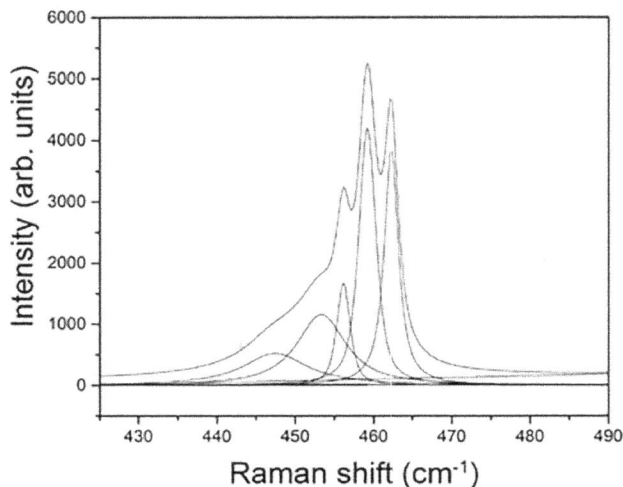

Curve Name	Centre	Width	Height	% Gaussian	Type	Area
Curve 1	459.206	2.59556	4211.55	35.068	Mixed	15230
Curve 2	462.268	2.20577	3847.01	9.44636	Mixed	12923.3
Curve 3	456.142	2.0278	1677.78	31.5991	Mixed	4799.84
Curve 4	453.326	7.4839	1157.4	9.15352	Mixed	13204.6
Curve 5	447.416	10.0746	523.534	0	Mixed	8284.97

FIGURE 1.13 Isotope effect on the ν_1 Raman band of CCl_4.

polarization of the various modes is measured using a polarizer. An incident laser is arranged to have vertical (V) polarization. Suppose a particular scattered mode is also vertically polarized, i.e., in that case, maximum intensity observed in the V–V configuration, and zero intensity in the crossed (H) polarization configuration V–H is called a polarized mode. On the contrary, if a band is found to be of maximum intensity in the V–H polarization and minimum in the V–V, it is called a depolarized mode. A_1 mode here is the symmetric stretch mode, and such modes are always found to be highly polarized. They are often of maximum intensity in Raman spectra and inactive in the IR.

Here we have ignored the fact that Cl isotopes reduce the symmetry of the molecules from T_d to C_{2v} or C_{3v}. The change in symmetry could result in differences in mode assignments [25].

The isotope effect shows itself spectacularly in the most intense A_1 (also called the ν_1) band since the different masses of the Cl isotopes lead to different frequencies [see, e.g., 26]. The relative abundance of ^{35}Cl is 75.77% while that of ^{37}Cl isotopes is 24.23%, and this leads to the percent distribution of the $^{12}C^{35}Cl_4$, $^{12}C^{35}Cl_3{}^{37}Cl$, $^{12}C^{35}Cl_2{}^{37}Cl_2$, $^{12}C^{35}Cl^{37}Cl_3$, and $^{12}C^{37}Cl_4$ species as 32.96, 42.16, 20.22, 4.31, and 0.35%, respectively. The A_1 band splits into five different, closely spaced peaks when the spectrum is recorded in a high-resolution instrument, with peak intensities proportional to the species' relative abundance (Figure 1.13).

1.1.5.2 Example 2

Next, we will carry out factor group analysis for a cubic crystal, $Zr(WO_4)_2$ (Figure 1.14). This is a well-known material that exhibits negative thermal expansion, with space group $P2_13(T^4)$, and the number of formula units in the primitive cell $Z=4$ [27]. This structure has an O atom associated with one of the two inequivalent WO_4 tetrahedra (O_4 in Figure 1.14) bonded to only one W atom and not bonded to any other metal atoms. Within the two tetrahedra, the corresponding W–O bonds

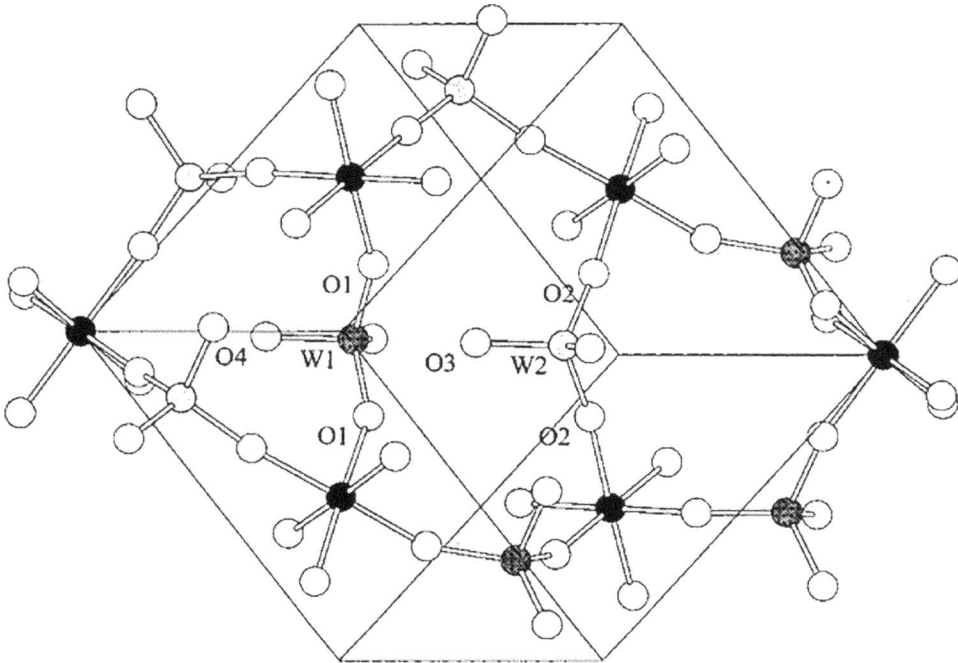

FIGURE 1.14 Cubic unit cell structure of $Zr(WO_4)_2$. (Reprinted with permission from American Physical Society [28].)

TABLE 1.2
Character Table of Point Group *T*

T	E	$4C_3$	$4C_3^2$	$3C_2$
A	1	1	1	1
E	1	ϵ	ϵ^*	1
	1	ϵ^*	ϵ	1
T	3	0	0	−1

(W_2–O_3 and W_1–O_4) are aligned along the <111> and three-fold axes of the cubic unit cell and point in the same direction for the nearest-neighbor WO_4 pairs (W_1–W_2 group) [28].

Step 1: The point group (~known as the factor group) of the space group T^4 is T, and its character table (Table 1.2) is given below:

Here $\epsilon = \exp(2\pi i/3)$ and $\epsilon^* = \exp(-2\pi i/3)$

Steps 2 and 3: Crystals like $Zr(WO_4)_2$, composed of polyatomic ions such as $(WO_4)^{4-}$, can be viewed as an interpenetrating lattice of the different entities. Interatomic forces in polyatomic units are distinctly stronger than those between the groups (e.g., between Zr and WO_4). The (Zr) translational and (WO_4) librational modes of the groups are called lattice modes or external modes. Internal modes of the polyatomic group arise from the $3n-6$ degrees of freedom, where '*n*' is the number of atoms in the polyatomic group. In the present case, the number of degrees of freedom is $(3 \times 5) - 6 = 9$. WO_4 units are tetrahedral like CCl_4, and it is shown earlier that these 9 degrees of freedom are distributed in four vibrational modes, $A_1 + E + 2F_2$.

The characters for the different types of atomic motion are [8]

$$\chi_j(\text{acoustic}) = 2\cos\varphi_j \pm 1 \tag{1.12}$$

$$\chi_j(\text{trans}) = m_j(s)(2\cos\varphi_j \pm 1) \tag{1.13}$$

$$\chi_j(\text{rot}) = m_j(s-v)(\pm 2\cos\varphi_j + 1) \tag{1.14}$$

$$\chi_j(\text{tot}) = m_j(2\cos\varphi_j \pm 1)$$

$$\chi_j(\text{internal}) = \chi_j(\text{tot}) - \chi_j(\text{rot}) - \chi_j(\text{trans}) \tag{1.15}$$

Here m_j is the number of atoms unshifted atoms under the various symmetry operations of the point group; $m_j(s)$ is the number of unshifted atomic groups, $m_j(s-v)$ is the number of unshifted polyatomic groups, where v is the number of monoatomic groups or linear groups (such as C≡N in $Zn(CN)_2$). The $\chi_j(\text{trans})$ in equation (1.13) includes acoustic modes, translations of Zr atoms, and WO_4 tetrahedra. Translations $\chi_j(\text{trans})$ and librations $\chi_j(\text{rot})$ constitute external modes. The notation 'rot' here denotes rotatory modes. Unlike fluid media, librations are restricted rotations since full rotations are generally not possible in solids.

Now the task is to determine m_j, $m_j(s)$, and $m_j(s-v)$ in the primitive cell of $Zr(WO_4)_2$ under the various symmetry operations in the above character table, viz., E, C_3, C_3^2, and C_2, and write these numbers in the corresponding columns. Since there are 44 atoms in the primitive cell ($Z=4$), all 44 atoms are unmoved under the identity operation. It is seen from Figure 1.14 that under the C_3 and C_3^2 operations, six atoms are unmoved. Under the C_2 operation, all atoms are moved, hence all $m_j=0$. Since there are two WO_4 units and one Zr in one formula unit, the total number of groups in the primitive cell is 12, and the number of polyatomic (WO_4) units is 8. Under the identity operation, all these groups of atoms are unmoved, but under all other operations, all groups are moved; hence $m_j(s)$, and $m_j(s-v)$ are zero under all three operations other than E. Table 1.2 can be updated (Table 1.2.1) as follows:

Using the magic formula equation (1.11) as illustrated for the case of CCl_4 above, $\Gamma_{(\text{tot})}(132) = \frac{1}{12}$ $[132A + 132E + 396F] = 11A + 11E + 33F$. We cross-check that since there are 44 atoms in the primitive

---

TABLE 1.2.1

Obtaining m_j Values and the Characters of Vibrational Motion for the Various Symmetry Operations

T	E	$4C_3$	$4C_3^2$	$3C_2$	Activity
A	1	1	1	1	Raman
E	1	ϵ	ϵ^*	1	Raman
	1	ϵ^*	ϵ	1	
F	3	0	0	−1	Raman, IR
m_j	44	6	6	0	
Total #groups $m_j(s)$	12	0	0	0	
Polyatomic $m_j(s-v)$	8	0	0	0	
$\chi_j(\text{acoustic}) = 2\cos\varphi_j \pm 1$	3	0	0	−1	
$\chi_j(\text{tot}) = m_j(2\cos\varphi_j \pm 1)$	132	0	0	0	
$\chi_j(\text{trans}) = m_j(s)(2\cos\varphi_j \pm 1)$	36	0	0	0	
$\chi_j(\text{rot}) = m_j(s-v)(\pm 2\cos\varphi_j + 1)$	24	0	0	0	

cell, we expect to get $3 \times 44 = 132$ degrees of freedom, which tallies with what we have obtained. At the Brillouin zone center, these degrees of freedom are distributed in 11 non-degenerate, 11 double degenerate, and 33 triply degenerate phonon modes. Thus, $11 + 11 + 33 = 55$ phonon modes are expected, including an acoustic mode that is triply degenerate: $\Gamma_{(acous)}(3) = \frac{1}{12}$ [(3–3)A + (3–3)E + (9+3)F] = F. All 54 optical modes are Raman active, and the F modes are also IR active. $\Gamma_{(trans)}(36) = \frac{1}{12}$ [36A + 36E + 108F] = 3A + 3E + 9F (including acoustic mode F); $\Gamma_{(rot)}(24) = \frac{1}{12}$ [24A + 24E + 72F] = 2A + 2E + 6T [29].

The structure of $Zr(WO_4)_2$ has turned out to be simple enough to be amenable to an unambiguous determination of the various m_j values. However, this is not always the case, and it may be difficult to determine the m_j values for structures with a large number of atoms. In such cases, the Tables of Adams and Newton [23] could be employed. We will do this exercise for $Zr(WO_4)_2$ and a few other crystal structures.

1.1.6   FACTOR GROUP ANALYSIS USING ADAMS AND NEWTON'S TABLES

1.1.6.1   Example 1: $Zr(WO_4)_2$

As mentioned earlier, we need to know the Wyckoff sites of all atoms to apply this method. Fractional coordinates for all atoms in $Zr(WO_4)_2$ have been reported, among others, in [28]; Wyckoff positions for the various atoms can be obtained from this information using International Tables for Crystallography [30] or its implementation in the Bilbao crystallographic website [31]. We find that Zr, W(1), W(2), O(3) and (O_4) atoms occupy the 4a site, and O(1) and O(2) occupy the 12b site. For space group #198, Adams and Newton's (AN) Table (Table 1.3) is as follows:

Translational modes: To determine Γ_{Trans}, the contributions of the sites of the center of gravity of the polyatomic and monoatomic species are summed (Table 1.3.1):

This Γ_{Trans} includes translations of Zr and WO_4 species, and also the acoustic modes.

Rotatory modes: WO_4 rotations are accounted for by the center of gravity of these tetrahedra, i.e., W ions (Table 1.3.2):

Tables for rotatory modes are given separately in the same set of Tables and can, in general, be different, but for space group 198, the corresponding characters are the same.

From the Character Tables for the point group T (Table 1.3 and 1.3.1), it is seen that $\Gamma_{Acoustic} = F$, since the acoustic vibrations transform under the symmetry operations of the point group as the translations of T_x, T_y, and T_z. We have obtained all these IRR without going through the task of determining the various m_j values, etc.

TABLE 1.3
Total Irreducible Representation of ZrW_2O_8
Obtained from Adams and Newton's Tables

#198	T	A	E	F
Atom	**Wyckoff Site**			
Zr	4A	1	1	3
O(1)	12B	3	3	9
O(2)	12B	3	3	9
W(1)	4A	1	1	3
W(2)	4A	1	1	3
O(3)	4A	1	1	3
O(4)	4A	1	1	3
Γ_{Total}	$\Sigma =$	11A	11E	33F

TABLE 1.3.1

Translational Modes of ZrW_2O_8 Obtained from Adams and Newton's Table

#198	T	A	E	F
Atom	Wyckoff Site			
Zr	4A	1	1	3
W(1)	4A	1	1	3
W(2)	4A	1	1	3
Γ_{Trans}	$\Sigma=$	3A	3E	9F

TABLE 1.3.2

Rotatory Modes of ZrW_2O_8 Obtained from Adams and Newton's Table

#198 Rotatory	T	A	E	F
Atom	Wyckoff Site			
W(1)	4A	1	1	3
W(2)	4A	1	1	3
Γ_{Rot}	$\Sigma=$	2A	2E	6F

TABLE 1.4

Total Irreducible Representation of $NaZr_2(PO_4)_3$ Obtained from Adams and Newton's Table

#167	D_{3d}	A_{1g}	A_{2g}	E_g	A_{1u}	A_{2u}	E_u
Atom	Wyckoff Site						
6Na	2B	0	0	0	1	1	2
12Zr	4C	1	1	2	1	1	2
18P	6E	1	2	3	1	2	3
36O(1), 36(O$_2$)	$2\times12F$	6	6	12	6	6	12
Γ_{Total}	$\Sigma=$	$8A_{1g}$	$9A_{2g}$	$17E_g$	$9A_{1u}$	$10A_{2u}$	$19E_u$

1.1.6.2   Example 2: $NaZr_2(PO_4)_3$

Next, we will do this exercise for the crystal structure of $NaZr_2(PO_4)_3$, a rhombohedral structure with space group $R\bar{3}c$ (D_{3d}^6), #167. From [32], we find that 6 Na atoms occupy the 6 b sites, 12 Zr atoms occupy the 12 c sites, 18 P atoms occupy the 18 e sites, 36 O(1) and 36 (O$_2$) atoms occupy the 2×36 f sites. We write down the AN Table (Table 1.4) for space group 167:

From the character table for D_{3d} point group, it is seen that $\Gamma_{Acoustic} = A_{2u} + E_u$; A_{1g} and E_g modes are Raman active, A_{2u} and E_u are IR active, A_{2g} and A_{1u} are optically inactive. 25 Raman bands and 27 IR bands are expected from this structure.

Translational modes Γ_{Trans} arise from the contributions of the translations of Na and Zr atoms and the center of gravity of the PO$_4$ tetrahedra (Table 1.4.1):

Rotatory modes: PO$_4$ rotations are accounted for by the center of gravity of these tetrahedra, i.e., P ions (Table 1.4.2):

TABLE 1.4.1

Translational Modes of $NaZr_2(PO_4)_3$ Obtained from Adams and Newton's Table

#167	D_{3d}	A_{1g}	A_{2g}	E_g	A_{1u}	A_{2u}	E_u
Atom	Wyckoff Site						
6Na	2B	0	0	0	1	1	2
12Zr	4C	1	1	2	1	1	2
18P	6E	1	2	3	1	2	3
Γ_{Trans}	$\Sigma=$	$2A_{1g}$	$3A_{2g}$	$5E_g$	$3A_{1u}$	$4A_{2u}$	$7E_u$

TABLE 1.4.2

Rotational Modes of $NaZr_2(PO_4)_3$ Obtained from Adams and Newton's Table

#167 Rotatory		A_{1g}	A_{2g}	E_g	A_{1u}	A_{2u}	E_u
Atom	Wyckoff Site						
18P	6E	1	2	3	1	2	3
Γ_{Total}	$\Sigma=$	A_{1g}	$2A_{2g}$	$3E_g$	$1A_{1u}$	$2A_{2u}$	$3E_u$

TABLE 1.5

Total Irreducible Representation of ZrO_2 Obtained from Adams and Newton's Table for the Pbca Space Group

#61	D_{2h}	A_g	B_{1g}	B_{2g}	B_{3g}	A_u	B_{1u}	B_{2u}	B_{3u}
Atom	Wyckoff Site								
Zr, O_1, O_2	8C	3	3	3	3	3	3	3	3
Γ_{Total}	$\Sigma=3\times$contrib.	$9A_g$	$9A_{2g}$	$9B_{2g}$	$9B_{3g}$	$9A_u$	$9B_{1u}$	$9B_{2u}$	$9B_{3u}$

All these IRRs tally with what is obtained from the Halford–Hornig site group method reported in [33].

1.1.6.3   Example 3: ZrO_2

The structure of ZrO_2 is monoclinic under ambient conditions. Above 4 GPa, it transforms into an orthorhombic structure called ortho-I. There was some ambiguity about the space group of this ortho-I structure, whether its space group is *Pbca* or *Pbcm*. The issue was resolved by us using spectroscopic methods in favor of *Pbcm* some years ago [34].

In *Pbca* (D_{2h}^{15}, #61, $Z=8$), all atoms (Zr, O_1 and O_2) sit at the 8c Wyckoff site [34]. The corresponding AN table is shown as Table 1.5 is

However, AN tables do not work for systems with the disorder: In the *Pbcm* structure (D_{2h}^{11}, #57, $Z=4$), 4 Zr atoms occupy the 4d Wyckoff sites, 4 O_2 atoms sit at 4c sites, and 4 O_1 atoms randomly occupy four of the 8e sites with equal probability [35]. AN table (Table 1.6) gives total IRR

which is incorrect. The correct IRR obtained by the original BV method is $\Gamma_{Total}=3A_g+5B_{1g}+5B_{2g}+5B_{3g}+3A_u+5B_{1u}+5B_{2u}+5B_{3u}$, including acoustic modes $B_{1u}+B_{2u}+B_{3u}$ [34].

1.1.6.4   Example 4: $Al_2(WO_4)_3$

At ambient conditions, $Al_2(WO_4)_3$ has an orthorhombic structure (space group *Pnca*, D_{2h}^{15}, #60, $Z=4$). 1 W_1 atoms occupy the Wyckoff 4c position, and 1 W_2 atom, 1 Al atoms, and 6 O atoms (i.e., a total of 8 atomic species) occupy the 8d site [36]. AN table (Table 1.7) gives total IRR

TABLE 1.6

Total Irreducible Representation of ZrO_2 Obtained from Adams and Newton's Table for the Disordered *Pbcm* Space Group Is Incorrect

#57	D_{2h}	A_g	B_{1g}	B_{2g}	B_{3g}	A_u	B_{1u}	B_{2u}	B_{3u}
Atom	Wyckoff Site								
4 O_2	4C	1	2	2	1	1	2	2	1
4 Zr	4D	2	2	1	1	1	1	2	2
4 O_1	8E	3	3	3	3	3	3	3	3
Γ_{Total}	Σ	$6A_g$	$7B_{1g}$	$6B_{2g}$	$5B_{3g}$	$5A_u$	$6B_{1u}$	$7B_{2u}$	$6B_{3u}$

TABLE 1.7

Total Irreducible Representation of $Al_2(WO_4)_3$ Obtained from Adams and Newton's Table

#60	D_{2h}	A_g	B_{1g}	B_{2g}	B_{3g}	A_u	B_{1u}	B_{2u}	B_{3u}
Atom	Wyckoff Site								
1 W_1	4C	1	2	1	2	1	2	1	2
1 W_2, 1 Al, 6 O	8D	24	24	24	24	24	24	24	24
Γ_{Total}	$\Sigma =$	$25A_g$	$26B_{1g}$	$25B_{2g}$	$26B_{3g}$	$25A_u$	$26B_{1u}$	$25B_{2u}$	$26B_{3u}$

TABLE 1.8

Total Irreducible Representation of $P\bar{4}3m$ Space Group of $Zn(CN)_2$ Obtained from Adams and Newton's Table

#215	T_d	A_1	A_2	E	F_1	F_2
Atom	Wyckoff Site					
2×Zn	2×(1A−B)	0	0	0	0	2
2×C/N	2×4E	2	0	2	2	4
Γ_{Total}	$\Sigma =$	$2A_1$	0	$2E$	$2F_1$	$6F_2$

which tallies with that reported in [37] using the original BV method.

1.1.6.5 Example 5: $Zn(CN)_2$

There was ambiguity in the space group of the ambient structure of $Zn(CN)_2$, whether it is an ordered structure or a disordered one. This can be resolved using spectroscopic methods using factor group analysis [17,18] and experimental Raman and IR spectra [6]. In the candidate ordered structure with space group $P\bar{4}3m$ (T_d^1, #215), Zn(1) atoms occupy Wyckoff 'a' sites, Zn(2) atoms are at Wyckoff 'b' sites, C and N atoms sit at distinct 4e sites. AN table (Table 1.8) gives total IRR, which tallies with the result obtained by BV and correlation methods [6].

In the other candidate, disordered structure with space group $Pn\bar{3}m$ (O_h^4, #224), Zn atoms occupy the Wyckoff 2a sites, and C and N atoms randomly occupy the 8e sites with a probability of 0.5 each. AN table (Table 1.9) gives total IRR

TABLE 1.9

Total Irreducible Representation of $Pn\bar{3}m$ Space Group of $Zn(CN)_2$ Obtained from Adams and Newton's Table

#224	O_h	A_{1g}	A_{2g}	E_g	F_{1g}	F_{2g}	A_{1u}	A_{2u}	E_u	F_{1u}	F_{2u}
Atom	Wyckoff Site										
Zn	A	0	0	0	0	1	0	0	0	1	0
C, N	8E	1	0	1	1	2	0	1	1	2	1
Γ_{Total}	$\Sigma=$	A_{1g}	0	E_g	F_{1g}	$3F_{2g}$	0	A_{2u}	E_u	$3F_{1u}$	F_{2u}

which tallies with the result in [6] obtained by the correlation method. It should be remarked that the correct IRR in [6] for this disordered structure was obtained after much deliberation, and we have obtained it in a straightforward manner using the AN Tables.

It is seen from the above examples that correct phonon irreducible representation can be obtained in most cases in a straightforward manner by using the AN Tables. There could be some exceptions in the case of disordered structures.

1.1.7 SPECIAL TECHNIQUES IN RAMAN SPECTROSCOPY

1.1.7.1 Resonance Raman Spectroscopy

When the exciting laser energy is close to an electronic energy gap in the material under investigation, resonant scattering of the laser photons occurs, accompanied by a correspondingly more significant number of Raman shifted photons.

If the laser frequency is denoted by ν_0 and the vibrational energy in question is ν_{mn}, the intensity of the Raman band corresponding to ν_{mn} will be [9,38]

$$I_{mn} = \text{Constant } I_0 \left(\nu_0 - \nu_{mn}\right)^4 \Sigma_{p\sigma} \left|\left(\alpha_{p\sigma}\right)_{mn}\right|^2 \tag{1.16}$$

where m and n denote the initial and final states, respectively, of the electronic ground state. I_0 is the incident laser intensity, and $\left(\nu_0 - \nu_{mn}\right)^4$ denotes the well-known ν^4 rule of Rayleigh scattering. $\left(\alpha_{p\sigma}\right)_{mn}$ represents the change in polarizability caused by the $m \rightarrow e \rightarrow n$ transition, where 'e' is an electronic energy level, ρ and σ are x, y and z components of the polarizability tensor. $\left(\alpha_{p\sigma}\right)_{mn}$ can be written as

$$\left(\alpha_{p\sigma}\right)_{mn} = \frac{1}{h} \Sigma_e \left(\frac{M_{me}M_{en}}{\nu_{em} - \nu_0 + i_e}\right) + \left(\frac{M_{me}M_{en}}{\nu_{en} + \nu_0 + i_e}\right) \tag{1.17}$$

where ν_{em} and ν_{en} are the frequencies corresponding to the energy differences between the subscribed states, and h is Planck's constant. The electronic transition moments such as M_{me} are given by

$$M_{me} = \int \psi_m^* \mu_\sigma \psi_e d\tau \tag{1.18}$$

where ψ_m and ψ_e are the wavefunctions of the corresponding states, and μ_σ is the σ component of the electric dipole moment. Γ_e is the bandwidth of the eth state, and $i\Gamma_e$ is a damping constant. The laser excitation frequency ν_0 is normally considerably less than the frequency ν_{em} corresponding to the electronic energy gap, $\nu_0 \ll \nu_{em}$. Then the Raman intensity is proportional to $(\nu_0 - \nu_{em})^4$. When ν_0 approaches ν_{em}, the denominator of the first term in equation (1.17) becomes very small, and this "resonance term" becomes so large that the intensity of the Raman band at $\nu_0 - \nu_{em}$ increases enormously. This phenomenon is called Resonance Raman Scattering (RRS).

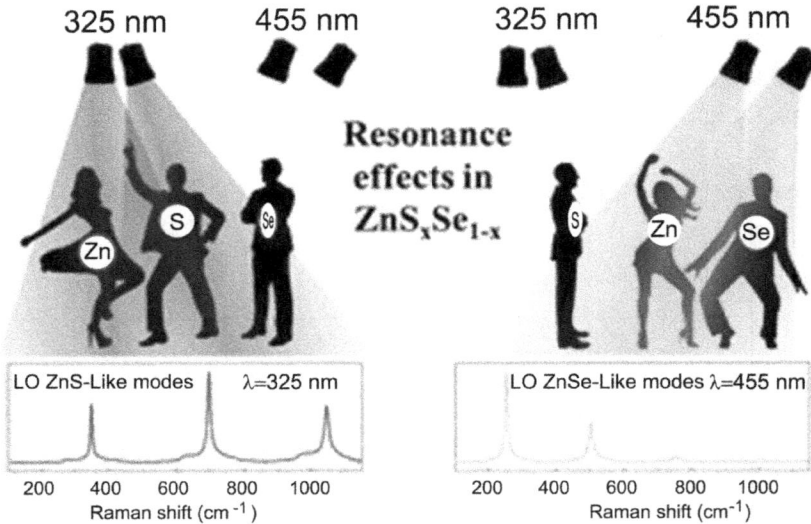

FIGURE 1.15 Enhancement in the Raman mode intensities of ZnSSe compounds under various resonant conditions. (Copyright © 2016 Royal Society of Chemistry [39].)

RRS has been observed in bulk solids as well as in low-dimensional materials, though it is highly popular in low-dimensional systems. A comprehensive Raman resonance scattering study of ZnS_xSe_{1-x}(ZnSSe) solid solutions over the whole compositional range ($0 \le x \le 1$) was carried out by Dimitrievska et al. [39] using 325 and 455 nm excitation wavelengths. The Raman scattering intensities of LO ZnS-like and ZnSe-like phonon modes, corresponding to pure S and Se vibrations, respectively, are revealed to be significantly enhanced when excited with 325 nm excitation in the case of S vibrations, and with 455 nm in the case of Se vibrations (Figure 1.15). This behavior is explained by the interaction of the excitation photons with the corresponding S or Se electronic states in the conduction band.

One of the most exciting examples of RRS is in single-walled carbon nanotubes (SWNT). Nanotubes are 1D systems with unique sets of electronic energy levels that arise because of the quantum confinement of electrons. We will briefly review the electronic DOS for bulk (3D), 2D, 1D, and 0D solids in this context.

1.1.7.1.1 Electronic Density of States in Crystals

To understand the effect of resonance Raman excitation on low-dimensional materials, we like to discuss the free electrons in 3D, 2D, and 1D with examples from bulk and 1D materials.

1.1.7.1.2 Free Electron Gas in Three Dimensions

Free particle Schrodinger equation in 3D can be written as [40]

$$-\frac{\hbar^2}{2m}\left(\frac{\partial^2}{\partial x^2}+\frac{\partial^2}{\partial y^2}+\frac{\partial^2}{\partial z^2}\right)\psi_k(\mathbf{r})=\in_k \psi_k(\mathbf{r}) \qquad (1.19)$$

If electrons are confined inside a cube of edge L, then the ψ can be envisaged as a standing wave with the solution:

$$\psi_n(\mathbf{r})=A\sin\left(\frac{n_x\pi x}{L}\right)\sin\left(\frac{n_y\pi y}{L}\right)\sin\left(\frac{n_z\pi z}{L}\right)$$

If the solution forms a standing wave inside the cube, then it is a stable solution. As the wave function is reflected from the boundary walls of the cube, the following conditions should be satisfied:

(i) the wave function should disappear outside the boundaries of the cube, and (ii) the standing wave's magnitude and phase should match at the boundary. These conditions imply that we can use periodic boundary conditions to describe the solutions at the boundary walls, and hence we can write:

$$\psi\left(x + L, \, y, z\right) = \psi\left(x, \, y, z\right), \text{ etc.}$$

Then $\psi_n\left(\boldsymbol{r}\right) = \exp\left(i\boldsymbol{k} \cdot \boldsymbol{r}\right)$, with $k_x = 0, \pm 2\pi/L, \pm 4\pi/L$, etc.

When we substitute $\psi_n\left(\boldsymbol{r}\right) = \exp\left(i\boldsymbol{k} \bullet \boldsymbol{r}\right)$ in equation (1.19), we get

$$\in_k = \frac{\hbar^2 \boldsymbol{k}^2}{2m} = \frac{\hbar^2}{2m}\left(k_x^2 + k_y^2 + k_z^2\right), \text{ where } k = 2\pi / \lambda$$

For a system of N free electrons in the ground state, the occupied states can be depicted as points in a sphere in the reciprocal, or $\boldsymbol{k}$ space, and the highest energy, which is the surface of the sphere, is called the Fermi energy expressed as $\in_F = \dfrac{\hbar^2 \boldsymbol{k}_F^2}{2m}$. Since $k_x = 0, \pm 2\pi/L, \pm 4\pi/L$, etc., there is one allowed wave vector per $(2\pi/L)^3$ volume element of $\boldsymbol{k}$ space. Hence, in a sphere of volume $\dfrac{4}{3}\pi \boldsymbol{k}_F^3$, the number of states can be obtained by

$$N = 2 \bullet \frac{\dfrac{4}{3}\pi \boldsymbol{k}_F^3}{\left(\dfrac{2\pi}{L}\right)^3} = \frac{V}{3\pi^2} k_F^3 \tag{1.20}$$

This means $k_F = \left(\dfrac{3\pi^2 N}{V}\right)^{1/3}$ depends only on N. The factor 2 in equation (1.20) comes from the two allowed values of spin quantum number m_s for each electron for each $\mathbf{k}$. Since $k_F^2 = \dfrac{2m \in_F}{\hbar^2}$, it follows that the Fermi energy can be obtained by writing $\dfrac{2m \in_F}{\hbar^2} = \left(\dfrac{3\pi^2 N}{V}\right)^{\frac{2}{3}}$, so that $\in_F = \dfrac{\hbar^2}{2m}\left(\dfrac{3\pi^2 N}{V}\right)^{\frac{2}{3}}$.

Knowing the Fermi energy, we can determine the number of states per unit energy range, $D(\in)$, called the density of states. The total numbers of states of energy $<\in$ are $N = \dfrac{V}{3\pi^2}\left(\dfrac{2m \in}{\hbar^2}\right)^{\frac{3}{2}}$, and

$$D(\in) = \frac{dN}{d \in} = \frac{V}{2\pi^2}\left(\frac{2m}{\hbar^2}\right)^{\frac{3}{2}} \in^{\frac{1}{2}}; \text{ i.e., } \in \propto D(\in)^2 \text{ (Figure 1.16)}.$$

1.1.7.1.3   Free Electron Gas in 2D

In analogy with the 3D case (equation 1.20) discussed above, when the electrons are free to move only in two dimensions and restricted in the third dimension, N can be written as $N = \dfrac{2\pi k_F^2}{\left(\dfrac{2\pi}{L}\right)^2} = \dfrac{L^2}{2\pi} k_F^2$

; so, $k_F^2 = \dfrac{2\pi N}{L^2}$; or, $\dfrac{2m \in_F}{\hbar^2} = \dfrac{2\pi N}{L^2}$; so that $\in_F = \dfrac{\hbar^2 \pi N}{mL^2}$. Hence the total number of states varies with

the energy as, $N(\in) = \dfrac{mL^2 \in}{\hbar^2 \pi}$; and the DOS is obtained as $D(\in) = \dfrac{dN}{d \in} = \dfrac{mL^2}{\hbar^2 \pi}$. That is, the DOS

in 2D is independent of energy. Also, the DOS per unit area is a constant, $D(\in) = \dfrac{1}{L^2}\dfrac{dN}{d \in} = \dfrac{m}{\hbar^2 \pi}$.

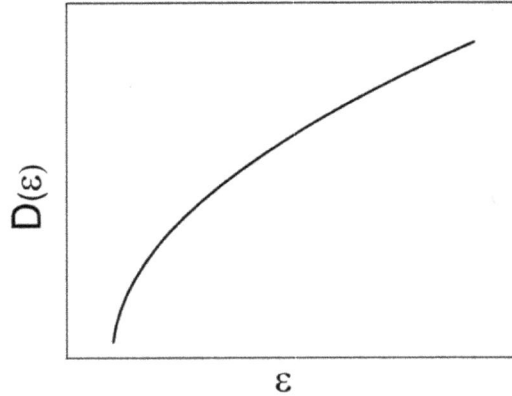

FIGURE 1.16   Free electron density of states of a bulk solid.

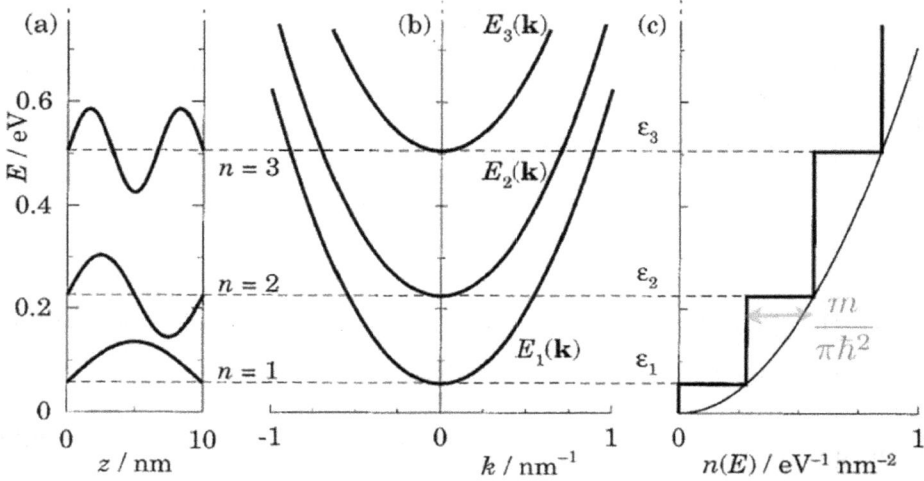

FIGURE 1.17   (a) Wavefunctions of electrons in an infinitely deep square well. (b) The corresponding energy subbands. (c) Step-like DOS of a 2D system superposed on the parabolic DOS of a 3D system of electrons. (From ref. [42].)

This description is true only for the ideal, infinitesimally thin 2D system with an infinite energy gap. In reality, however, a 2D system has a finite thickness, and the energy gap between the different bands is finite. The number of the carrier states in a conduction band is written with respect to the conduction band edge energy ϵ_n as [41], $N(\epsilon) = \sum_{n \text{ (filled states)}} \dfrac{mL^2(\epsilon - \epsilon_n)}{\hbar^2 \pi}$, and the DOS is obtained as $\dfrac{dN}{d\epsilon} = \dfrac{mL^2}{\hbar^2 \pi} \sum_n \theta(\epsilon - \epsilon_n)$, where $\theta = (\epsilon - \epsilon_n) = \epsilon_n$ for integer values of n, and zero otherwise (Figure 1.17).

Examples of such systems are thin-film quantum well structures that are quasi-2D (Figure 1.18). The effects of quantum confinement are seen when the quantum well thickness becomes comparable to the de Broglie wavelength of electrons, leading to discrete energy levels called energy subbands. These subbands evolve from the energy levels of an electron in an infinite potential well,

Quantum Well

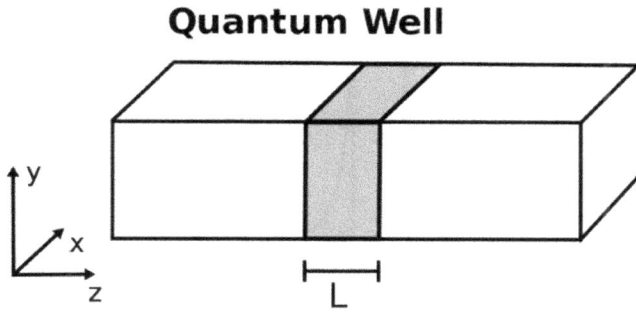

FIGURE 1.18 Quantum well structure; $L \sim 10\text{–}100$ nm. (From ref. [43].)

when a large (of the order of Avogadro number) of electrons are brought together in a quantum well structure.

1.1.7.1.4 Free Electron Gas in 1D

When the electrons are confined in two dimensions and free to move only in one direction, like in a carbon nanotube, we get a quasi-1D system. The number of states, in this case,

is, $= N = 2 \cdot \dfrac{k}{\frac{2\pi}{L}} = \dfrac{kL}{\pi}$. Since $k = \sqrt{\dfrac{2m \in_F}{\hbar^2}}$, we can write $N(\in) = \dfrac{L}{\pi\hbar}\sqrt{2m \in}$; and the DOS,

$\dfrac{dN}{d\in} = \dfrac{L}{\pi\hbar}\sqrt{2m} \cdot \dfrac{1}{2\sqrt{\in}} = \dfrac{L}{2\pi\hbar}\sqrt{\dfrac{2m}{\in}}$. So that we get $\dfrac{1}{L}\dfrac{dN}{d\in} = \dfrac{1}{2\pi\hbar}\sqrt{\dfrac{m}{\in}} \cdot \sqrt{2} = \dfrac{1}{\pi\hbar}\sqrt{\dfrac{m}{2\in}}$. In anal-

ogy with the 2D case above, we can write $D(\in) = \dfrac{1}{\pi\hbar}\sqrt{\dfrac{m}{2(\in - \in_n)}}$ for an ideal cylinder of infinite

length and an infinitesimally small radius, or $D(\in) = \sum_n \dfrac{1}{\pi\hbar}\sqrt{\dfrac{m}{2(\in - \in_n)}}\theta(\in - \in_n)$, for a real 1D

material with a finite radius (Figure 1.19).

The sharp peaks in the DOS are called Van Hove singularities.

Being a 1D system, the DOS of SWNT exhibits such singularities (Figure 1.20). Nanotubes are specified by the chiral vector $C_h = na_1 + ma_2 = (n, m)$ in the hexagonal honeycomb lattice contained in the vector C_h. Thus, the notation (n, m) indicates the atomic coordinates for the 1D unit cell of the nanotube [44]. For $n \neq m \neq 0$, the tube has chiral symmetry. "Zigzag" and "armchair" nanotubes have $(n, 0)$ and (n, n), respectively, and they are achiral (Figure 1.21).

When a laser of certain energy, such as 2.4 eV is incident on a sample of SWNTs with different diameters and chiralities, we obtain spectra (Figure 1.22) due to resonance of this laser excitation with several SWNTs with matching the energy separations in their DOS (Figure 1.23).

1.1.7.2 UV Raman Spectroscopy

We discussed earlier in this chapter some of the advantages of UV Raman spectroscopy, mainly to avoid fluorescence. UV Raman spectroscopy is ideally suited to the analysis of carbon phases containing both sp^2 and sp^3 bonding that also fluoresce. When 514 nm excitation is used, the scattering cross section for sp^2 carbon is considerably larger than that of sp^3 carbon due to a resonance enhancement [47]. Thus, even if small amounts of sp^2-bonded carbon are present in the sample, it can be difficult to observe sp^3 bonded carbon under visible laser excitation. The scattering cross sections for sp^2 and sp^3 carbon are nearly equal when UV excitation is used instead of visible excitation.

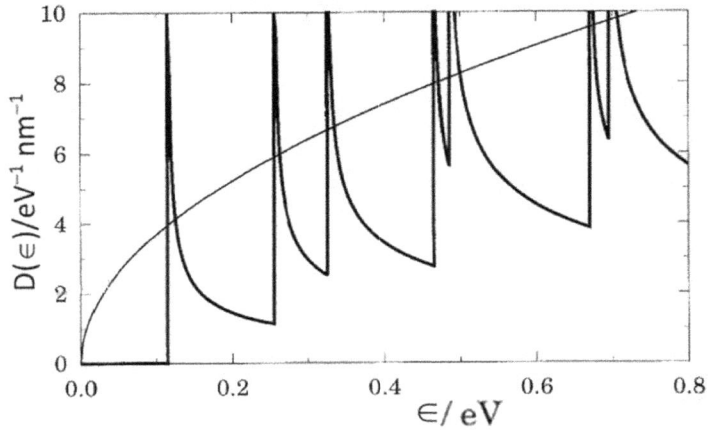

FIGURE 1.19   DOS for unconfined electrons in 3D (parabola) and 1D (with sharp peaks). (From ref. [42].)

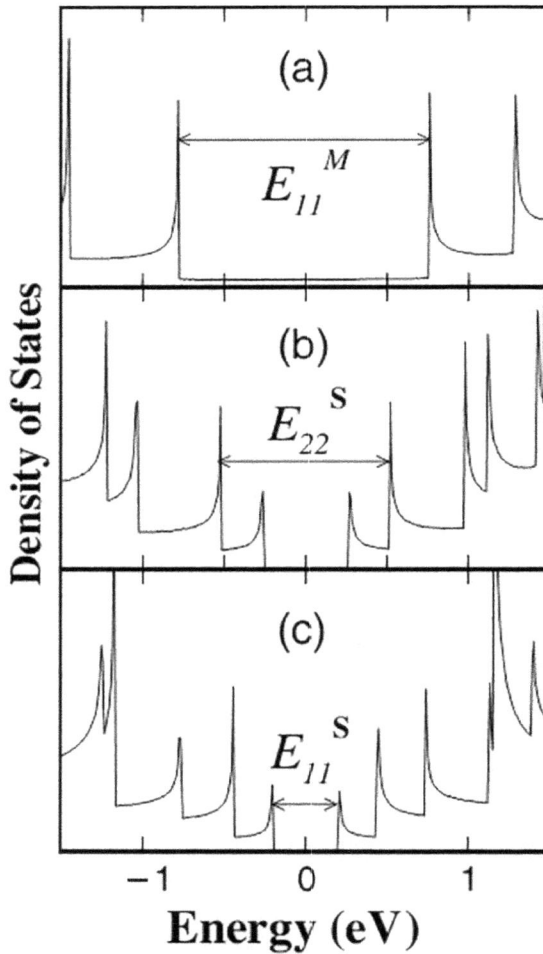

FIGURE 1.20   DOS for an (a) armchair (10,10) SWNT, (b) chiral (11,9) SWNT, and (c) zigzag (22,0) SWNT obtained with the tight-binding model. (From ref. [44]. © Deutsche Physikalische Gesellschaft. Reproduced by permission of IOP Publishing. All rights reserved.)

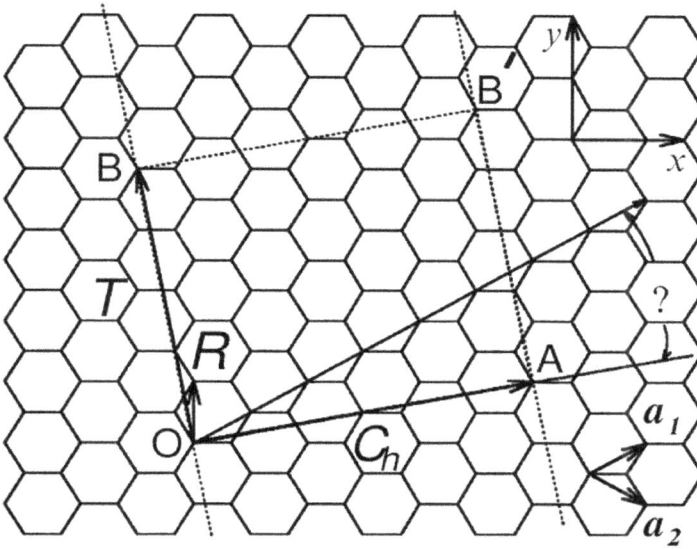

FIGURE 1.21   The unrolled honeycomb lattice of a nanotube. When we connect lattice sites O and A, and sites B and B′, a nanotube can be constructed. $\overrightarrow{OA}$ and $\overrightarrow{OB}$ define the chiral vector $\mathbf{C}_h$ and the translational vector $\mathbf{T}$ of the nanotube, respectively. The rectangle OABB′ defines the unit cell for the nanotube. The figure is constructed for an $(n, m) = (4, 2)$ nanotube. (From ref. [45], reproduced with permission from World Scientific Publishing Company.)

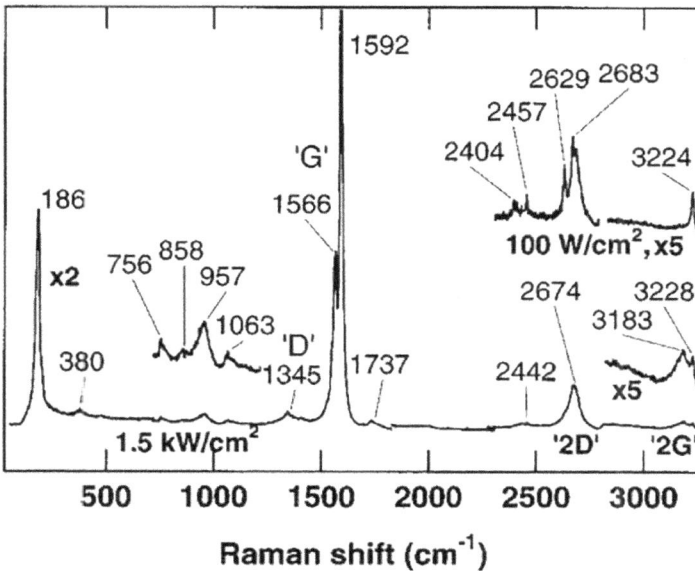

FIGURE 1.22   Raman spectrum of SWNT with 2.4 eV excitation. Features in the vicinity of "G" peaks are typical of semiconducting tubes. (Reprinted with permission from ref. [46]. Copyright © 2001 American Chemical Society.)

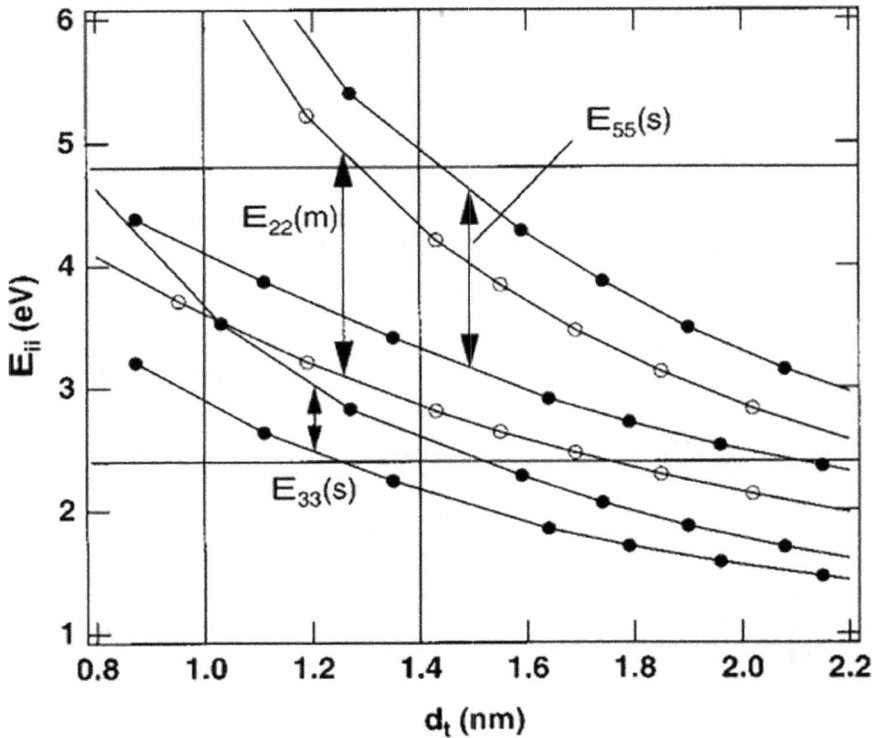

FIGURE 1.23 Energy separations ΔE_{ii} for semiconducting (closed circles) and metallic (open circles) nanotubes of various diameters d_t. Only the extremes of each energy band (populated by zigzag tubes) are shown. Values between the vertical lines correspond to the diameters of SWNTs in the present sample. Horizontal lines are drawn at the laser excitation energies 2.4 and 4.8 eV. The ΔE_{22} band of metallic nanotubes overlaps with the ΔE_{55} semiconducting band. (Reprinted with permission from ref. [46]. Copyright © 2001 American Chemical Society.)

For example, a C_{60} sample non-hydrostatically compressed to 30 GPa and retrieved to ambient conditions, when excited by 514.5 nm laser exhibited Raman spectra that were featureless due to large background fluorescence. However, deep UV (257 nm) laser excitation of this sample showed a strong Raman band at 1,336 cm^{-1} (Figure 1.24) which is characteristic of the A_{1g} mode of diamond [48]. The observation was only the second report of the synthesis of the diamond from carbon at high pressure and ambient temperature.

Other special techniques, such as SERS and its variant, tip-enhanced Raman spectroscopy (TERS), are dealt with in Chapters 5 and 6, respectively, in this book.

1.2 ACOUSTIC PHONON CONFINEMENT

Phonons are collective excitations in a periodic and elastic arrangement of atoms (or molecules). A phonon is a quasiparticle arising from the quantization of vibrational modes in solids. Phonons can be envisaged as quantized sound waves analogous to photons for the quantized light waves.

A coherent movement of lattice atoms from their equilibrium positions is named acoustic phonons. For long wavelengths, the frequency of acoustic phonon, showing a linear relationship with phonon wavevector, is close to zero. An anomaly in the measured specific heat of metal nanoparticles [49,50] led to a theoretical study of lattice dynamics of small particles [51,52]. In this regard, the vibrational modes of a homogeneous elastic sphere were studied by Lamb [53]. Two types of

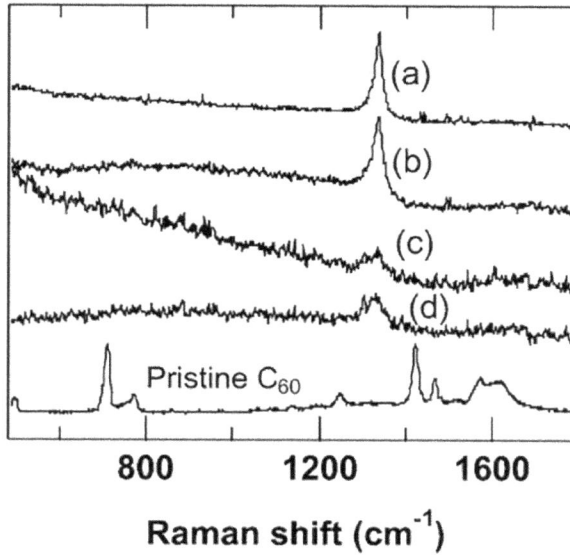

FIGURE 1.24 UV Raman spectra of C_60 uniaxially compressed to 30 GPa and retrieved to ambient conditions, at different points (a, b, c, and d) on the sample, exhibiting a diamond peak. Raman spectrum of pristine C_60 is included for comparison. (Reprinted with permission from ref. [48]. Copyright © 2002 Elsevier Publishers.)

vibrational modes, spheroidal and torsional, were found from the stress-free boundary condition of a spherical shape. Tamura et al. [52] invoked the Lamb theory to understand the anomaly in the specific heat measurement for metal nanoparticles, considering the effects of matrix, porosity, and shape. Duval et al. [54] reported the first observation of a very low-frequency vibrational mode, corresponding to the eigenmode for small spinel crystallites using Raman spectroscopy. This study indicated the possibility that acoustic phonon confinement of small particles could contribute to the excess amount of specific heat observed in small metal particles [49,50].

1.2.1 LAMB'S EQUATION AND ITS SOLUTIONS

Low-frequency Raman modes from a small crystallite originate from the elastic vibrations of the crystallite. Lamb [53] derived the equation of motion of a 3D elastic body under stress-free boundary conditions as

$$\rho \bullet \partial^2 D / \partial t^2 = (\lambda + \mu)\nabla(\nabla \bullet D) + \mu\nabla^3 \bullet D \qquad (1.21)$$

where D is the displacement and λ and μ are Lame's constants. Equation (1.21) can be solved by a scalar and a vector potential. The scalar potential solution of the wave equation is

$$\phi_s \propto Z_l(\mathrm{hr})P_l^m(\cos\theta)_{\sin m\varphi}^{\cos m\varphi} \ \exp(-i\omega t) \qquad (1.22)$$

where Z_l is a spherical Bessel function and $h = \omega/c$ (c being the velocity of light). The displacement derived from ϕ_s is $D_s = \nabla\phi_s$.

The vector potential is set as $A = (r\,\psi_v, 0, 0)$:

$$\psi_v \propto Z_l(\mathrm{kr})P_l^m(\cos\theta)_{\sin m\varphi}^{\cos m\varphi} \ \exp(-i\omega t) \qquad (1.23)$$

For $k=\omega/v_t$, there are two kinds of displacement:

$$\boldsymbol{D}_{v1} = \nabla \times \boldsymbol{A} \text{ and } \boldsymbol{D}_{v2} = \nabla \times \nabla \times \boldsymbol{A} = \boldsymbol{D}_{v1} = \nabla \times \boldsymbol{D}_{v1}$$

Lamb obtained two types of modes at the spherical surface under stress-free boundary conditions. The spheroidal modes are calculated from the condition of the disappearance of the stress derived for $\boldsymbol{D}_s$ and $\boldsymbol{D}_{v2}$ at $r=R$ (radius of the crystallite).

The eigenvalue is derived as

$$
\begin{aligned}
&[\eta^2 + (l-1)(l+2)\{\eta j_{l+1}(\eta)/j_l(\eta) - (l+1)\}]\zeta j_{l+1}(\zeta)/j_l(\zeta) - \tfrac{1}{2}\eta^4 \\
&+ (l-1)(2l+1)\eta^2 + \{\eta^2 - 2l(l-1)(l+2)\}\eta j_{l+1}(\eta) j_l(\eta) = 0
\end{aligned}
\tag{1.24}
$$

where $\xi = hR = \omega R/v_l$ and $\eta = kR = \omega R/v_t$.

Equation (1.24) can be solved by setting the longitudinal (v_l) and transverse (v_t) wave velocity, with $j_l(\eta)$ being the first kind of Bessel function.

The torsional mode can be derived by considering the stresses deduced from $\boldsymbol{D}_{v1}$ resulting in the eigenvalue equation

$$j_{l+1}(\eta) - (l-1) j_l(\eta)/\eta = 0 \quad (l \geq 1) \tag{1.25}$$

As discussed earlier, the displacement vectors of the two types of vibrational modes of torsional, $\boldsymbol{D}_t$, and spheroidal, $\boldsymbol{D}_s$ can be explained further by one scalar, ϕ_0 and two vector potentials, ϕ_1 and ϕ_2. $\boldsymbol{D}_t = \nabla \times \phi_1$ and $\boldsymbol{D}_{v2} = \nabla \phi_0 + \nabla \times \nabla \times \phi_2$, where $\phi_i = (r\phi_l, 0, 0)$; ϕ_0, ϕ_1, and ϕ_2 are scalar functions proportional to spherical harmonics $Y_l^m(\theta, \phi)$ [52,53]. $Y_l^m(\theta, \phi)$ is even for even l values and odd for odd values of l. Therefore spheroidal mode having $l=0, 2, 4, \ldots$ and torsional modes having $l=1, 3, 5, \ldots$ can be Raman active [55]. However, Duval [56] argued that only spheroidal modes are allowed: unlike spheroidal mode, the eigenvalues η_{nl}^T do not depend on the material as the torsional mode has null displacement. In practice, however, the torsional modes are weak.

1.2.2 Acoustic Modes and Determination of Size and Shape

Low-frequency Raman studies can be performed to determine the size of clusters. Confined surface acoustic phonons in metallic or semiconductor nanoclusters give rise to low-frequency modes in the vibrational spectra of the materials. The Raman active spheroidal motions are associated with dilation and strongly depend on the cluster material through v_t, the transverse and v_l, the longitudinal sound velocities. These modes are characterized by two indices l and n, where l is the angular momentum quantum number and n is the branch number; $n=0$ represents surface modes. The surface quadrupolar mode $l=2$, eigenfrequency η_2^s appears in Raman scattering geometries, whereas the surface spherical mode $l=0$, eigenfrequency ξ_0^s appears only in the polarized geometry. The surface quadrupolar mode frequencies corresponding to $l=0$, and 2 are given by $\omega_0^s = \xi_0^s v_l/Rc$ and $\omega_2^s = \eta_2^s v_t/Rc$.

With the experimentally observed frequencies of the Raman modes and the known sound velocities in the materials under study along the transverse and longitudinal directions and eigenfrequencies derived from equation (1.24), one can find the size of the particle. The shape factor and the matrix effect can also be estimated by calculating the eigenfrequencies under different boundary conditions as prescribed by Tamura et al. [52] and later studied extensively by Dhara et al. [57,58]. The depolarization ratio corresponding to $l=2$ mode (ratio of intensities in the polarized and unpolarized conditions) close to 0.75 is predicted for the spherical clusters [59]. A large deviation from the estimated value indicates the elongated shape (elliptical or rod shape) of the particle.

1.2.3 RADIAL BREATHING MODE (RBM)

Radial breathing mode (RBM) is extensively studied for single-walled carbon nanotube (SWNT). The phonon dispersion relations for an SWNT can be determined by folding a graphene layer and discussed in Section 1.1.7.1.4 in some detail. In the case of carbon nanotube (CNT), $6N$ phonon dispersion branches for the 3D vibrations of atoms can be realized as there are $2N$ carbon atoms in the unit cell. The corresponding 1D phonon energy dispersion relation for the CNT is given by

$$\omega^{\mu\lambda}(q) = \omega_{\text{graphite}}^{\lambda}\left(q\frac{\boldsymbol{K}_2}{|\boldsymbol{K}_2|} + \mu\boldsymbol{K}_1\right) \tag{1.26}$$

where $\lambda = 1, \ldots 6$ represents the polarization, $\mu = 0, \ldots N-1$ denotes the azimuthal quantum numbers, and $-\pi/T < q \leq \pi/T$ limits the phonon wavevectors. However, the zone-folding method does not always give the correct dispersion relation for a CNT (rolled graphene), especially in the low-frequency region. For example, the out-of-plane tangential acoustic modes of a graphene sheet do not give zero energy at the Brillouin zone center ($q=0$) of a CNT. At the zone center, all the carbon atoms of the CNT move radially in and out-of-plane radial acoustic vibration, which corresponds to a breathing mode (RBM) with a non-zero frequency [60]. To avoid these difficulties, one can directly diagonalize the dynamical matrix. Fundamental phonon polarizations in CNTs are radial, transverse, and longitudinal. Zone center phonons, also referred to as Γ-point phonons, can belong to the various acoustic and optical modes along with the RBM. The longitudinal optical (LO) phonon branch near the Γ-point has an energy of ≈ 190 meV. On the contrary, the energy of the RBM phonon branch (≈ 28 meV/d_{CNT}) is inversely proportional to the diameter of CNT (d_{CNT}).

Radial vibrational mode is reported in the ultrathin-walled anatase TiO_2 nanotube of 2–5 nm wall thickness [61]. Large surface energy in the finite-size nanotubes energetically favored only the $B_{1g}(\nu_4)$ mode having unidirectional molecular vibrations in the parallel configuration out of all other allowed Raman modes with molecular vibration normal to the radial modes. The electron-phonon coupling in the resonance Raman spectroscopy of ultrathin-wall nanotubes might be the origin of the observed radial vibrational mode. Observed enhanced amount of specific heat with increasing temperatures in these nanotubes compared to that reported for nanoparticles of diameter (2–5 nm) similar to the nanotube wall thickness was associated with the presence of the prominent radial mode along with other energetic phonons [61].

1.2.4 BOSONIC MODE (SPIN WAVE)

Spin waves are low-energy (few meV) collective excitations observed in magnetic lattices having continuous symmetry. In the quasiparticle picture, spin waves, which are bosonic modes of the spin lattice that correspond to the phonon excitations of the nuclear lattice, are also called magnons. Spin waves are experimentally observed by a few methods (i) inelastic neutron scattering, (ii) inelastic light scattering techniques of Raman and Brillouin scattering, inelastic X-ray scattering, and (iii) inelastic electron scattering of spin-resolved electron energy loss spectroscopy, and ferromagnetic (spin-wave) resonance. In the first methodology, the energy loss of a beam of neutrons exciting a magnon is measured as a function of scattering vector, temperature, and external magnetic field. As in the case of phonons, inelastic neutron scattering measurements can determine the dispersion curve for magnons. Raman and Brillouin scattering measure the energy loss of photons for the magnetic material. On the contrary, ferromagnetic resonance measures the absorption of incident microwaves by spin waves, as a function of angle, temperature, and applied field. Spin-polarized electron energy loss spectroscopy can demonstrate very high energy surface magnons allowing one to probe the dispersion of magnons in the ultrathin ferromagnetic films.

The simplest way to understand spin wave is to consider a 1D spin ½ Heisenberg system:

$$H_0 = J \sum_j S_j \cdot S_{j+1} \tag{1.27}$$

where J (>0) is the exchange interaction between neighboring spins (S). The Tomonaga–Luttinger (TL) liquid with spinon excitation helps in explaining the low-energy physics of this system [62]. No Raman peak is expected without additional perturbation, as the Hamiltonian H_0 and the corresponding Raman operator commute with each other. A trivial perturbation v_0 owing to spin-lattice coupling leads to the bond dimerization [63] and is represented as

$$v_0 = \sum_j J (-1)^j u S_j \cdot S_{j+1} \tag{1.28}$$

where u is the distortion. In the presence of the perturbation, the effective Hamiltonian H' suits an exactly solvable Sine-Gordon (SG) model:

$$H' = H_0 + \int \frac{dx}{a_0} u d \sin\left(\sqrt{2\pi}\varphi\right) \tag{1.29}$$

where $x = ja_0$, a_0 is the lattice spacing, and φ is the canonical part of Bosonic fields.

There are three kinds of spin excitations; soliton (S; spinon), antisoliton (S'; antispinon), and breather modes (B_n), which are the soliton-antisoliton (spinon-antispinon) bound states. The nth breather's excitation, $E_n = 2E_s \sin[n\pi/(8/K-2)]$ for $n = 1, ..., [4/K-1]$. In the SU(2)-symmetry, the soliton mass, E_s (soliton, antisoliton, and the first breather forming a spin-triplet), and the singlet second breather mass, E_2 can be calculated as

$$E_s \sim 1.5 u^{2/3} J, \text{ and } E_2 = \sqrt{3} E_s \tag{1.30}$$

In a recent study, a new collective mode in the low-frequency Raman spectra of insulating phases of VO_2 has been reported. The reported frequency is the singlet spin excitation of breather mode about a spin-Peierls dimerized 1D spin-½ Heisenberg chain and is phenomenologically consistent with the superexchange coupling strength between V spin ½ moments in the insulating phases of VO_2 [64,65]. The observation of spin excitations resulting from the 1D Heisenberg spin-½ chain finally resolved the long-standing debate whether structural phase transition (SPT) or metal-insulator transition (MIT) occurs first? In the case of VO_2, the shift in the frequency of spin wave and simultaneous increase in the transition temperature in the absence of any structural change confirm that SPT does not prompt MIT in VO_2. At the same time, the presence of a spin wave confirms the perturbation due to spin-Peierls dimerization leading to SPT.

REFERENCES

1. W. Abney, E. R. Festing, *Philos. Trans. Roy. Soc.* **172**, 887–918 (1881).
2. W. W. Coblentz, *Investigations of Infrared Spectra Part 1*. Publication No. 35, Carnegie Institute of Washington, Washington (1905).
3. T. Rojalin, L. Kurki, T. Laaksonen, T. Viital, J. Kostamovaara, K. C. Gordon, L. Galvis, S. Wachsmann-Hogiu, C. J. Strachan, M. Yliperttula, *Anal. Bioanal. Chem.* **408**, 761 (2016).
4. American Chemical Society International Historic Chemical Landmarks, *The Raman Effect*, http://www.acs.org/content/acs/en/education/whatischemistry/landmarks/ramaneffect.html (accessed April 30, 2021).
5. C. V. Raman, K. S. Krishnan, *Nature* **121**, 3048 (1928).
6. T. R. Ravindran, A. K. Arora, T. N. Sairam, *J. Raman Spectrosc.* **38**, 283 (2007).
7. W. G. Fateley, F. R. Dollish, N. T. McDevitt, F. F. Bentley, *Infrared and Raman Selection Rules for Molecular and Lattice Vibrations*, Wiley-Interscience, New York (1972).
8. G. Turrell, *Infrared and Raman Spectra of Crystals*, Academic Press, London (1972).

9. J. R. Ferraro, K. Nakamoto, C. W. Brown, *Introductory Raman Spectroscopy*, 2nd ed., Elsevier, San Diego, USA (2003).
10. *Infrared and Raman Spectroscopy: Methods and Applications*, edited by B. Schrader, VCH Publishers, Weinheim, Federal Republic of Germany (1995).
11. T. R. Ravindran, A. K. Arora, *J. Raman Spectrosc.* **42**, 885 (2011).
12. T. R. Ravindran, J. V. Badding, *J. Mater. Sci.* **41**, 7145 (2006).
13. *Roger Crosby's Internet Web page*, http://www.chm.bris.ac.uk/webprojects1997/RogerEC/welcome.htm.
14. P. R. Bartholomew, *Geostand. Geoanal. Res.* **37**, 353 (2012).
15. http://www.fen.bilkent.edu.tr/~gulseren/phys545/pdf/phonons-chain.pdf.
16. K. Wu, E. Torun, H. Sahin, B. Chen, X. Fan, A. Pant, D. P. Wright, T. Aoki, F. M. Peeters, E. Soignard, S. Tongay, *Nature Commun.* **7**, 12952 (2016).
17. S. Bhagavanta, T. Venkatarayudu, *Proc. Indian Acad. Sci.* **9A**, 224 (1939).
18. S. Bhagavantam, T. Venkatarayudu, *Theory of Groups and Its Application to Physical Problems*, 2nd ed., Bangalore Press, Bangalore (1951).
19. R. S. Halford, *J. Chem. Phys.* **14**, 8–15 (1946).
20. H. Winston, R. S. Halford, *J. Chem. Phys.* **17**, 607–616 (1949).
21. D. F. Hornig, *J. Chem. Phys.* **16**, 1063–1076 (1948).
22. D. M. Adams, D. C. Newton, *J. Chem Soc. (A)* **2822**, (1970); https://doi.org/10.1039/J19700002822.
23. D. M. Adams, D. C. Newton, *Tables for Factor Group Analysis*, Beckman-RIIC Ltd, Croydon (1970).
24. J. B. Williams, C. S. Trevisan, M. S. Schoffler, T. Jahnke, I. Bocharova, H. Kim, B. Ulrich, R. Wallauer, F. Sturm, T. N. Rescigno, A. Belkacem, R. Dorner, Th. Weber, C. W. McCurdy, A. L. Landers, *J. Phys. B: At. Mol. Opt. Phys.* **45**, 194003 (2012).
25. T. Chakraborty, A. L. Verma, *Spectrochimica Acta A* **58**, 1013 (2002).
26. T. Chakraborty, S. N. Rai, *Spectrochimica Acta A* **62**, 438 (2005).
27. T. R. Ravindran, A. K. Arora, T. A. Mary, *Phys. Rev. Lett.* **84**, 3879 (2000).
28. J. D. Jorgensen, Z. Hu, S. Teslic, D. N. Argyriou, S. Short, J. S. O. Evans, A. W. Sleight, *Phys. Rev. B* **59**, 215 (1999).
29. T. R. Ravindran, A. K. Arora, T. A. Mary, *J. Physics: Condens. Matter* **13**, 11573 (2001).
30. *International Tables for Crystallography, Vol. A: Space-Group Symmetry*, 5th ed., Kluwer AcademicPress, Dordrecht, The Netherlands (2002), http://it.iucr.org/.
31. M. I. Aroyo, J. M. Perez-Mato, C. Capillas, E. Kroumova, S. Ivantchev, G. Madariaga, A. Kirov, H. Wondratschek, *Zeitschriftfuer Kristallographie* **221**, 15–27 (2006).
32. L. O. Hagman, P. Kierkegaard, *Acta Chem. Scand.* **22**, 1822–1826 (1968).
33. K. Kamali, T. R. Ravindran, C. Ravi, Y. Sorb, N. Subramanian, A. K. Arora, *Phys. Rev. B* **86**, 144301 (2012).
34. T. R. Ravindran, K. Yadav, *Eur. Phys. J B* **88**, 24 (2015).
35. H. Arashi, M. Ishigame, *Phys. Stat. Sol. A* **71**, 313 (1982).
36. P. Villars (Chief Editor), *Pauling File In: Inorganic Solid Phases*, Springer Materials (Online Database), Springer, Heidelberg. SpringerMaterials $Al_2(WO_4)_3$ ($Al_2[WO_4]_3$) TaP Crystal Structure https://materials.springer.com/isp/crystallographic/docs/sd_1503954sd_1503954 (Springer-Verlag GmbH, Heidelberg, © 2016).
37. T. R. Ravindran, V. Sivasubramanian, A. K. Arora, *J. Phys. Condens. Matter* **17**, 1 (2005).
38. *Biological Applications of Raman Spectroscopy*, edited by T. G. Spiro, **Vols. 1–3**, John Wiley, NewYork (1987–88).
39. M. Dimitrievska, H. Xie, A. J. Jackson, X. Fontane, M. E. Rodriguez, E. Saucedo, A. P. Rodriguez, A. Walsh, V. I. Roca, *Phys. Chem. Chem. Phys.* **18**, 7632 (2016).
40. *Introduction to Solid State Physics*, 7th ed., edited by C. Kittel, John Wiley & Sons, Inc (1996).
41. https://alan.ece.gatech.edu/ECE6451/Lectures/StudentLectures/King_Notes_Density_of_States_2D1D0D.pdf.
42. http://www.ehu.eus/nezabala/Irakaskuntza/Low%20dimensional%20systems_files/2.ELECTRONIC%20STATES%20AND%20QUANTUM%20CONFINES%20SYSTEMS%20.pdf.
43. https://commons.wikimedia.org/wiki/File:Quantum_well.svg.
44. A. Jorio, M. A. Pimenta, A. G. S. Filho, R. Saito, G. Dresselhaus, M. S. Dresselhaus, *New J. Phys.* **5** 139 (2003).
45. R. Saito, G. Dresselhaus, M. S. Dresselhaus, *Physical Properties of Carbon Nanotubes*, Imperial College Press, London (1998).
46. T. R. Ravindran, B. R. Jackson, J. V. Badding, *Chem. Mater.* **13**, 4187 (2001).

47. V. I. Merkulov, J. S. Lannin, C. H. Munro, S. A. Asher, V. S. Veerasamy, W. I. Milne, *Phys. Rev. Lett.* **78**, 4869 (1997).
48. T. R. Ravindran, J. V. Badding, *Solid State Commun.* **121**, 391 (2002).
49. V. Novotny, P. P. M., Meincke, J. H. P. Watson, *Phys. Rev. Lett.* **28**, 901 (1972).
50. V. Novotny, P. P. M., Meincke, *Phys. Rev. B* **12**, 2520 (1975).
51. R. Lautenschlager, *Soild State Commun.* **16**, 1331 (1975).
52. A. Tamura, K. Higeta, T. Ichinokawa, *J. Phys. Chem. C* **15**, 4957 (1982).
53. H. Lamb, *Proc. London Math. Soc.* **13**, 187 (1882).
54. E. Duval, A. Boukenter, B. Champagnon, *Phys. Rev. Letter.* **5**, 2052 (1986).
55. M. Fujii, T. Nagareda, S. Hayashi, K. Yamamoto, *Phys. Rev. B* **44**, 6243 (1991).
56. E. Duval, *Phys. Rev. B* **46**, 5795 (1992).
57. S. Dhara, R. Kesavamoorthy, P. Magudapathy, M. Premila, B. K. Panigrahi, K. G.M. Nair, C. T. Wu, K. H. Chen, L. C. Chen, *Chem. Phys. Lett.* **370**, 254 (2003).
58. S. Dhara, B. Sundaravel, T. R. Ravindran, K. G.M. Nair, C. David, B. K. Panigrahi, P. Magudapathy, K. H. Chen, *Chem. Phys. Lett.* **399**, 354 (2004).
59. B. Palpant, H. Portales, L. Saviot, J. Lerme, B. Prevel, M. Pellarin, E. Duval, A. Perez, M. Broyer, *Phy. Rev. B* **60**, 17107 (1999).
60. R. A. Jishi, L. Venkataraman, M. S. Dresselhaus, G. Dresselhaus, *Chem. Phys. Lett.* **209**, 77 (1993).
61. R. P. Antony, A. Dasgupta, S. Mahana, D. Topwal, T. Mathews, S. Dhara, *J. Raman Spectrosc.* **46**, 231 (2015).
62. T. Giamarchi, *Quantum Physics in One Dimension*, Oxford University Press, New York (2004).
63. M. Sato, H. Katsura, N. Nagaosa, *Phys. Rev. Lett.* **108**, 237401 (2012).
64. R. Basu, A. Patsha, S. Chandra, S. Amirthapandian, Raghavendra K. G., A. Dasgupta, S. Dhara, *J. Phys. Chem. C* **123**, 11189 (2019).
65. R. Basu, V. Srihari, M. Sardar, S. Srivastava, S. Bera, S. Dhara, *Sci. Rep.* **10**, 1977 (2020).

2 Plasmonic and Optical Confinement

Binaya Kumar Sahu, Pratap K. Sahoo, and Sandip Dhara

CONTENTS

2.1 INTRODUCTION TO SURFACE PLASMON RESONANCE

The quanta assigned for collective oscillation of conduction electrons in a metal is coined as a plasmon [1–3]. In the metal, the free electron cloud can be imagined with a positive lattice core. The incident electromagnetic field repels the electron cloud in its positive cycle, resulting in a movement of these clouds toward the surface. However, as the electron is confined to the metal, pushing away these electrons cloud from the positive core result in an electric dipole in a particular direction. However, the positive core will try to pull back the electron cloud to its equilibrium condition, and hence a restoring force is originated. Upon removal of the incident field, the dipole continues to oscillate with a certain frequency due to the restoring force [4–6]. Importantly, for a certain frequency of the incident field, the conduction electrons oscillate collectively to form the resonance. The quanta of such oscillation are known as surface plasmon (SP), and the resonance frequency is called surface plasmon resonance (SPR). The SPR frequency is given by the following equation [1]:

$$\omega_p = \sqrt{\frac{ne^2}{m_{\text{eff}}\varepsilon_0}} \tag{2.1}$$

In the above equation n, e, and m_{eff} are the density, charge, and effective mass of the electron. Whereas ε_0 stands for permittivity of the free space. Metals like Au, Ag, Cu, and Pt are in higher demand as their SPR frequency falls in the range of ultraviolet–visible (UV-Vis) range [3,6].

DOI: 10.1201/9781003248323-2

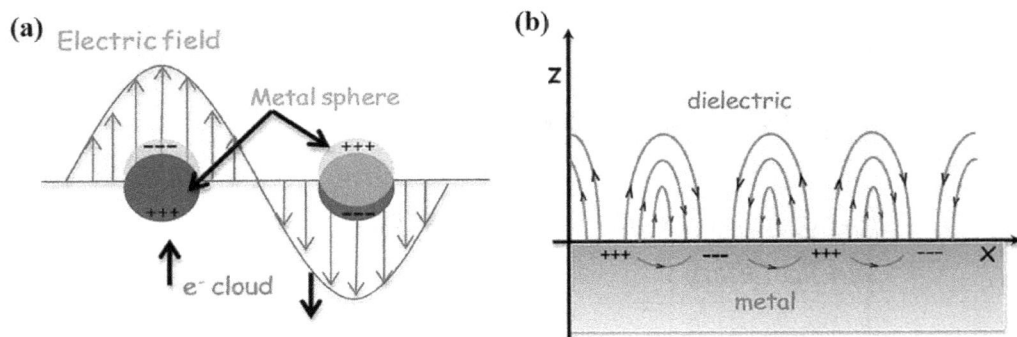

FIGURE 2.1 Schematic diagram for (a) LSPR and (b) PSPR. (J. Jana et al., *RSC Adv.* 2016, 6, 86174. Creative Commons [7].)

Further, depending on the dimensionality of the metal, SPR can be localized or propagating and classified into (Figure 2.1) [1,7]:

1. Localized surface plasmon resonance (LSPR)
2. Propagating surface plasmon resonance (PSPR)

2.1.1 LOCALIZED SURFACE PLASMON RESONANCE

The dimension of the metallic nanostructure is lower than the wavelength of the incident light or comparable to it for a sustainable LSPR. Such excitation causes collective but localized SP on the metallic nanostructure [3,8]. It may be noted that such oscillation happens for a specific wavelength, otherwise called a resonance condition. In return, a strong electromagnetic field is generated due to the strong confinement of the incident field. This feature of LSPR is utilized for applications like surface-enhanced Raman spectroscopy (SERS) [9,10]. A brief detail regarding the SERS study is presented in the later section of this chapter. Both the intensity and the position of SPR strongly depend on the size, shape, and composition of the nanostructures and the dielectric properties of the surrounding environment. The frequency of LSPR (ω_{LSPR}) and the dielectric value of the surroundings are correlated by the following equation [11]:

$$\omega_{\mathrm{LSPR}} = \frac{\omega_{\mathrm{p}}}{\sqrt{\left(1 + 2\varepsilon_{\mathrm{diel}}\right)}} \tag{2.2}$$

The above equation suggests a change in the LSPR frequency due to a change in the surrounding dielectric condition. This concept is also used for sensing the local environment by observing the shift in the LSPR peak position.

When the electromagnetic radiation incidents on the plasmonic nanoparticles, the photons can be absorbed by the metal surface and converted into heat and can be scattered in every direction. These processes can be quantified via the absorption and scattering cross section, the effective area over which the nanoparticle can absorb or scatter the photons of light falling [12–14]. Mie's theory effectively explains the relation between scattering cross section and absorption from the spherical plasmonic nanoparticle. This is derived by solving Maxwell's equation using the series expansion method for the spherical harmonics wave equation. For the particle of radius R and dielectric constant $\varepsilon(\omega) = \varepsilon_1(\omega) + i\varepsilon_2(\omega) = n_p^2$ (ε_1 and ε_2 are real and imaginary dielectric values, respectively) in the medium of permittivity $\varepsilon_m = n_m^2$ (n_m is the refractive index) and the wavelength of the incident light λ_0, the respective expression for scattering and extinction cross section are as follows [12]:

$$\sigma_{\text{sca}} = \frac{24\pi^3 V^2 \varepsilon_m^2}{\lambda_0^4} \frac{(\varepsilon_1 - \varepsilon_m)^2 + \varepsilon_2^2}{(\varepsilon_1 + 2\varepsilon_m)^2 + \varepsilon_2^2} \tag{2.3}$$

$$\sigma_{\text{abs}} = \frac{18\pi V \varepsilon_m^{\frac{3}{2}}}{\lambda_0} \frac{\varepsilon^2}{(\varepsilon_1 + 2\varepsilon_m)^2 + \varepsilon_2^2} \tag{2.4}$$

where V is the volume of the particle ($=4/3\pi R^3$). The resonance between the photons of incoming light and scattered light from nanoparticles occurs when $\varepsilon_1 = -2\varepsilon_m$ for small ε^2. Absorption is dependent on V, and scattering is on V^2.

In the case of metallic nanorods, as a solution to Maxwell equations, Mie-Gans predicted that for small ellipsoidal nanoparticles with dipole approximation, the SR mode would split into two distinct modes due to the surface curvature and geometry of the ellipsoid nanoparticles. These small rods have commonly been treated as ellipsoids when explaining the optical properties of these rod-shaped particles. Therefore, the Gans (or Mie-Gans) theory can be applied to describe the optical behavior of metallic nanorods, and according to the Gans formula, the excitation cross section C_{ext} for metallic nanorods can be calculated as follows [12,15]:

$$C_{\text{ext}} = \frac{2\pi V N \varepsilon_m^{3/2}}{3\lambda} \sum_j^3 \frac{(1/P_j^2)\varepsilon_i}{(\varepsilon_r + ((1-P_j)/P_j)\varepsilon_m)^2 + \varepsilon_i^2} \tag{2.5}$$

where P_j is the depolarized factor. The depolarized factor for the elongated particles may be described as follows [16,17]:

$$P_{\text{length}} = \frac{1-\xi^2}{\xi^2}\left[\frac{1}{2\xi}\ln\left(\frac{1+\xi}{1-\xi}\right) - 1\right] \tag{2.6}$$

$$P_{width} = \frac{1 - P_{length}}{2} \tag{2.7}$$

where ξ is the ellipticity, given by: $\xi^2 = 1-(\text{length/width})^{-2}$; the last term being the aspect ratio of the rod. The LSPR occurs when $\varepsilon_r = -[(1-P_j)/P_j]\varepsilon_m$, where $P_j = P_{\text{width}}$ for the transverse plasmon resonance. Absorption does not depend upon the absolute dimensions of the rod but only on the aspect ratio. Due to a strong dependency on the aspect ratio, nanorods are significantly more sensitive to the dielectric of the medium. Because of this property, nanorods are promising candidates for LSPR sensing.

Besides the anisotropic shape, the electromagnetic field was found to be high for non-uniform structures like nanotips, interparticle gaps, or particle-substrate nanogaps. Such points are known as 'hotspots'. The molecule that resides at the 'hotspot' experiences a strong electromagnetic field. These hotspots can have the following category [18]:

1. First-generation hotspot: Single nanoparticle with shell structure, nano tip, ridges or hinges form single nanostructure.
2. Second-generation hotspot: Coupled nanostructure with controlled interparticle nanogap.
3. Dimers or trimers are a few examples of such hotspots.
4. Third-generation hotspot: Hybrid structure of particle of plasmonic nanostructure and the substrate where hybridization of the electromagnetic field is formed between the scattering field from nanostructure and reflected field from the surface.

2.1.2 PROPAGATING SURFACE PLASMON RESONANCE

Propagating surface plasmons are raised from the oscillation of the conduction electron at the metal-dielectric surface upon excitation of the electromagnetic field [19,20]. Unlike LSP, PSP propagates on the interface before it decays. The suitable condition to excite PSP is given by the following equation [1]:

$$k_{spp} = \frac{\omega}{c} \sqrt{\frac{\varepsilon_{metal}\varepsilon_{diel}}{\varepsilon_{metal} + \varepsilon_{diel}}} \qquad (2.8)$$

However, the momentum of the oscillating charge wave is always greater than the incident photon. Hence, a sustainable PSP cannot be generated directly by indenting light. Hence, a special arrangement is required to gain extra momentum. Exciting light through a prism also known as the Kretschmann geometry, the use of a grating meets the required condition.

For the particular condition when the real part of the dielectric constant is negative, equation (2.8) will take the form [1]:

$$\omega_{SPP} = \frac{\omega_P}{\sqrt{(1 + \varepsilon_{diel})}} \qquad (2.9)$$

Due to the correlation of ω_{SPP} and ε_{diel}, as discussed in equation (2.9), PSP can be used in the sensor as a transducer. The change in the dielectric environment is likely to shift the PSP (equation 2.9) frequency and can be taken as sensor input.

2.1.3 RELAXATION PROCESS OF SURFACE PLASMON EXCITATIONS

Light-induced SP modes gradually undergo relaxation by (i) non-radiative or (ii) radiative mechanisms [21,22]. First of all, non-radiative decay of SP results in an electron-hole, electron-electron, or electron-phonon excitations. Interestingly, this non-radiative decay happens on different time scales and contributes to different physical phenomena. The first kind of excitation, i.e., electron-hole, is generated due to multiple photon absorption from the SP field and is of the scale of 100 fs. Hence, it introduces damping in the SP field, known as the Landau damping, which causes a non-thermalized electron distribution in the system (Figure 2.2) [22]. Subsequently, the adsorption of a sufficient amount of photon energy causes a large multiphoton emission. In this lineup, electron-electron excitation arises, also known as Auger transitions. This type of transition happens on a 100 fs to 1 ps scale and initiates a quasi-equilibrium in the system by the creation of a hot electron. Such hot electrons come down on a time scale of 1–10 ps by electron-phonon coupling by the Ohmic loss process.

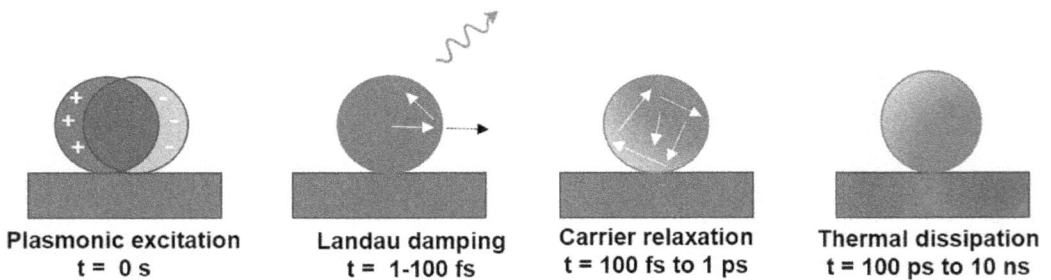

| Plasmonic excitation | Landau damping | Carrier relaxation | Thermal dissipation |
| t = 0 s | t = 1-100 fs | t = 100 fs to 1 ps | t = 100 ps to 10 ns |

FIGURE 2.2 Relaxation process of surface plasmon in different processes in different time scales.

During the process of hot-electron generation, intense light incident on the metal causes the excitation of the electron to a higher level above the Fermi level of metal by inter/intraband transition. Subsequently, these hot electrons can be directly injected into the conduction band close to the semiconductor. Hot-electron generation in the plasmonic metal and transfer of the same to the nearby system (mostly to a suitable semiconductor) regulates various plasmonic-based applications such as photocatalysis and photovoltaic [11,23,24].

2.2 APPLICATION OF SPR

2.2.1 PLASMON ENHANCED-OPTICAL SENSOR

2.2.1.1 Surface-Enhanced Raman Spectroscopy-Based Sensor

Surface-enhanced Raman spectroscopy demonstrates an enormous scientific validation owing to its capability of ultra-trace detection down to single-molecule level and has become a prodigious tool in the field of analytical chemistry. The primary mechanism of SERS is enhancing the weak inelastic Raman signal from the analyte molecule of interest, which is kept close to plasmonic nanostructures under suitable excitation. Raman signal, from the molecule, being the fingerprint of the analyte molecule, provide sufficient information even in ultra-low concentration during SERS. Mostly, SERS finds its origin due to (i) electromagnetic enhancement or (ii) charge transfer enhancement [18,25]. The former is the most accepted mechanism which is regulated by plasmonic interaction. Electromagnetic enhancement occurs due to the confinement of the electromagnetic field to suitable exciton on the metal nanostructure to the plasmonic property. The confined light, in return, boosts the Raman signal by (i) local field enhancement (light being incident in the system) and (ii) radiation enhancement (light being scattered back from the system). Because of the combined effect, a massive improvement in the Raman signal is probable. Hence obtained, the net enhancement (EF_{EM}) is represented in equation (2.10) [18,26]:

$$EF_{EM}\left(\omega_0, \omega_R, r_m\right) \approx \left|\frac{E_{Local}\left(\omega_0, r_m\right)}{E_0\left(\omega_0, r_m\right)}\right|^4 \tag{2.10}$$

The equation is well known as the E^4 approximation for the EF_{EM} of SERS. However, the finest enhancement is dependent on various parameters such as the morphology of plasmonic materials, nature of excitation, bonding nature of analytes with metals, and others, and is detailed in a dedicated Chapter 5. In a real-life scenario, SERS is becoming a popular tool in various fields, including environmental monitoring, health diagnosis, explosive detection, food safety, etc. [18,27]. The popularity of this technique is mainly attributed to a minimum or no sample preparation and a small amount of sample requirement. The probe material, i.e., a metallic nanostructure, can be used in colloidal form or on any solid chip based on the requirement in a SERS-based sensor. As an instance, for a few examples of environmental monitoring, explosive detection, food safety, and in vivo and in vitro determination of cancerous cell plasmonic probes can be injected through colloidal form and later on can be guided to the area of interest for detection [6,28]. In contrast, the other way of the signal collection can be done by drop-casting the analyte of interest on a prepared SERS substrate. Generally, in biological applications, certain molecular labels are used during SERS experiments. The function of these label molecules is to get attached to the area of interest [29], leading to an enhancement in the Raman signal from these labels to provide the required information. Additionally, SERS imaging is an important tool to give a spatial distribution of the spectra [30,31]. Chapter 5 exclusively deals SERS related phenomena.

2.2.1.2 Plasmonic Sensor Based on Kretschmann Configuration

This mode of sensing is closely associated with PSPR, which can be excited on a planar plasmonic substrate. Interestingly the PSPR is dependent on the dielectric constant or, in other words, refractive index of the medium as expressed in equation (2.9). This fact is utilized to monitor the change in the surrounding refractive index. As discussed in Section 2.1.2, additional momentum is necessary to sustain the PSPR, possibly by a prism or grating. This special geometry is known as the Kretschmann configuration (Figure 2.3a) [7]. This geometry demonstrates both angles as well as time-resolved measurement through a Glass prism. In this setup for a particular angle, the refractive index is minimum and known as the SPR angle. However, due to changes in the refractive index, the SPR angle changes and is engaged as the transducer for the sensor [32]. The wavelength of the SPR peak varies linearly with the refractive index of the surrounding medium according to the Drude model.

In a modified structure, the planar metal film is generally replaced with a large-area periodic nano-array pattern (Figure 2.3b), such as a nano-sphere array, nano-disk array, or nano-triangle array which supports both a stronger and propagating SPR.

2.2.1.3 Colloidal Particle-Based Plasmonic Sensor

Nanoparticles such as Au, Ag, and Cu exhibit shape- and size-dependent LSPR absorption and scattering bands, which have been utilized to construct plasmonic sensors. There are two types of plasmonic sensors based on the LSPR peak shift: (i) the LSPR peak wavelength shifts when an analyte binds to the surface of a nanoparticle, changing the local refractive index, and (ii) the plasmonic fields of multiple nanoparticles are coupled when an analyte brings the nanoparticles into proximity, causing a shift of the LSPR and thereby a color change. Among the various plasmonic nanoparticles, colloidal Au nanoparticles are the most commonly used plasmonic transducer because of their chemical stability and visual color change to the naked eye [33].

2.2.2 Plasmon-based Photocatalysis

During photocatalysis, absorbed photon creates active electrons and holes in a semiconductor to reduce the toxic chemicals mostly from the water medium [34]. This technique has a wide range of applications in various fields, including wastewater treatment, air purification, and water splitting. Two important factors that need to be improved for enhancing the photocatalytic properties are (i) delaying the recombination rate of electrons and holes and (ii) increasing the photon absorption

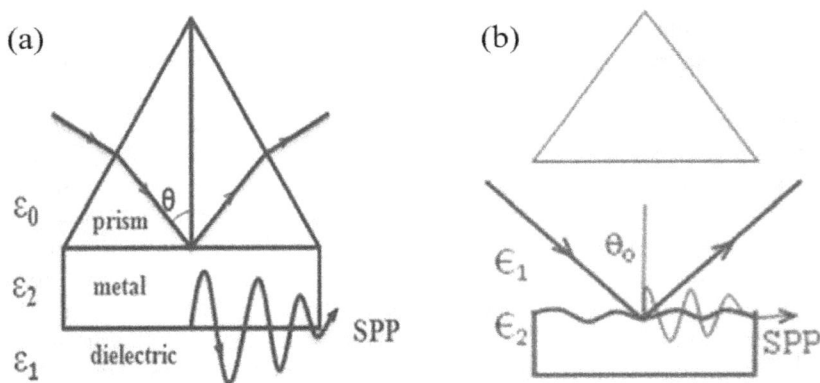

FIGURE 2.3 Schematic for (a) Kretschmann and (b) grating configuration. (J. Jana et al., *RSC Adv.* 2016, 6, 86174. Creative Commons [7].)

mostly in the visible range [11]. In a normal semiconductor, the excited electron undergoes a random walk and is prone to recombine with holes. Further, semiconductors like TiO_2 and SnO_2 have a wide bandgap that can only absorb UV light, and semiconductors like CdS and Fe_2O_3 do not maintain the photocatalytic property for a long time [35–38].

The inclusion of plasmonic material in photocatalysis demonstrates an excellent solution to solve these issues. It involves the dispersal of noble metal nanoparticles (mostly Au and Ag, in the sizes of tens to hundreds of nanometers) into semiconductor photocatalysts and obtains drastic enhancement of photoreactivity under the irradiation of UV and a broad range of visible light [11,39].

1. Schottky junction builds up a space charge region and forces both electron and hole to move in different directions to delay the recombination process.
2. Besides, the presence of metal provides a better medium for charge transfer.
3. The most brilliant benefits of plasmonic-based photocatalysis come from the LSPR of the plasmonic material. The resonance wavelength for Au and Ag nanoparticles can be tailored to fall in the visible range or the near-UV range, depending on the size, the shape, and the surrounding environment. Visible light absorption initiates photocatalysis in the presence of plasmonic nanostructures and semiconductors.
4. Finally, LSPR can create an intensive local electric field that favors the photocatalytic reaction by exciting an abundance of electrons and holes, increasing the heat generation, and increasing the redox reaction rate and mass transfer.

The following schematic (Figure 2.4) shows the role of Au in TiO_2 for plasmonic-based catalysis.

2.2.3 Plasmonic in Photovoltaic Application

The inclusion of plasmonic material in photovoltaic cells improves photon absorption and directly enhances the photon-to-electron efficiency. As described, SPR is nothing but the collective oscillations of conduction or free electrons in metal at the metallic and dielectric substrate [24,40].

a. Interestingly, the addition of plasmonic materials in the photovoltaic cell increases the far-field scattering and decreases the reflection to enhance the probability of absorption.
b. Another important factor is a near-field enhancement that happens due to the strong confinement of electromagnetic field at plasmonic substrate compared to the bare substrate. This confinement increases the absorption by many folds.
c. Besides, hot-electron injection from the plasmonic structure into a nearby semiconductor under suitable condition additionally increases the free carrier, which is further beneficial for photovoltaic applications.

2.2.4 Plasmonic Photothermal Application

Plasmonic photothermal therapy (PTT) involves the generation of heat from the plasmonic nanostructure after exposure to near-infrared light, which is then utilized for localized thermal therapy of cancerous tissues [10,41]. Conventionally, the therapeutic procedure is known as photothermal therapy or PTT. PTT is a minimally invasive therapeutic procedure where a selectively thermal dose is delivered to cancerous tissues by means of introducing apoptosis or directly through rapid necrosis as the local temperature increases beyond 42°C–47°C [42]. Further, simultaneous PTT along with conventional therapeutic protocols, like chemotherapy, radiotherapy, and surgery, augments the effective therapeutic benefits. Figure 2.5 schematically shows the working principle of PTT [43].

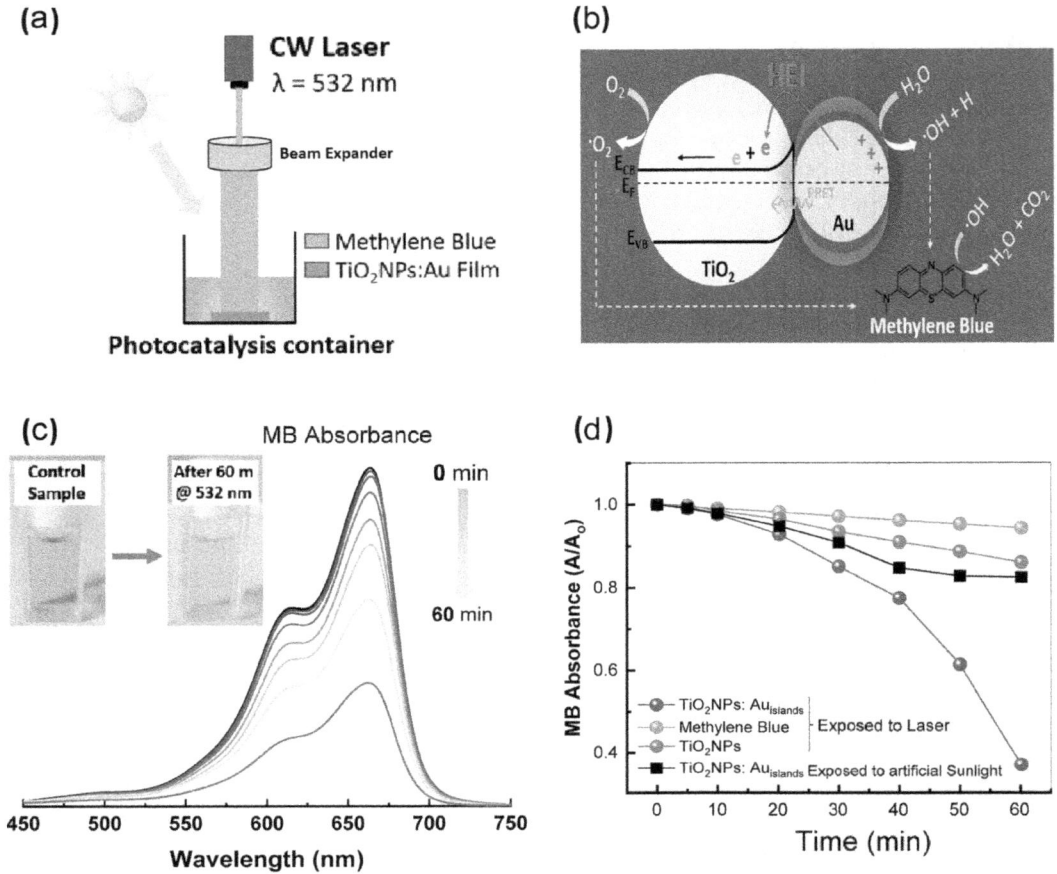

FIGURE 2.4 Schematic for (a) Photocatalysis setup of TiO$_2$:Au system, (b) model showing plasmonic absorption and hot-electron injection from Au to TiO$_2$, (c) methylene blue degradation from UV-Vis, and (d) degradation kinetics. (M. A. Ibrahem, *RSC Adv.* 2020, 10, 22324. Creative Commons [39].)

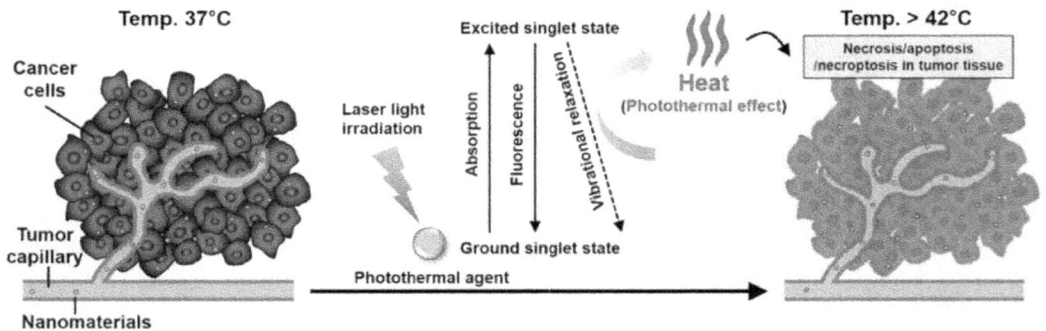

FIGURE 2.5 Schematic showing the mechanism of nanoparticle-mediated PTT within tumor microenvironment. Due to enhanced permeation and retention within the leaky tumor vasculature, nanoparticles accumulate within the solid tumors. The nanoparticles have strong absorbance in the NIR window and this leads to light-to-heat conversion, which then results in thermal ablation of the tumor cells (for temperature > 42°C). (H. S. Han, K. Y. Choi, *Biomedicines* 2021, 9, 305. Creative Commons [43].)

In PTT, photothermal agents, which are plasmonic nanostructures, are utilized for light-heat conversion. On irradiation by a photon source of a specific wavelength, the PTT agents absorb energy and undergo excitations from the ground singlet state to an excited singlet state, followed by non-radiative vibrational relaxation, thereby returning to the ground state by collisions between the excited PTT agents and the surrounding molecules [43]. This leads to an increase in the temperature of the tumor micro-environment [43]. For an efficient PTT, nanoparticles with a large absorption cross section within biological windows (700–1,400 nm) are beneficial, as the absorption and scattering of incident light by body tissue are significantly lower [44,45]. Apart from high photothermal conversion efficiency, excellent biocompatibility, good photostability, and solubility within a biocompatible liquid medium are essential characteristics of an efficient PTT agent. Of late, facile functionalization of various PTT agents has been performed to impart tumor targeting capability [46].

Generally, photothermal agents are divided into the following categories: (i) Au nanostructures, (ii) carbon-based nanomaterials (carbon nanotubes and graphene oxides), (iii) organic dyes, (iv) organic nanoparticles or polymeric nanoparticles, and (v) quantum dots, other inorganic nanoparticles, and metal oxide nanoparticles. Among these, Au nanostructures have been extensively explored as PTT agents due to the possibility of tunable absorbance spectra with size, chemical inertness, and reasonable biocompatibility [10,23]. The US Food and Drug Administration (FDA) approved two devices for PTT of high-grade glioma in the late 2000s, viz., Visualase Thermal Therapy (150 W, 980-nm laser) and NeuroBlate Laser Ablation System (12 W, 1,064-nm laser) [43]. Four clinical studies based on these PTT platforms were reported by Nanospectra Biosciences using 150 nm Au nano-shells. The Au nano-shells comprised of a 120 nm silica dielectric core, 15 nm Au shell, and a polyethylene glycol (PEG) layer for imparting colloidal stability. The clinical trials involving metastatic lung tumors and head-neck tumors were completed [47]. PTT is a comparatively modern therapeutic protocol that has shown immense potential due to the possibility of cancer therapy sans significant side effects. However, the feasibility of the treatment protocol in clinical settings needs further studies before practical adaptation.

2.3 CONCLUSION

The current chapter describes the basics of plasmon with their classification, which includes localized and propagating surface plasmons. Moving forward, different relaxation procedure of plasmon is discussed. Finally, various plasmonic applications are briefly studied. An important role of plasmonic in applications in various sensors, photocatalysis, solar cell, and PTT is elucidated. Indeed, plasmonic-based applications have come a long way not only in research but also in real-life applications. The concurrent theoretical understanding and advance in the experimental field open up a widespread possibility and research interest in the area of plasmonic.

REFERENCES

1. Li, M.; Cushing, S. K.; Wu, N., Plasmon-enhanced optical sensors: A review. *Analyst* 2015, *140* (2), 386–406.
2. Jain, P. K.; Eustis, S.; El-Sayed, M. A., Plasmon coupling in nanorod assemblies: Optical absorption, discrete dipole approximation simulation, and exciton-coupling model. *J. Phys. Chem. B* 2006, *110* (37), 18243–18253.
3. Stockman, M. I., Nanoplasmonics: Past, present, and glimpse into future. *Opt. Exp.* 2011, *19* (22), 22029–22106.
4. Orendorff, C. J.; Sau, T. K.; Murphy, C. J., Shape-dependent plasmon-resonant gold nanoparticles. *Small* 2006, *2* (5), 636–639.
5. Wang, X.; Huang, S.-C.; Hu, S.; Yan, S.; Ren, B., Fundamental understanding and applications of plasmon-enhanced Raman spectroscopy. *Nat. Rev. Phys.* 2020, *2*, 253–271.
6. Langer, J.; Jimenez de Aberasturi, D.; Aizpurua, J.; Alvarez-Puebla, R. A.; Auguié, B.; Baumberg, J. J.; Bazan, G. C.; Bell, S. E.; Boisen, A.; Brolo, A. G., Present and future of surface-enhanced Raman scattering. *ACS Nano* 2020, *14* (1), 28–117.

7. Jana, J.; Ganguly, M.; Pal, T., Enlightening surface plasmon resonance effect of metal nanoparticles for practical spectroscopic application. *RSC Adv.* 2016, *6* (89), 86174–86211.

8. Nehl, C. L.; Hafner, J. H., Shape-dependent plasmon resonances of gold nanoparticles. *J. Mater. Chem.* 2008, *18* (21), 2415–2419.

9. Sahu, B. K.; Dwivedi, A.; Pal, K. K.; Pandian, R.; Dhara, S.; Das, A., Optimized Au NRs for efficient SERS and SERRS performances with molecular and longitudinal surface plasmon resonance. *Appl. Surf. Sci.* 2021, *537*, 147615.

10. Pan, S.; Sahu, B. K.; Amirthapandian, S.; Dhara, S.; Das, A., Growing AuNRs in a single step with NIR plasmon for superior SERS and plasmonic photothermal performance. *J. Phys. Chem. Solids* 2022, *161*, 110421.

11. Zhang, X.; Chen, Y. L.; Liu, R.-S.; Tsai, D. P., Plasmonic photocatalysis. *Rep. Prog. Phys.* 2013, *76* (4), 046401.

12. Olson, J.; Dominguez-Medina, S.; Hoggard, A.; Wang, L.-Y.; Chang, W.-S.; Link, S., Optical characterization of single plasmonic nanoparticles. *Chem. Soc. Rev.* 2015, *44* (1), 40–57.

13. Smitha, S.; Gopchandran, K.; Ravindran, T.; Prasad, V., Gold nanorods with finely tunable longitudinal surface plasmon resonance as SERS substrates. *Nanotechnol* 2011, *22* (26), 265705.

14. Rycenga, M.; Cobley, C. M.; Zeng, J.; Li, W.; Moran, C. H.; Zhang, Q.; Qin, D.; Xia, Y., Controlling the synthesis and assembly of silver nanostructures for plasmonic applications. *Chem. Rev.* 2011, *111* (6), 3669–3712.

15. Ros, I.; Placido, T.; Amendola, V.; Marinzi, C.; Manfredi, N.; Comparelli, R.; Striccoli, M.; Agostiano, A.; Abbotto, A.; Pedron, D., SERS properties of gold nanorods at resonance with molecular, transverse, and longitudinal plasmon excitations. *Plasmonics* 2014, *9* (3), 581–593.

16. Link, S.; Mohamed, M.; El-Sayed, M. A., Simulation of the optical absorption spectra of gold nanorods as a function of their aspect ratio and the effect of the medium dielectric constant. *J. Phys. Chem. B* 1999, *103* (16), 3073–3077.

17. Kuwata, H.; Tamaru, H.; Esumi, K.; Miyano, K., Resonant light scattering from metal nanoparticles: Practical analysis beyond Rayleigh approximation. *Appl. Phys. Lett.* 2003, *83* (22), 4625–4627.

18. Ding, S.-Y.; You, E.-M.; Tian, Z.-Q.; Moskovits, M., Electromagnetic theories of surface-enhanced Raman spectroscopy. *Chem. Soc. Rev.* 2017, *46* (13), 4042–4076.

19. Stockman, M. I., Nanofocusing of optical energy in tapered plasmonic waveguides. *Phys. Rev. Lett.* 2004, *93* (13), 137404.

20. Barnes, W. L.; Dereux, A.; Ebbesen, T. W., Surface plasmon subwavelength optics. *Nature* 2003, *424* (6950), 824–830.

21. Link, S.; El-Sayed, M. A., Optical properties and ultrafast dynamics of metallic nanocrystals. *Ann. Rev. Phys. Chem.* 2003, *54* (1), 331–366.

22. Brongersma, M. L.; Halas, N. J.; Nordlander, P., Plasmon-induced hot carrier science and technology. *Nat. Nanotechnol.* 2015, *10* (1), 25–34.

23. Ali, M. R.; Wu, Y.; El-Sayed, M. A., Gold-nanoparticle-assisted plasmonic photothermal therapy advances toward clinical application. *J. Phys. Chem. C* 2019, *123*, 15375–15393.

24. Jang, Y. H.; Jang, Y. J.; Kim, S.; Quan, L. N.; Chung, K.; Kim, D. H., Plasmonic solar cells: From rational design to mechanism overview. *Chem. Rev.* 2016, *116* (24), 14982–15034.

25. Fraire, J. C.; Sueldo Ocello, V. N.; Allende, L. G.; Veglia, A. V.; Coronado, E. A., Toward the design of highly stable small colloidal SERS substrates with supramolecular host–guest interactions for ultrasensitive detection. *J. Phys. Chem. C* 2015, *119* (16), 8876–8888.

26. Xu, H.; Aizpurua, J.; Käll, M.; Apell, P., Electromagnetic contributions to single-molecule sensitivity in surface-enhanced Raman scattering. *Phys. Rev. E* 2000, *62* (3), 4318.

27. Sahu, B. K.; Dwivedi, A.; Pal, K. K.; Pandian, R.; Dhara, S.; Das, A., Optimized Au NRs for efficient SERS and SERRS performances with molecular and longitudinal surface plasmon resonance. *Appl. Surf. Sci.* 2020, *537*, 147615.

28. Huang, X.; El-Sayed, I. H.; Qian, W.; El-Sayed, M. A., Cancer cells assemble and align gold nanorods conjugated to antibodies to produce highly enhanced, sharp, and polarized surface Raman spectra: A potential cancer diagnostic marker. *Nano Lett.* 2007, *7* (6), 1591–1597.

29. Faulds, K.; Barbagallo, R. P.; Keer, J. T.; Smith, W. E.; Graham, D., SERRS as a more sensitive technique for the detection of labelled oligonucleotides compared to fluorescence. *Analyst* 2004, *129* (7), 567–568.

30. Lahr, R. H.; Vikesland, P. J., Surface-enhanced Raman spectroscopy (SERS) cellular imaging of intracellulary biosynthesized gold nanoparticles. *ACS Sustainable Chem. Eng.* 2014, *2* (7), 1599–1608.

31. Yu, H.; Shan, X.; Wang, S.; Chen, H.; Tao, N., Plasmonic imaging and detection of single DNA molecules. *ACS Nano* 2014, *8* (4), 3427–3433.
32. Law, W.-C.; Yong, K.-T.; Baev, A.; Prasad, P. N., Sensitivity improved surface plasmon resonance biosensor for cancer biomarker detection based on plasmonic enhancement. *ACS Nano* 2011, *5* (6), 4858–4864.
33. Lee, J. S.; Han, M. S.; Mirkin, C. A., Colorimetric detection of mercuric ion (Hg2+) in aqueous media using DNA-functionalized gold nanoparticles. *Angew. Chem. Int. Ed.* 2007, *46* (22), 4093–4096.
34. Sahu, B. K.; Juine, R. N.; Sahoo, M.; Kumar, R.; Das, A., Interface of GO with SnO_2 quantum dots as an efficient visible-light photocatalyst. *Chemosphere* 2021, *276*, 130142.
35. Sahu, B. K.; Kaur, G.; Das, A., Revealing interplay of defects in SnO_2 quantum dots for blue luminescence and selective trace ammonia detection at room temperature. *ACS Appl. Mater. Interf.* 2020, *12* (43), 49227–49236.
36. Huy, T. H.; Kang, F.; Wang, Y.-F.; Liu, S.-H.; Thi, C. M.; You, S.-J.; Chang, G.-M., SnO_2/TiO_2 nanotube heterojunction: The first investigation of NO degradation by visible light-driven photocatalysis. *Chemosphere* 2019, *215*, 323–332.
37. Reddy, C. V.; Ravikumar, R.; Srinivas, G.; Shim, J.; Cho, M., Structural, optical, and improved photocatalytic properties of CdS/SnO_2 hybrid photocatalyst nanostructure. *Mater. Sci. Eng. B* 2017, *221*, 63–72.
38. Jiang, J.; Olsen, J. S.; Gerward, L.; Mørup, S., Enhanced bulk modulus and reduced transition pressure in γ-Fe_2O_3 nanocrystals. *Europhys. Lett.* 1998, *44* (5), 620.
39. Ibrahem, M. A.; Rasheed, B. G.; Mahdi, R. I.; Khazal, T. M.; Omar, M. M.; O'Neill, M. J., Plasmonic-enhanced photocatalysis reactions using gold nanostructured films. *RSC Adv.* 2020, *10* (38), 22324–22330.
40. Yuan, Z.; Wu, Z.; Bai, S.; Xia, Z.; Xu, W.; Song, T.; Wu, H.; Xu, L.; Si, J.; Jin, Y.; Sun, B., Hot-electron injection in a sandwiched TiO_x–Au–TiO_x structure for high-performance planar perovskite solar cells. *Adv. Energy Mater.* 2015, *5* (10), 1500038.
41. Liu, Y.; Ai, K.; Liu, J.; Deng, M.; He, Y.; Lu, L., Dopamine-melanin colloidal nanospheres: An efficient near-infrared photothermal therapeutic agent for in vivo cancer therapy. *Adv. Mater.* 2013, *25*, 1353–1359.
42. Kumar, P.; Srivastava, R., IR 820 dye encapsulated in polycaprolactone glycol chitosan: Poloxamer blend nanoparticles for photo immunotherapy for breast cancer. *Mater. Sci. Eng. C* 2015, *57*, 321–327.
43. Han, H. S.; Choi, K. Y., Advances in nanomaterial-mediated photothermal cancer therapies: Toward clinical applications. *Biomedicines* 2021, *9*, 305.
44. Smith, A. M.; Mancini, M. C.; Nie, S., Bioimaging: Second window for in vivo imaging. *Nat. Nanotechnol.* 2001, *4*, 710–711.
45. Weissleder, R., A clearer vision for in vivo imaging: Progress continues in the development of smaller, more penetrable probes for biological imaging. *Nat. Biotechnol.* 2001, *19*, 316–317.
46. Jaque, D.; Martínez Maestro, L.; Del Rosal, B.; Haro-Gonzalez, P.; Benayas, A.; Plaza, J. L.; Martín Rodríguez, E.; García Sole, J., Nanoparticles for photothermal therapies. *Nanoscale* 2014, *6*, 9494–9530.
47. Rastinehad, A. R.; Anastos, H.; Wajswol, E.; Winoker, J. S.; Sfakianos, J. P.; Doppalapudi, S. K.; Carrick, M. R.; Knauer, C. J.; Taouli, B.; Lewis, S. C.; Tewari, A. K.; Schwartz, J. A.; Canfield, S. E.; George, A. K.; West, J. L.; Halas, N. J., Gold nanoshell-localized photothermal ablation of prostate tumors in a clinical pilot device study. *Proc. Natl. Acad. Sci. USA* 2019, *116*, 18590.

3 Dielectric and Metallodielectric Nanophotonics and Optical Confinement

Sushil Mujumdar, Rabisankar Samanta, and Sandip Mondal

CONTENTS

DOI: 10.1201/9781003248323-3

3.1 INTRODUCTION TO PHOTONIC BANDGAP

3.1.1 PHOTONIC CRYSTAL

Photonic crystals (PCs) are special materials where the refractive index of the material changes periodically in space. It is analogous to electronic crystals. In electronic crystals, atoms or molecules are arranged in a periodic manner. This periodicity affects the transport of electrons in various ways. At some particular energy of the electrons, transport happens without any scattering. Also, some energies are not allowed to exist in the system. With the progress of quantum mechanics, these phenomena were well understood with the concept of electronic wavefunctions. Similar to the electronic wavefunction, the electromagnetic waves are manipulated in PCs. The photonic bandgap is the most exciting property of PCs. At the photonic bandgap frequencies, photons are not allowed to propagate in the system. Later, we will discuss the origin of the photonic band gaps in detail.

One-dimensional (1D) PCs are the simplest ones. In 1D PCs, the refractive index varies only in one direction, while the other two directions are homogeneous. Similarly, in two-dimensional (2D) and three-dimensional (3D) PCs, the refractive index varies in all directions. A schematic of three different types of PCs is shown in Figure 3.1. The concept of PCs was first proposed by E. Yablonovitch [1] and S. John [2]. In the following years, many literature was published on PCs and photonic bandgaps [1,3–5]. Many fundamental physics phenomena have been studied along with their potential applications.

Low-dimensional PCs are widely used in controlling and manipulating light. For instance, 1D crystals are used as reflecting mirrors, reflection coatings on lenses, optical filters, and others. 2D PCs, appropriately modified, are used to guide light and are used as waveguides. Higher dimensional crystals are also becoming more and more applicable in sensing devices, nonlinearity, and others [6].

3.1.2 ORIGIN OF PHOTONIC BANDGAP

In the simplest case for 1D photonic crystal, multiple layers of two different dielectric media are stacked together. For propagation in the perpendicular direction to the interface, when the wavelength is half of the periodicity of the crystal, the forward and backward propagating Bloch waves interfere constructively and give rise to a standing wave. From the figure, it is evident that the standing wave can be arranged with two different configurations without breaking the symmetry of the system. For the first case, the maxima of the standing wave reside at the centers of the higher refractive-index material. In the second case, the maxima reside at the lower refractive material.

In order to understand the origin of the bandgap, we have to look at the energy distribution of the modes [6]. The energy is higher for the mode, where the optical density is concentrated in higher

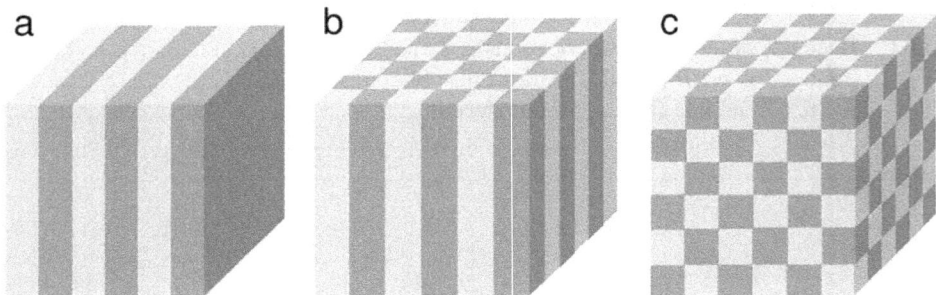

FIGURE 3.1 Photonic crystals (PC). Schematic of different types of photonic crystals. (a) One-dimensional. (b) Two-dimensional. (c) Three-dimensional. Here, two different colors indicate different refractive indices.

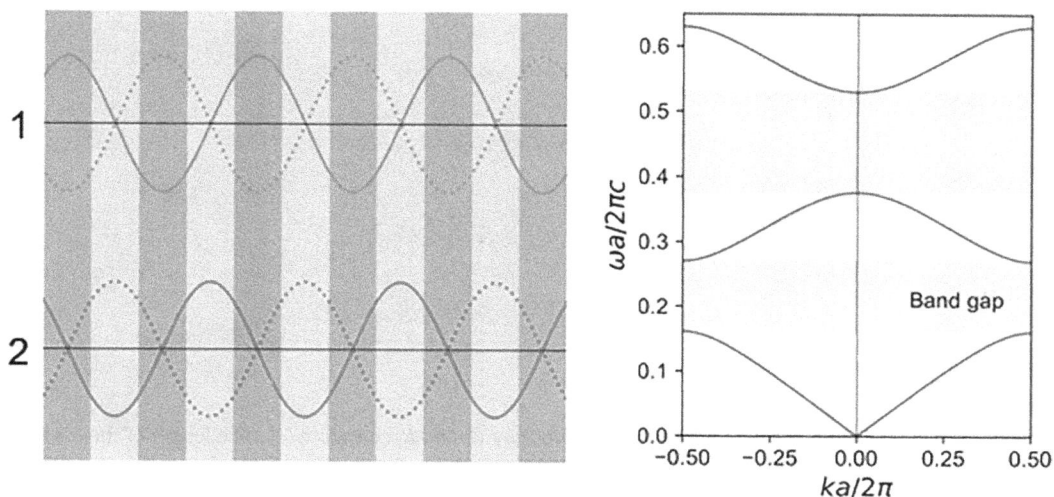

FIGURE 3.2 Photonic bandgap. Left Panel: The field distribution in a 1D photonic crystal. For the green (1) curve, the maxima of the fields are at the higher refractive index material, but for the red (2) curve, the maxima are located at the lower refractive index material. Right Panel: Band diagram of the 1D photonic crystal.

refractive-index material, as shown in Figure 3.2 (green curve, 1). Conversely, the energy is lower for the second mode, where optical density is higher in lower refractive-index material (red curve, 2). These two modes have different energies associated with different frequencies, leading to gaps in the frequency. The counter-propagating waves interfere destructively in the bandgap frequencies and hinder the propagation. The band structure for a 1-D photonic crystal (multiple layers of two materials with different refractive indices) is shown in Figure 3.2.

3.2 CONFINEMENT OF LIGHT BY MICRO- AND NANO-CAVITIES

3.2.1 Cavity Confinement of Light

The simplest form of the cavity is two parallel mirrors, otherwise called the Fabry-Perot cavity. Here, two counter-propagating waves from the mirrors interfere and form a standing wave. These cavities are widely used in conventional lasers for their tunability. Figure 3.3 depicts the schematic of a Fabry-Perot cavity.

In 1878, Lord Rayleigh observed that a whisper could be heard near the wall around the gallery but not in the dome's center in St. Paul's Cathedral. He explained this phenomenon by the wave theory of sound, where the wall reflects sound without any loss, and thus, the sound waves are confined inside the dome. This type of confinement is observed in other waves, such as vibrational modes and photonic modes. These modes are called whispering gallery (WG) modes. These WG modes for light waves within a micro-cavity can be very long lived as they are supported by total internal reflection. This type of micro-cavities can be of different shapes and sizes. Figure 3.4 shows two cavities with circular (Figure 3.4a) and star (Figure 3.4c) shapes along with the intensity patterns (Figure 3.4b and d) of the supported WG modes [7]. One of the many applications of WG modes is micro-lasers. These lasers are very small in size and ideal for biosensing [8].

Defect cavities based on band gaps in PC are robust and have very high quality. Significant research has been done in this area and its application in integrated optics. Various device configurations based on PC cavities have been proposed and demonstrated experimentally, mostly based on planar 2D air hole lattice PC slab structures. An example of the PC cavity mode of a square lattice with air holes (white circles) in the dielectric slab is shown in Figure 3.5.

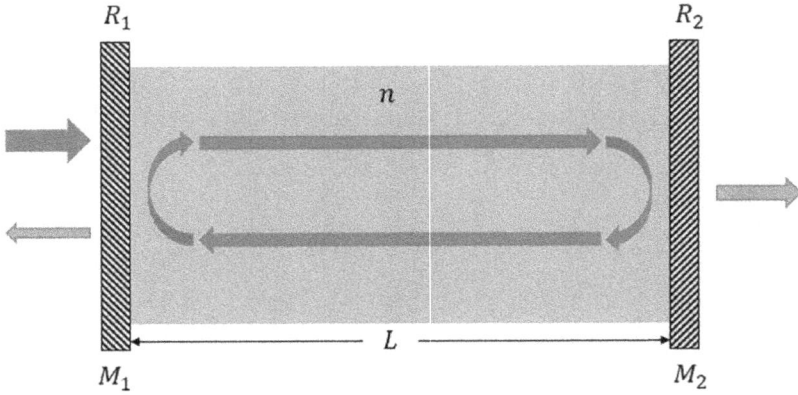

FIGURE 3.3 M_1 and M_2 are two mirrors with reflectivity R_1 and R_2, respectively. The length of the cavity is L, with a dielectric medium of refractive index n inside.

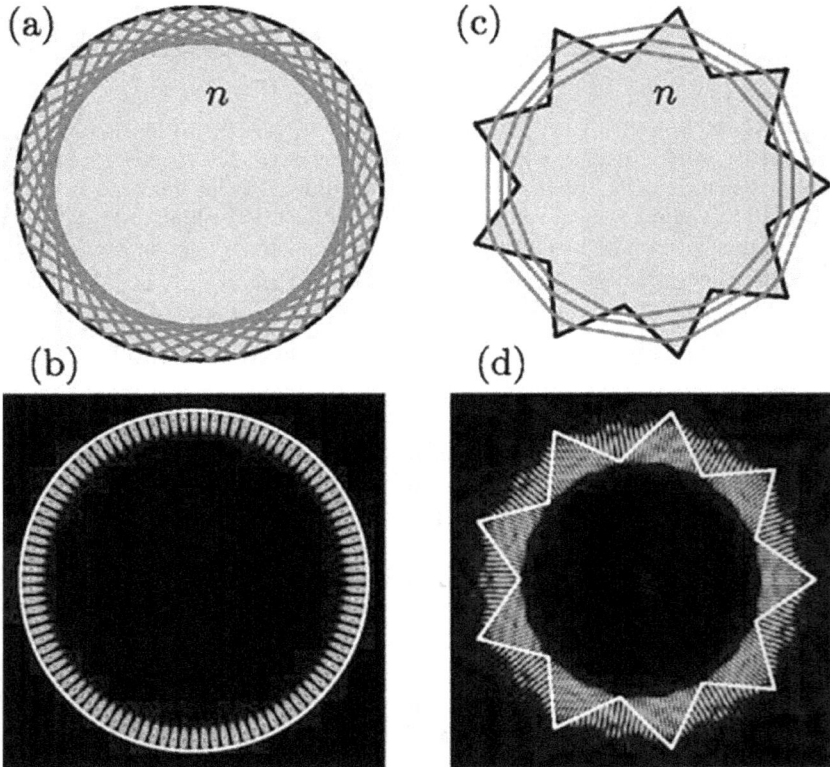

FIGURE 3.4 (a) and (c) are ray diagrams of whispering gallery modes in two different-shaped micro-cavities. (b) and (d) are corresponding intensity patterns of the modes. (This figure is reproduced with permission from Ref. [7], American Physical Society, under the Creative Commons Licence.)

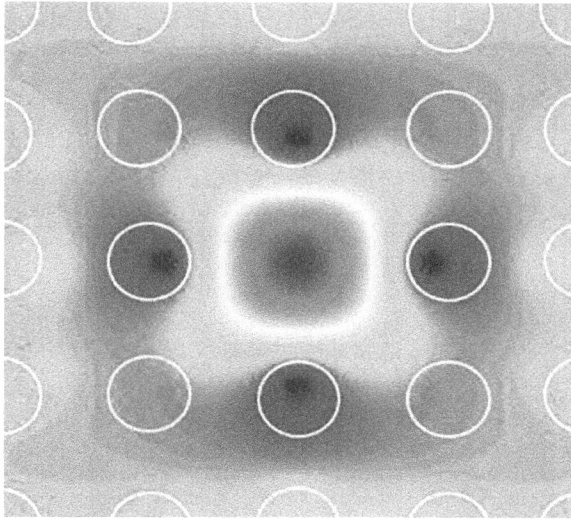

FIGURE 3.5 Simulated photonic crystal cavity mode (electric field) based on a square lattice.

3.2.2 Anderson Localization of Light

The scattering of light in disordered media is ubiquitous in nature. The blue or red sky, the clouds, the fog, and white color of milk are all results of the scattering of broadband light. The photons (electromagnetic waves) scatter from random particles in all these media and diffuse with time. The scattering strength of the medium is determined by the number density of particles and the refractive index of the scattering particles. The scattering strength can be quantified with a single parameter called the mean free path $\ell^* = 1/n\sigma$, where n is the number density and σ is the scattering cross section of the particle. For very strong scattering strength when $k\ell^* \leq 1$, the transport of waves can get localized due to interference between the scattered wavelets. This phenomenon was first proposed for disordered metals, where electronic wavefunction gets localized, and the transport is hindered due to the scattering of the waves from random potentials giving rise to the metal-insulator transition. The theory of this disorder-induced localization was proposed by P. W. Anderson and is hence known as the Anderson localization. The wavefunction of the Anderson localized mode can be written as, $\psi(\vec{r}) \propto e^{-\vec{r}/\xi}$, where ξ is the localization length.

Being a wave phenomenon, it was soon realized that electromagnetic waves or photons are the perfect candidates to observe and measure Anderson localization. In 3D systems, to observe Anderson localization, a critical disorder strength has to be reached [9], whereas in low-dimensional systems (1D, 2D, and quasi-2D) [10–14], all modes can be localized if the system size is large enough. In low-dimensional systems, the modes are directly accessible for measurement, while in 3D systems, the signature of Anderson localization can only be measured from transmission data, which can sometimes be corrupted by absorption. The figure below shows a systematic transition of a delocalized state to a localized state. Figure 3.6a shows a scanning electron microscope (SEM) image of a disordered structure. The system consists of air holes lithographically written on a GaAs membrane. Figure 3.6b shows an extended mode in a periodic structure. However, the modes are localized with increasing disorder strength, as seen in Figure 3.6c and d. More on Anderson localization of light can be found in some recent reviews [15,16].

3.2.3 Chiral Light-Matter Interactions in Optical Resonators

A molecule that cannot be superimposed with its mirror image is called a chiral molecule. An example of a chiral molecule [17] is illustrated in Figure 3.7. Chiral molecules behave differently

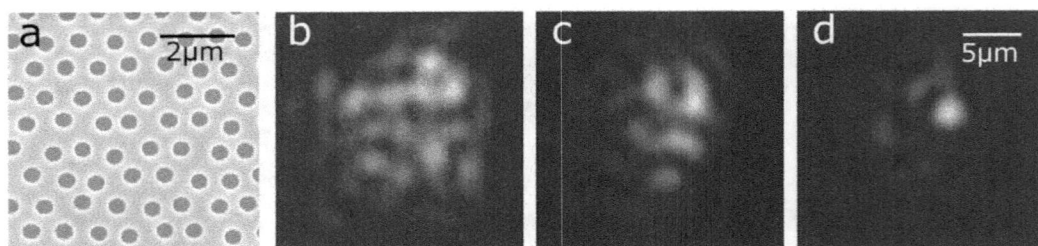

FIGURE 3.6 Anderson localization of light in 2D disorder. (a) Scanning electron micrograph image of a 2-D disordered sample where nanoholes are lithographically written on a GaAs membrane. (b) An extended mode in a periodic structure. (c) Weakly localized mode in a low disordered system. (d) A strongly localized mode in a high-disordered configuration.

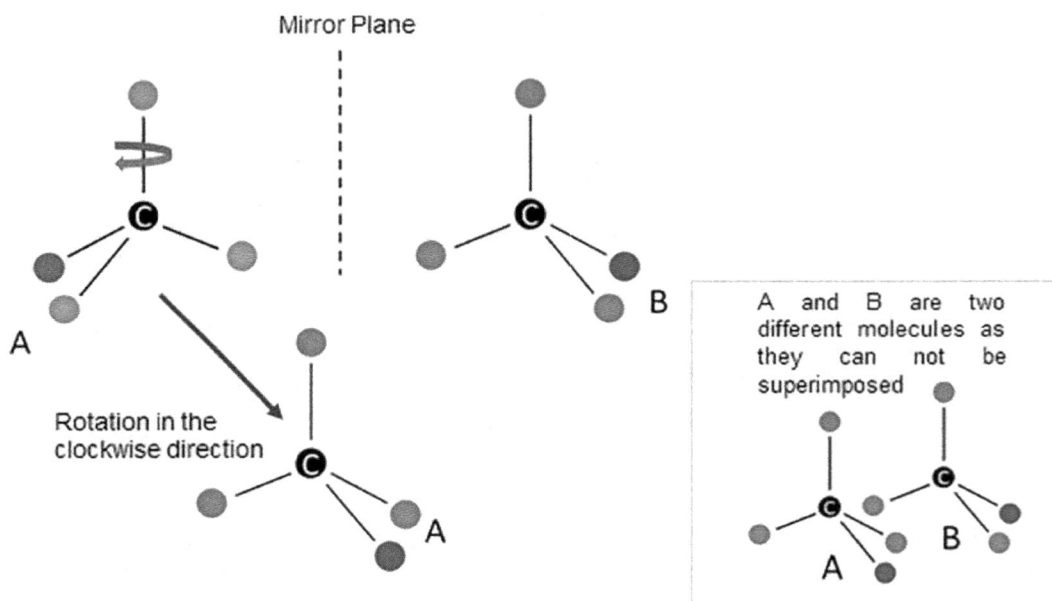

FIGURE 3.7 Schematic of a chiral molecule.

with left and right circular polarized light. Chiroptical signals are essential to determine the properties of a chiral molecule. However, these signals are very weak in nature. There have been many studies to enhance the chiroptical signal. One of the methods to enhance the chiroptical signal is to locally vary the electromagnetic field, for which nanostructures have been used. However, the mechanism of enhancement of spontaneous decay of chiral molecules with respect to the local density of states is not very well known. On the other hand, the Purcell effect successfully describes the enhancement of spontaneous decay in the presence of an optical resonator in the case of achiral molecules. This idea of the Purcell factor can be extended to chiral molecules also.

Yoo and Park [18] reported numerical results of the enhanced fluorescence detected circular dichroism of a mirror cavity with different reflectivities (Figure 3.8a). The refractive index of the chiral sample inside the cavity was 1.5. They even proposed a quite realistic structure for the chiral Purcell effect. Their structure looked like a double fishnet (Figure 3.8b) where a dielectric layer (35 nm hydrogen silsesquioxane) was sandwiched between two 30 nm thick Ag layers. The whole structure was deposited on a glass plate while the chiral molecule could stay inside the square

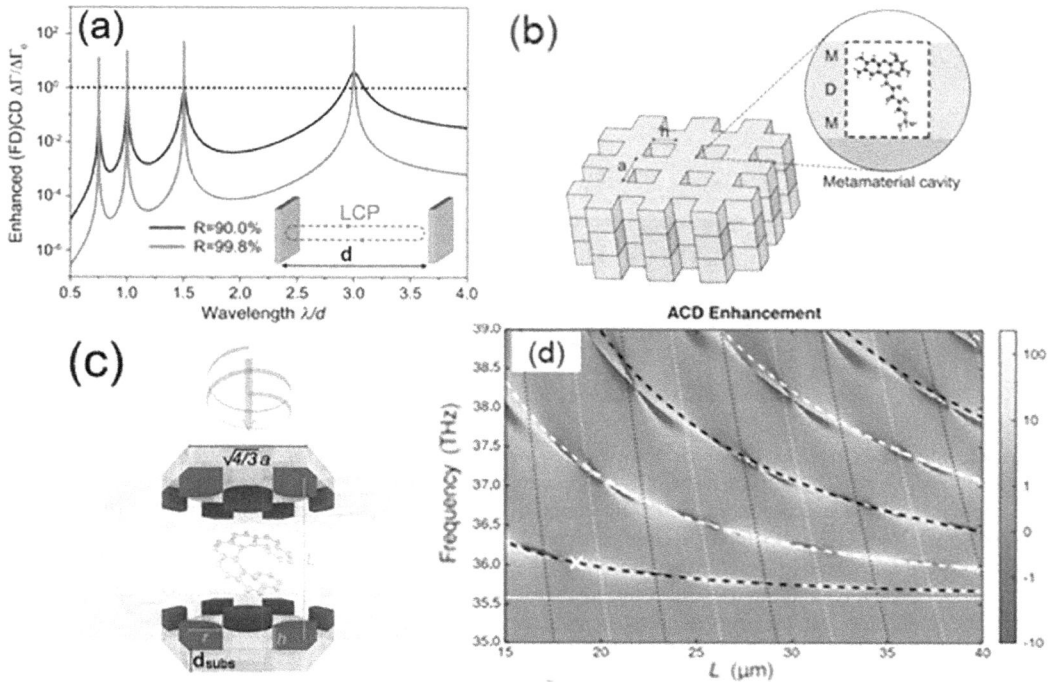

FIGURE 3.8 (a) Enhanced fluorescence detected circular dichroism of a cavity with different mirror reflectivities. (b) Proposed experimental structure: a double fishnet comprised of metal-dielectric-metal multilayers metamaterial where chiral molecules stay inside the cavity. (c) Structure of a cavity with Si discs and a substrate of $\epsilon_{subs} = 2.14$. (d) Absorbance circular dichroism as a function of the length of the cavity (L) and frequency. (Panels (a), (b) and (c, d) are adapted with permission from Refs. [18,19], respectively, American Physical Society.)

cavities. Another cavity structure composed of Si nanodiscs (Figure 3.8c) was proposed by Feis et al. [19]. They claimed that their proposed structure could enhance more than two orders of magnitude circular dichroism signals of chiral molecules. Figure 3.8d illustrates the computed enhanced absorbance circular dichroism as a function of frequency and cavity length.

3.3 CONFINEMENT OF LIGHT BY HYBRID NANOSTRUCTURES

3.3.1 Sub-Wavelength Confinement in Dielectric-Metal Wedges and Grooves

Surface plasmon polaritons (SPPs) confine light at the nanoscale beyond the diffraction limit, which is desirable in many applications like on-chip photonic operation, nanofocusing, biosensing, and many more. SPPs are the electromagnetic waves that propagate along the metal-dielectric interface, and they involve collective oscillations of electrons in the metal. Recent advances in fabrication technology have scaled the production of sub-wavelength architectures. Many structures have been proposed, like slot waveguides, V-shaped grooves, and wedges, among others. However, the main drawback of plasmonic waveguides is that they possess huge losses (mostly ohmic loss and generate heat). Recently, hybrid metal-dielectric (HMD) structures are proposed that can concentrate light at the nanoscale and suffer from low transmission loss. Among all structures, here we will discuss grooves and wedge waveguides.

Metal with V-shaped grooves supports channel plasmon polaritons (CPPs) which tightly confine the light at the sub-wavelength region and also help in long-distance propagation [20]. The field of

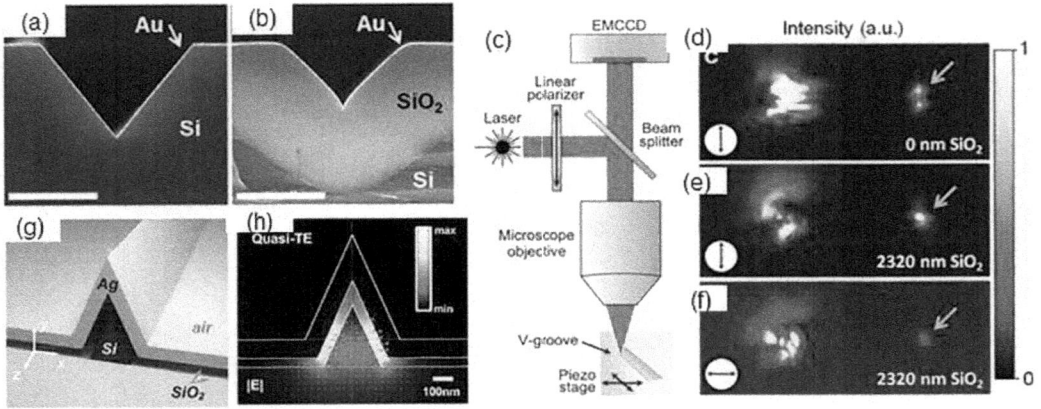

FIGURE 3.9 Hybrid metal-dielectric grooves and wedges. (a) Cross section of a V-groove of Si covered by a thin layer of Au. (b) A thick layer of SiO$_2$ introduced by thermal oxidation step on top of Si groove. (c) Schematic setup of the experiment. (d–f) Out-coupled radiation pattern from the other end of the V-groove as seen on the EMCCD for different system configuration and input polarization. (g) Hybrid wedge structure where a triangular Si wedge structure is covered by a subsequent silica buffer layer and metallic layer (Ag). (h) Local field enhancement at the low-index buffer region. (Panels (a–f) are adapted with permission from Ref. [20], American Chemical Society; Panels (g–h) are adapted with permission from Ref. [22], Optica Publishing Group.)

the CPP can be tightly localized if the depth of the groove is high enough. However, the field confinement at the bottom of the groove strongly depends on the smoothness of the metal surface. Any kind of roughness on the metal surface creates a sufficient amount of scattering loss. Unfortunately, fabricating a very smooth metal surface is challenging, if not impossible. Researchers have come up with hybrid plasmonic grooves that can tightly confine the field and even have low propagation loss to cope with the problem. Generally, two methods are followed to make hybrid groove structures. A V-shaped groove is first milled onto the metal surface in one process. Then, subsequently, low- and high-index dielectric materials are deposited on the metal groove [21]. The other method follows opposite steps. Here, the groove is first created on the high-index dielectric, and subsequently, low-index material and metal are coated on the structure. Figure 3.9a shows an SEM image of a V-shaped groove of Si, which is covered by a thin layer of Au.

In another configuration, a thick layer of SiO$_2$ (thickness 2,320 nm) is introduced between Si and Au by the thermal oxidation method (Figure 3.9b). Figure 3.9c depicts the schematic of the experimental setup. A laser diode of 811 nm wavelength was illuminating the sample, and the out-coupled light from the groove was imaged on an EMCCD. A pair of intensity peaks were observed at the opposite end of the groove with no SiO$_2$ layer and when the laser was focused on the surface plane (Figure 3.9d). With the introduction of the SiO$_2$ layer, a strong single peak was observed, which resembles the CPP mode (Figure 3.9e). When the polarization of the input light was rotated by an angle of 90°, the relative intensity of the single peak was lowered by a factor of 10 (Figure 3.9f). The phenomenon suggests that the mode no longer supports the electric field of the CPP.

Another type of structure is a triangular-shaped metal wedge waveguide that supports wedge plasmon polaritons (WPPs). This structure can simultaneously sustain sub-wavelength modal volume and guide light over a long distance. They are even sometimes found superior to the channel plasmonic waveguides. Thus, WPPs have great potential in many sub-wavelength scale applications. Theoretically, it has been shown that wedge plasmonic waveguides with small wedge angles and sharp tips can confine light tightly. However, it is very challenging to fabricate these structures.

Different hybrid wedge structures have been introduced to realize sub-wavelength confinement with low loss, like metal wedges covered with a thick layer of low dielectric claddings or both the combination of low- and high-index dielectric layers. The high refractive-index contrast forces the mode to confine tightly around the metal wedges with the enhanced local field. Moreover, the mode size can be tuned by changing the thickness of the dielectric layers and even changing the shape of the metallic wedges. Different types of hybrid dielectric layer wedge plasmon polariton waveguides (HDLWPPWs) and their properties are discussed elaborately in Ref. [23]. Bian et al. [22] proposed another structure where a triangular wedge of Si is covered by a low-index buffer layer of silica and, on top of that, a thin Ag layer (Figure 3.9g). The simulated mode profile is depicted in Figure 3.9h. At the low-index buffer region, one could observe a significant enhancement in the local electric field.

3.3.2 Exciton–Plasmon Interactions in Metal-Semiconductor Nanostructures

Due to quantum size effects in metal and semiconductor nanoparticles, many exotic properties are observed that are quite different from their bulk properties. Controlled synthesis of these nanostructures has been already exploited in many applications. For instance, metal nanostructures have huge applications in localized surface plasmon (LSP) resonance-based sensing, catalysis, near-infrared photothermal therapy and many more [24]. On the other hand, semiconductor nanostructures also find extensive applications in solar cells [25], production of photovoltaic and optoelectronic devices [26], chemical- and biosensing [27], and others. When electromagnetic fields fall on metal nanostructures, coherent delocalized electrons in the conduction band oscillate collectively along the interface of the metal, and the nonconductive environment is essentially called localized surface plasmons. At the resonance frequency, LSPs show enhanced amplitude at the near-field, and this resonance frequency can be tuned by changing the size and shape of the nanostructures.

Similarly, semiconductor nanostructures show excitations at optical frequencies. Here, discrete electronic levels in the conduction and valence bands can be engineered by changing the shape and size [28]. When the semiconductor captures a photon with energy larger than the bandgap, an electron in the valence band is excited to the conduction band and creates an electron-hole pair. The bound state of the electron and hole is called an exciton. Recently, advanced fabrication techniques have enabled the production of metal-semiconductor heterostructures. Here, the coupling of exciton dipole and the electromagnetic field of surface plasmon shows better promises in nanoscale energy transfer [29], and enhanced optical nonlinearity [30] and different new types of excitations.

Generally, Ag and Au are most often treated as suitable metals for excitons-plasmons nanostructure. Sometimes Ag is preferred most for its low intrinsic loss. In semiconductor nanostructures, indirect bandgap materials are chosen most of the time, although Si-based nanostructures (a direct bandgap material) have been already established in exciton–plasmon interactions. Usually, metal and semiconductor nanostructures are fabricated separately and then mixed together using sophisticated techniques. In recent times, different types of metal-semiconductor nanostructures have been realized, like core-shell structures [31], dumbbells [32], dots-on-rod [33,34], and others. These kinds of assembly are very stable, and due to the absence of spacer layers between metal and semiconductor, exciton–plasmon interactions are improved.

Thus, engineered metal-semiconductor nanostructures can improve specific optical properties and offer exciton-plasmon-based new optical phenomena. Apart from the fundamental aspects, these nanostructures have already been explored in sensing, photovoltaics, quantum information, and other areas [35]. Like in photovoltaics, the metallic part localizes the impinged light while the semiconductor part absorbs that. In this way, a large amount of light can be absorbed, which, in turn, enhances the performance of photovoltaic devices [36–39].

3.3.3 Hybrid Metal-Dielectric Nanostructures

In the last two decades, with the advancement of fabrication technology, light-matter interaction at the nanoscale has attracted lots of interest for various applications and also for exploring new physical phenomena. Metallic nanostructures help in strong light confinement at the nanoscale, and they are used in biosensing, solar cell, microscopy, and many other applications. However, metallic nanostructures suffer from inherent ohmic losses and subsequent heating effects. As a result, they reduce the performance efficiency of many plasmonic devices. On the other hand, high refractive-index dielectric nanostructures have very low absorption in the visible and near-infrared spectral ranges. Moreover, they show magnetic and electric resonances for different nanoparticle shapes and strong directionality.

In recent years, researchers have combined these nanostructures and made HMD nanostructures where the metal part confines the electromagnetic energy and the dielectric part reduces the loss and scatters strongly in a specific direction [40]. Different types of HMD nanostructures have been realized, like metal-dielectric core-shell nanoparticles, metal nanoparticles on dielectric substrates or vice versa, metallic nanoparticles on dielectric PCs, or metallic nanoparticles in dielectric cavities, [41] among others.

In the last few years, researchers have observed enhanced nonlinear effects like a second [42–44] and a third harmonic [45,46] generation (THG) in HMD nanostructures. Renaut et al. [42] observed enhanced second-harmonic generation (SHG) from barium titanate (BTO)-Au nano dimer compared to a bare BTO nanoparticle. The enhancement is due to the overlapping between BTO Mie resonant higher order modes and LSP modes of the Au nanoparticle at visible frequencies. Shibanuma et al. [45] showed enhanced THG from a Si nanodisc surrounded by a Au nanoring.

3.3.4 FAR-FIELD OPTICAL HYPERLENS FOR THE MAGNIFICATION OF NANOSTRUCTURES IN THE SUB-DIFFRACTION REGIME

Getting a perfect focus using a lens is a long-standing problem in imaging science and lithography technology. According to Abbe's 'diffraction limit', any optical system cannot resolve tiny features of an object smaller than half the wavelength of light. When we illuminate an object with a lens, the scattered light comprises both the propagating and evanescent components. The evanescent fields that contain the sub-wavelength information of the object die down exponentially. As a result, at the far-field, we only get an imperfect image realized by the propagating components.

It took almost 100 years from the Abbe's diffraction limit to get some idea about perfect focusing through a negative-index material, which was a proposal by Vesalago [47]. The first realization of a negative-index material became possible after Pendry's seminal works [48–50]. After that, many groups started to make experimental negative-index materials that could achieve a perfect focus. When an object is illuminated with a beam of light, the scattered light carries all the information about the object. Large features of the object are encoded with the propagating waves, while small details are carried by the evanescent waves. If one uses a conventional lens to gather the scattered light, only propagating components can reach the image plane while the evanescent parts die down (Figure 3.10a). As a result, a diffraction-limited image is formed at the image plane. Researchers have come up with a new type of lens called 'superlens' made up of negative-index material to combat this problem. If we place a slab of superlens near the object (Figure 3.10b), also called near-field superlens, the evanescent field can be enhanced [51–53]. However, this is only limited to the near-field. For the far-field sub-diffraction-limited imaging (far-field superlens, see Figure 3.10c), researchers [54] have put many additional nanoscale features (like a designed surface scatterer) on the slab superlens of Ag. Here, the evanescent fields are not only enhanced but become propagating into the far-field.

In an alternative approach, instead of a slab geometry, anisotropic metamaterials with curved multilayer stacks were proposed [55,56]. Here, a far-field image is formed in the cylindrical geometry. The most interesting aspect of such a lens is that it can immediately convert the evanescent fields into propagating ones inside the lens system (Figure 3.10d). Their large transverse wavevectors subsequently transform into smaller ones such that they become true propagating waves to make sub-diffraction-limited images in the far-field. This kind of lens system is called a 'hyperlens'. The

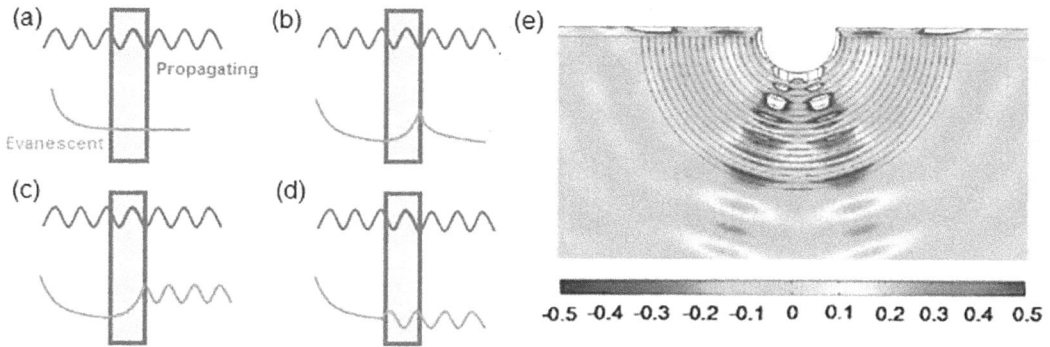

FIGURE 3.10 Optical superlens and hyperlens. When light falls on an object, the scattered light carries the object's information in the form of propagating and evanescent components. The scattered light can be collected by four different types of lenses: (a) Conventional lens, (b) Near-field superlens, (c) Far-field superlens, (d) Hyperlens, and (e) Example of a hyperlens structure. Eight alternative layers of Ag (35 mm) and Al$_2$O$_3$ (35 nm) are designed on a curved quartz mold. The color plot depicts the magnetic field distribution where 365 nm light illuminates two line objects on the curved surface. (Panel (e) is adapted with permission from Ref. [57], Optica Publishing Group.)

term 'hyper' comes from the lens material's hyperbolic dispersion. A typical structure of a hyperlens was first proposed by Jacob et al. [55]. The structure consists of a cylindrical geometry with a hollow core surrounded by 160 alternating layers of metal ($\epsilon = -1 + 0.01\,i$) and dielectric ($\epsilon = 1.1$), where the thickness of each layer was 10 nm. The inner and outer radii of the hollow core were 250 and 1,840 nm, respectively. Two point sources separated by 100 nm were placed inside the hollow core, and the operating wavelength was 300 nm. Clearly, the separation between the point sources was lower than the resolution limit. The simulated intensity map depicted that two output beams are well separated at the boundary, even seven times more than the distance between the point sources. Lee et al. [57] fabricated a different type of hyperlens structure containing eight alternative layers of Ag (35 mm) and Al$_2$O$_3$ (35 nm) inside a half-cylindrical quartz mold. Figure 3.10e is the simulated magnetic field distribution of a hyperlens structure where two lines of 50 nm depth, separated by 150 nm, are drawn on the inner surface (50 nm thick Cr layer). The wavelength of the illuminated light was 365 nm. Clearly, a magnified and well-separated image was formed in the far-field. Later on, many different configurations were used for far-field hyperlens imaging [58,59].

3.4 THE EFFECT OF DIELECTRIC CONTRAST IN THE OPTICAL CONFINEMENT

The dielectric contrast or the refractive-index contrast is a measure of the difference in refractive index between two dielectric media. Here the refractive-index contrast is defined as n_1 / n_2, where n_1 is, in the current context, the higher refractive index. The dielectric contrast plays a crucial role in an optical confinement. The phenomenon can be understood with a simple example of a straight dielectric waveguide, as illustrated in Figure 3.11. The waveguide is embedded in the air ($n_2 = 1$) and excited by a continuous point source at one end of the waveguide. Figure 3.11a and b shows the electric field distribution within the waveguide with refractive index (n_1) 2 and 4. The electric field is confined within the waveguide and propagates to the other end. As can be seen from the figure, the field is extended well outside the waveguide for $n_1 = 2$, whereas the field in the waveguide with $n_1 = 4$ is tightly confined within the waveguide. The boundary loss is higher because of the low-refractive-index contrast (Figure 3.11a), and modes leak through the waveguide. Figure 3.11c and d shows the time-averaged power distribution in the same waveguides as in Figure 3.11a and b. Evidently, more power is concentrated in the waveguide with a higher refractive index.

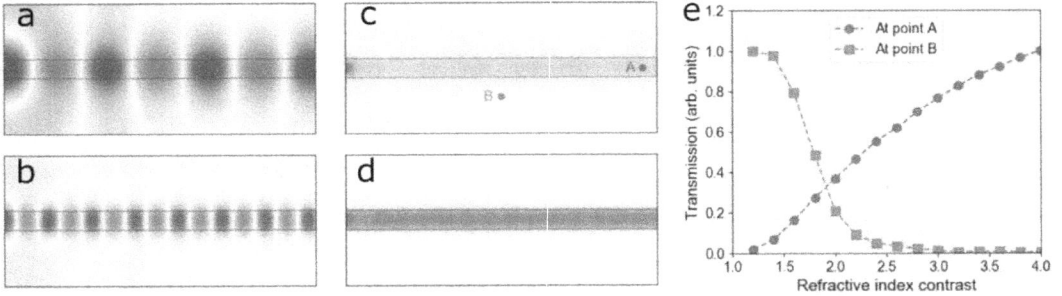

FIGURE 3.11 Electric field distribution in a waveguide with refractive index 2 (a) and 4 (b). (c) and (d) are power distributions in the same waveguides mentioned earlier. (e) Transmitted power at positions A (blue discs) and B (green squares).

To quantify the effect of dielectric contrast, the transmission is measured at two locations named as A (within and at the other end of the waveguide) and B (outside the waveguide) while changing the refractive index of the waveguide from 1.2 to 4 (Figure 3.11e). The blue discs and green squares depict the transmission at locations A and B, respectively. As the refractive index of the waveguide increases, thus increasing the dielectric contrast, the transmitted power at A increases. The exactly opposite behavior is seen at B. With increasing refractive index contrast, the boundary loss decreases, and the fields get more and more confined within the waveguide, thus reducing the loss or transmission outside the waveguide. Similar to this example, in an optical cavity, with increasing dielectric contrast (the contrast between the cavity and surrounding medium), the loss or the leakage decreases, and this increases the lifetimes of the cavity modes. This is also seen in complex systems such as Anderson localization systems. In disordered systems, the scattering strength determines the strength of the disorder, and the scattering strength depends on the refractive index contrast.

In Figure 3.12, we present the results of Vanneste and Sebbah [60], where they numerically showed how the variation in dielectric contrast leads to the transition from weakly scattering to disorder-induced Anderson localization in a 2D disordered medium. The system was composed of parallel dielectric cylinders (radius = 60 nm, infinite height) positioned randomly in the air. The volume fraction of those cylinders was kept at 40%. For lower refractive indices ($n = 1.05, 1.25$), the modes in the system are extended. As the refractive index of the cylinders increases, the modes start to localize. At high-refractive-index contrast ($n = 2$), the modes are confined to a small region. Insets in all sub-figures illustrate the phase probability distribution ($P(\phi)$). It is clearly observed that $P(\phi)$ is almost flat for low-index medium. As the index contrast increases, $P(\phi)$ starts to peak at 0 and π.

3.5 CONFINEMENT OF LIGHT IN AN OPTICAL FIBER WITH A SUB-WAVELENGTH AIR CORE

Since the late 1970s, human civilization has witnessed tremendous uses of optical fiber in modern telecommunications, modern-day internet, medicine, or industry. An optical fiber is typically made of silica glass. It comprises mainly two elements, the central part is called the core and the outer part is called cladding which encloses the core. The core has a slightly higher refractive index than the cladding due to the doping of different materials. This refractive index contrast keeps the light confined within the fiber. Despite many applications, optical fiber finds some disadvantages. For instance, it has a limited bend radius. It suffers from a large dispersion which means the light pulse gets broadened over the fiber length. Another aspect is that modern optical fiber finds problems in the transmission of high-power laser pulses, which get self phase-modulated via the

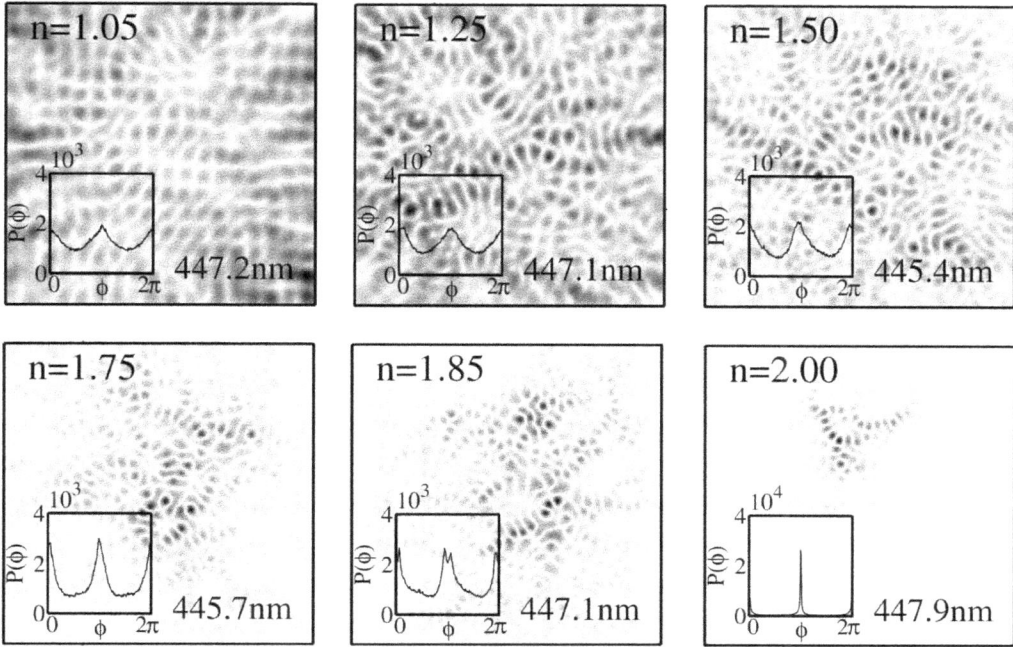

FIGURE 3.12 Mode structures of a complex system for different refractive index contrast values. Inset figures depict the phase probability distribution which peaks around 0 and π for localized modes ($n = 2$). (This figure is adapted with permission from Ref. [60], American Physical Society.)

intensity-dependent Kerr effect. Similarly, doping causes many nonlinear effects, which can create a nuisance in telecommunications.

Several efforts have been made to combat these problems. Yeh et al. [61] gave the first theoretical concept of a Bragg fiber, where light travels in a hollow core, and the core is surrounded by multiple alternative concentric layers of glass and polymer (Figure 3.13a). These layers are arranged in a 1D periodic manner and act as a 'Bragg reflector'. The experimental realization came another 20 years later [62]. In the late 1980s, John [2] and Yablonovitch [1] individually gave the idea of a photonic bandgap, just like electrons form an electronic bandgap in condensed matter physics. With this new idea, Knight et al. [63] predicted a new kind of fiber called 'photonic crystal fiber' (PCF). Several air holes are arranged in a periodic manner in the cladding section, and these holes run through the fiber length. Depending upon the core material, the PCF can be divided into two types. One type has a solid core, and the outer holes are arranged in such a manner that the effective index of the cladding becomes lower compared to the core (Figure 3.13b). We denote this type of PCF as 'index-guiding photonic crystal fiber' (IG-PCF). The refractive index contrast guides the light through the fiber. The other kind of fiber uses an air core, while the outer part has periodic structures of air holes that act as a bandgap material (Figure 3.13c) [64]. This type of fiber is termed as a 'holey fiber'.

Although PCFs enable strong light confinement in a small core that is often useful in several nonlinear interactions and atomic manipulation [65,66], they ultimately suffer diffraction [67]. As a result, light spreads away from the core region. In 2003, Michal Lipson's group, for the first time, proposed a slot waveguide structure embedded in a silica wafer [68]. The structure has a nanometer-wide low refractive-index region sandwiched between two rectangular size high refractive-index regions. They numerically showed that this type of slot waveguide strongly confines intense light with minimal loss in the nanometer-wide low-index region, which is not achievable in conventional waveguides. Subsequently, they fabricated the structure and experimentally showed light confinement in the slot waveguide structure [69]. This idea was later translated within a PCF by

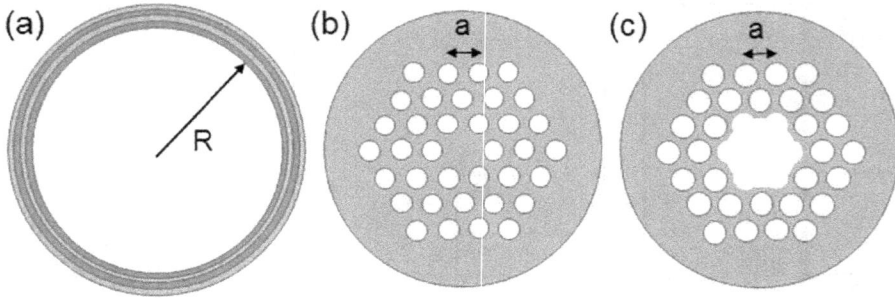

FIGURE 3.13 Photonic crystal fibers (PCFs). (a) Optical fiber with hollow cylindrical core surrounded by several alternating sub-micron thick high- and low-refractive-index concentric layers (called "Bragg fiber"). (b) PCF with solid silica core and cladding is composed of air-silica based two-dimensional periodic structure (called "index-guiding Photonic crystal fiber"). Here light is confined by the index guiding. (c) Basic design of a "holey fiber" where light is confined in the hollow region by the photonic band gap.

Wiederhecker et al. [70], which opened a new way to transmit light through a PCF. Later on, many other groups reported different types of PCFs with sub-wavelength air-core designs [71–73].

3.6   OPTICAL METAMATERIALS

3.6.1   INTRODUCTION

The last two decades have seen an enormous growth of a new type of materials called 'metamaterials' that are not found in nature. The prefix 'meta' comes from the Greek word '$\mu\varepsilon\tau\alpha$', which means 'beyond'. That illustrates the meaning of metamaterials beyond conventional materials. The properties of these materials depend on their structure rather than the material properties they are made of. Unlike other materials, metamaterials have simultaneously negative values of electric susceptibility (ϵ) and magnetic permeability (μ). Based on the values of ϵ and μ, all materials can be depicted in parameter space (Figure 3.14). The first quadrant ($\epsilon, \mu > 0$) represents most of the dielectric materials. The second quadrant is for those materials (metals, doped semiconductors, ferroelectric materials) whose $\epsilon < 0$ at certain frequencies (below plasma frequency). Few ferrite materials show negative μ below microwave frequency. So, they fall in the fourth quadrant. The most interesting region is the third quadrant, where $\epsilon, \mu < 0$. Unfortunately, no such material exists in nature.

Metamaterials generally consist of specially designed sub-wavelength elements (also called meta-atoms) in a periodic manner or in a special way to get some unique electromagnetic responses. In the early days of metamaterials, researchers were able to produce metamaterials at microwave frequencies. Later on, metamaterials at visible frequencies were invented. Among different types of metamaterials, 2D counterparts, also called metasurfaces, have been created for many applications, including wavefront shaping (WFS), meta-lens, pulse shaping, optical cloak, and holography [74,75]. The following sections discuss the fundamentals and recent applications of optical metamaterials.

3.6.2   ELECTRIC METAMATERIALS

According to the Drude model, the permittivity of can be written as

$$\epsilon(\omega) = 1 - \frac{\omega_p^2}{\omega(\omega + i\Gamma)} \tag{3.1}$$

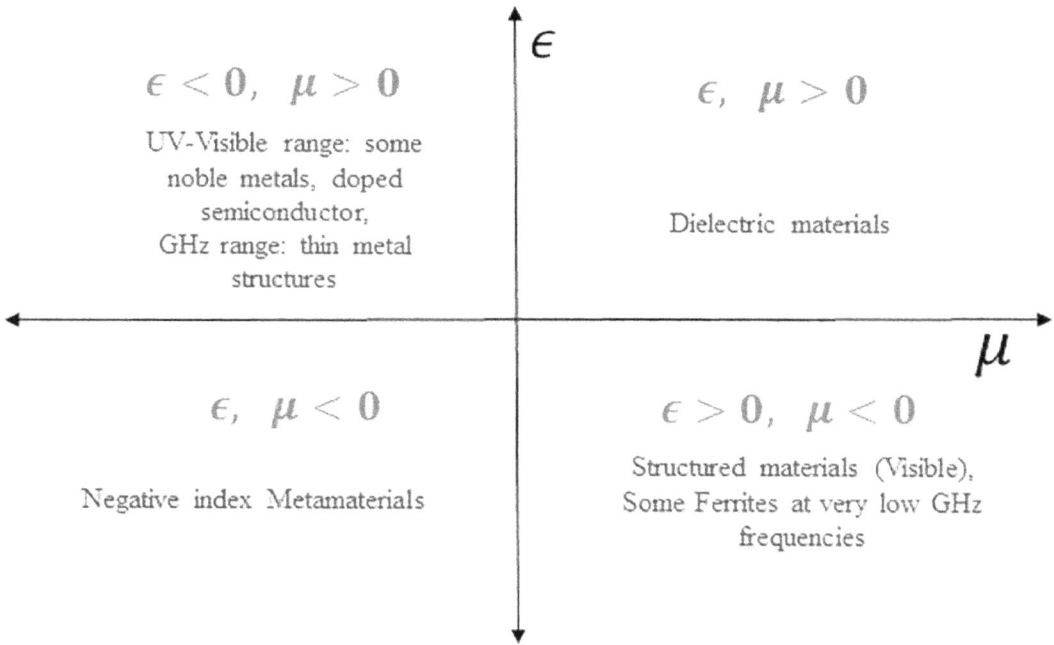

FIGURE 3.14 Parameter space of material based on the values of electric permittivity (ϵ) and magnetic permeability (μ).

where $\omega_p = \dfrac{ne^2}{\epsilon_0\, m}$ is the plasma frequency, n is the density of electrons, and m is the mass of electrons. Γ is the damping constant. In many metals, plasma frequency is in the UV and visible range, and the damping constant is also negligible compared to ω_p. So, the permittivity becomes negative below plasma frequency in this frequency range. Now, as metals suffer from losses at lower frequencies (GHz range), plasma frequency becomes comparable with Γ. The loss, in turn, restricts metals from negative permittivity at lower frequencies.

3.6.3 Magnetic Metamaterials

It is well known that many noble materials show negative electric permittivity (ϵ) at optical frequencies below their plasma frequency. In nature, however, no material shows negative magnetic permeability (μ) at optical frequencies. Moreover, μ is always close to the free space value, i.e., $\mu \approx \mu_0$ in all materials existing in nature due to the absence of magnetic response at optical frequencies. The main reason for the negligible magnetic response is that the light that couples with the atom of any material has a very weaker magnetic component than the electric one. As a result, only the electric component interacts with the material. Unlike electric response, magnetic response due to electronic spin states starts to vanish above a few GHz frequencies. Again, there is no concept of magnetic plasma frequency, unlike electron plasma frequency, due to the absence of magnetic monopoles. So, effectively it has always been challenging work to get negative μ. Nevertheless, with the growth of metamaterials research, researchers have invented many artificial sub-wavelength nanostructures like split-ring resonators, nano-clips, nanorings, and nanoplates, which can provide negative μ at optical frequencies. Here, the magnetic response does not depend on the electronic spin states of the atom. Rather, by designing a special structure of the 'meta-atoms', i.e., the functional unit of the metamaterial, magnetism can be accomplished. In Figure 3.15a, we present the

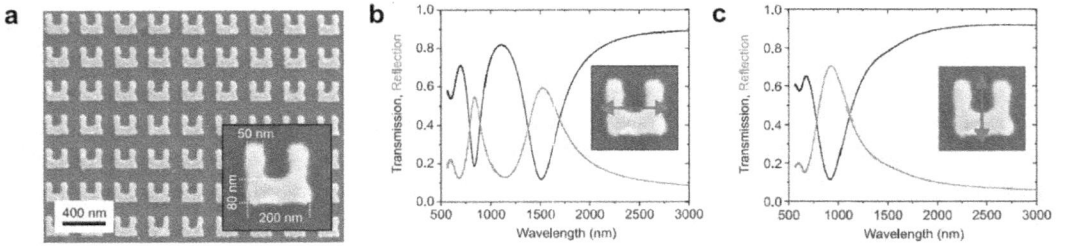

FIGURE 3.15 Magnetic metamaterials. (a) SEM image of split-ring resonator arrays made of Au. Inset shows the dimension of a single resonator. (b) and (c) are measured normal-incidence transmission and reflection spectra for two different input polarizations. (This figure is adapted with permission from Ref. [76], American Physical Society.)

SEM image of Au-based split-ring resonator arrays. This structure has been regarded as the first experimentally reported magnetic metamaterials at telecom wavelengths [76]. Figure 3.15b and c depicts the measured transmission and reflection spectra for horizontal and vertical input polarized light, respectively. It was observed that for horizontal polarization, fundamental magnetic mode at 1.5 μm wavelength could be coupled. Later on, other groups showed magnetic metamaterials at visible wavelengths [77,78].

3.6.4 NEGATIVE-INDEX METAMATERIALS (NIMS)

Maxwell's EM theory reads (for free charge and current)

$$\vec{\nabla} \times \vec{E} = -\frac{\partial \vec{B}}{\partial t}; \vec{\nabla} \times \vec{H} = -\frac{\partial \vec{D}}{\partial t} \tag{3.2}$$

Now, for a plane monochromatic wave we can label $\vec{E}$ and $\vec{H}$ as $\vec{E} = \vec{E}_0 e^{i(\vec{k} \times \vec{r} - wt)}$ and $\vec{H} = \vec{H}_0 e^{i(\vec{k} \times \vec{r} - wt)}$. Again, we know that $\vec{D} = \epsilon \ \vec{E}$ and $\vec{B} = \mu \ \vec{H}$. Putting all these in equation (3.2), we obtain

$$\vec{k} \times \vec{E} = \mu \omega \vec{H}; \vec{k} \times \vec{H} = -\epsilon \omega \vec{E} \tag{3.3}$$

In a normal dielectric where $\epsilon, \mu > 0, \vec{k}, \vec{E}$, and $\vec{H}$ vectors construct a right-handed triplet, while in metamaterials ($\epsilon, \mu < 0$), those three vectors form a left-handed triplet.

Vesalago first discussed the concept of having simultaneously negative ϵ and μ in his seminal paper [47], where he showed that a negative index could affect refraction, the Doppler effect, and the Cerenkov radiation. But the practical realization of negative-index material took another two decades when Pendry and co-workers demonstrated special materials structure that can show negative permittivity and permeability [48–50]. Smith et al. [79] experimentally showed metamaterials and a negative refractive index for the first time.

3.6.5 NONLINEAR OPTICS WITH METAMATERIALS

From the invention of the laser, the field of nonlinear optics has witnessed tremendous applications in modern photonic technology, material science, chemistry, biology, and industry. Apart from the applications, many new physical phenomena have emerged due to the photon–photon interactions. In the macroscopic domain, the induced polarization $\vec{P}$ can be written as a Taylor series in an electric field $\vec{E}$ as

$$P_i = \epsilon_0 \left(\sum_j \chi_{ij}^{(1)} E_j + \sum_{jk} \chi_{ijk}^{(2)} E_j E_k + \sum_{jkl} \chi_{ijkl}^{(3)} E_j E_k E_l + ... \right) \tag{3.4}$$

where ϵ_0 is the permittivity in a vacuum, $\chi^{(1)}$ is the linear optical susceptibility, $\chi^{(2)}$ and $\chi^{(3)}$ are the second- and third-order optical susceptibility, respectively. The first term denotes the linear effects like reflection and refraction. Generally, the value of $\chi^{(2)}$ can be an order of a few pm/V. The $\chi^{(3)}$ has orders of magnitudes lower values like 10^{-12} of $\chi^{(2)}$. So, a high E value is required for the higher order terms in order to give sufficient contribution in P and hence various nonlinear effects. These nonlinear effects again depend on the symmetry of the bulk crystals. Like, centrosymmetric materials lack the $\chi^{(2)}$ term where $\chi^{(2)}$ is responsible for many nonlinear effects like SHG, sum frequency generation (SFG), difference frequency generation (DFG), spontaneous parametric downconversion (SPDC), parametric amplification, three- and four-wave mixing, and many more. The $\chi^{(3)}$ term does not require any symmetry constraint, and it can initiate THG, Kerr effects, and four-wave mixing. Other higher order terms in the expression of P generate higher harmonics.

Frequency conversion is one of the most interesting topics of research among all nonlinear effects. Extensive efforts have been put to increase the conversion efficiency of the bulk samples. Engineered chemical synthesis and artificial materials are a few of them. However, most bulk nonlinear materials suffer from phase-matching conditions, which decrease the conversion efficiency. In the last decade, researchers have focused on the 2D nonlinear metasurfaces where sub-wavelength structures support strong light-matter interaction, almost beating phase-matching conditions. Generally, the overall efficiency of the nonlinear metasurface depends on the material composition, arrangement, and geometry of the meta-atoms.

However, metasurface structures still require noncentrosymmetric nature. To combat this problem, many different shapes of plasmonic nanostructures were introduced like L-shaped [80], G-shaped [81], split-ring resonators [82], and so on, where the symmetry is broken by its constituent structures. In Figure 3.16a, we show a metasurface consisting of L-shaped meta-atoms of Au [80].

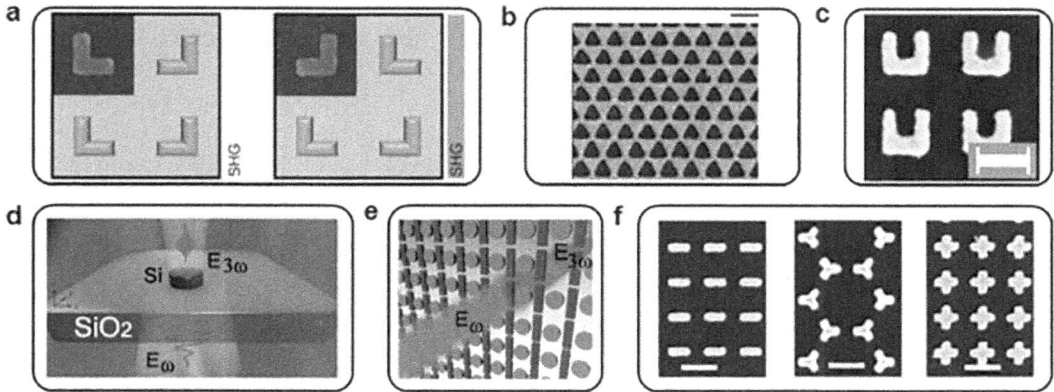

FIGURE 3.16 Nonlinear photonic metasurfaces. (a) Changing of mutual ordering in meta-atoms causes distinct behavior in SHG response. (b) Example of plasmonic nonlinear metasurfaces with threefold rotational symmetry. (c) Example of a split-ring resonator that generates second-harmonic light. (d) Third harmonic generation (THG) from Si nanoparticles. (e) Fano-resonance assisted THG enhancement in a Si metasurface. (f) THG from plasmonic nonlinear crystals with different rotational symmetries. (Panel (a) is adapted with permission from Ref. [80], American Chemical Society; Panel (b) is adapted with permission from Ref. [84], American Physical Society; Panel (c) is adapted with permission from Ref. [82], American Physical Society; Panel (d) is adapted with permission from Ref. [83], American Chemical Society; Panel (e) is adapted with permission from Ref. [86], American Chemical Society; Panel (f) is adapted with permission from Ref. [87], American Physical Society.)

The metasurface was found to exhibit distinct second-harmonic efficiency when the mutual ordering in the meta-atoms was changed. Figure 3.16b is the SEM image of a plasmonic metasurface having threefold rotational symmetry. This metasurface could generate a counter-circularly polarized second-harmonic light when a circularly polarized fundamental beam was made incident on the metasurface [84]. In Figure 3.16c, we present a scanning electron micrograph image of a split-ring resonator-based meta-atoms made of Au metal. An SEM image of plasmonic nonlinear crystals having different rotational symmetries and generating THG is depicted in Figure 3.10f. Although plasmonic structures enhance the nonlinear process with the help of magnetic resonances, they still suffer from high dissipation. Moreover, they are not suitable for high-power laser applications. On the other hand, all-dielectric metasurfaces show magnetic resonances and can survive at high laser intensities. Dielectric nanoparticles with a high refractive index like crystalline Si, GaAs, and Ge are preferred most. At visible frequencies, they can show magnetic and electric resonances tuned for nanoscale light manipulation. GaAs is known for its high $\chi^{(2)}$ – nonlinearity ($d_{36} = 360$ pm/V, [85] (Boyd, 2008)). On the other hand, Si and Ge have higher third-order optical nonlinearity ($\chi^{(3)}_{Si} = 2.8 \times 10^{-18}$ m^2/V^2, $\chi^{(3)}_{Ge} = 5.6 \times 10^{-19}$ m^2/V^2). Figure 3.16d illustrates strong THG from Si nanodiscs deposited on a silica substrate. Enhanced THG was observed in the proximity of magnetic dipolar resonances. Another type of structure that shows enhanced THG is illustrated in Figure 3.16e [86]. Here, strong nonlinearity was observed with the assistance of the Fano resonance of the Si metasurface. Figure 3.16f is the SEM image of plasmonic nonlinear crystals with different rotational symmetries generating strong third harmonic light.

3.6.5.1 BaTiO$_3$ Nanoparticles-based Nonlinear Metasurfaces

Recently, there has been a growing interest in perovskite-based nanomaterials. They show broadband nonlinear effects that span from near UV to visible. As discussed earlier, the enhancement of nonlinear effects at the nanoscale has always been favorable. Although metal-based plasmonic nanostructures are able to do that, they possess high loss and are difficult to fabricate. Recently, Timpu et al. [88] have observed Mie resonance-enhanced SHG in a single BaTiO$_3$ nanoparticle. Here, the enhanced SHG was four orders of magnitude larger within the same nanoparticle at the visible range. Interestingly, these nanoparticles are non-toxic, biocompatible, photo-stable, and non-bleaching, making them suitable for various applications like bioimaging, biosensing, and others. Moreover, BaTiO$_3$ has high $\chi^{(2)}$ nonlinearity, high refractive index, noncentrosymmetric structure, and the second harmonic are generated from the bulk, further increasing its nonlinear conversion efficiency. Most recently, the group of Prof Rachel Grange has fabricated BaTiO$_3$-based metasurfaces that show enhanced SHG in the UV-Vis range [89]. The BaTiO$_3$ layer was deposited on a transparent glass surface using the pulsed laser deposition (PLD) method.

Today, the field of nonlinear metamaterials has traveled a long path. It is not just limited to the enhanced frequency conversion; several advanced applications have been made possible using nonlinear metamaterials. Almeida et al. [90] reported nonlinear metamaterial holograms where the phase and amplitude information of an optical element were stored in the V-shaped Au nanoantennae. Moreover, these holograms create background-free, high-resolution holographic images at the third harmonic of the input laser.

3.6.6 Recent Advances and Applications

With the experimental realization of metamaterials at optical frequencies, researchers, with the help of advanced fabrication technologies, have invented different types of new metamaterials to implement in different applications. Some of them even paved the paths of new physical phenomena. Here, we will briefly mention some of the current applications and future directions of optical metamaterials. For more details about the current trends and applications, readers are advised to follow these references [91–94].

3.6.6.1 Wavefront and Ultrafast Pulse Shaping

Wavefront shaping is a widely used technique in many applications mainly related to disordered media. The optical wavefront is manipulated for imaging inside biological tissue and focusing inside or through disordered media. In most cases, liquid crystal-based spatial light modulators (SLM) or digital micromirror devices (DMD) are used for WFS. Recently, WFS and disordered media have been exploited to get the perfect focusing of a lens system. In another study, researchers have succeeded in coupling evanescent and propagating fields with the help of WFS and disordered medium to focus light at the near-field. Jang et al. [95] demonstrated that a specially designed disordered metasurface can alter the disordered medium and can get rid of the complete input-output calibration. Generally, a metasurface is composed of many sub-wavelength structures that can change the phase of the input beam. This type of metasurface is capable of having a wide field of view, high transmission, large optical memory effect, and sharp focusing. Divitt et al. [96] performed Fourier transform pulse shaping using a pair of diffraction gratings, a pair of parabolic mirrors, and a dielectric metasurface segmented in N superpixels.

3.6.6.2 Tunable Optical Metamaterials

There are some types of metamaterials whose optical properties like reflectance, transmittance, and absorbance, among others, can be tuned. The external factors responsible for the change in the properties are temperature, magnetic field, strain [97], pressure, voltage, and others. If the substrate of a metamaterial is somehow deformed, then the nanostructures' arrangement, as well as mutual couplings, will be changed. This, in turn, will affect the optical properties. In other cases, if the surrounding medium is filled with temperature/voltage/magnetic field sensitive liquid, that can also change the optical properties [98,99]. Dielectric metasurfaces integrated with nematic liquid crystals (LC) have changed transmittance upon the introduction of voltage [100]. Here, the external voltage realigns the LC, changing the electric and magnetic dipolar resonances.

3.6.6.3 Cloaking

Making an object invisible or undetectable has always been a dream of a human being, especially a child. The concept of invisibility has been nurtured in many myths, stories, science fiction, video games, and TV series. The fascination with invisibility would have revolved around myth and stories only until metamaterials were invented. Metamaterials have enabled people to start thinking of ways to make a practical cloaking device. Many schemes about electromagnetic cloaking have been proposed in the last few years. Among them, the coordinate transformation technique has attracted the most as it deals with macroscopic objects to be cloaked. Using transformation optics, researchers can design curved optical spaces that can give unusual shapes to the flow of light. In 2007, Cai et al. [91] first proposed an optical cloaking device. A few other groups [101–103] have also experimentally shown cloaking effect at optical frequencies. Although researchers are coming up with many new structures of optical cloaking devices, there is still a long way to go before one can convert the dreams about invisibility into reality.

Besides many applications, researchers have observed many topological effects like robust edge states and Möbius symmetry in metamaterials [74,104–107].

REFERENCES

1. Yablonovitch, E. (1987). Inhibited spontaneous emission in solid-state physics and electronics. *Phys. Rev. Lett.*, *58*, 2059.
2. John, S. (1987). Strong localization of photons in certain disordered dielectric superlattices. *Phys. Rev. Lett.*, *58*, 2486.
3. Yablonovitch, E., & Gmitter, T. J. (1989). Photonic band structure: The face-centered-cubic case. *Phys. Rev. Lett.*, *63*, 1950.

4. Ho, K. M., Chan C. T., & Soukoulis C. M. (1990). Existence of a photonic gap in periodic dielectric structures. *Phys. Rev. Lett.*, *65*, 3152.

5. Yablonovitch E. Gmitter T.J., & Leung K. M. (1991). Photonic band structure: The face-centered-cubic case employing nonspherical atoms. *Phys. Rev. Lett.*, *67*, 2295.

6. Joannopoulos J. D., Johnson S. G., Winn J. N., & Meade R. D. (2008). *Photonic Crystals: Molding the Flow of Light*. Princeton University Press, Princeton, USA.

7. Kullig J, Jiang X.,Yang L., & Wiersig J. (2020). Microstar cavities: An alternative concept for the confinement of light. *Phys. Rev. Res.*, *2*, 012072(R).

8. Toropov N., Cabello G., Serrano, M. P., Gutha R. R., Rfti M., & Vollmer F. (2021). Review of biosensing with whispering-gallery mode lasers. *Light: Sci. Appl.*, *10*(42), 1–19.

9. Wiersma D. S., Bartolini P., Lagendijk A., & Righini R. (1997). Localization of light in a disordered medium. *Nature*, *390*, 671–673.

10. Schwartz T., Bartal G., Fishman S., & Segev M. (2007). Transport and Anderson localization in disordered two-dimensional photonic lattices. *Nat. Photon.*, *446*, 52–55.

11. Segev M., Silberberg Y., & Christodoulides D. N. (2013). Anderson localization of light. *Nat. Photon.*, *7*, 197–204.

12. Kumar R., Balasubrahmaniyam, M., Alee K. S., & Mujumdar S.(2017). Temporal complexity in emission from Anderson localized lasers. *Phys. Rev. A*, *96*, 063816.

13. Mondal S., Kumar R., Kamp M., & Mujumdar S. (2019). Optical Thouless conductance and level-spacing statistics in two-dimensional Anderson localizing systems. *Phys. Rev. B*, *100*, 060201(R).

14. Kumar R., Mondal S., Balasubrahmaniyam, M., Kamp M., & Mujumdar S. (2020). Discrepant transport characteristics under Anderson localization at the two limits of disorder. *Phys. Rev. B*, *102*, 220202(R).

15. Mafi, A. (2015). Transverse Anderson localization of light: A tutorial. *Adv. Opt. Photon.*, *7*(3), 459–515.

16. Giordani T., Schirmacher W., Ruocco G., & Leonetti M. (2021). Transverse and quantum localization of light: A review on theory and experiments. *Front. Phys.*, *9*, 715663.

17. https://chem.libretexts.org/Bookshelves/Organic_Chemistry/Map%3A_Organic_Chemistry_ (Vollhardt_and_Schore)/05._Stereoisomers/5.1%3A_Chiral__Molecules.

18. Yoo, S., & Park, Q.-H. (2015). Chiral light-matter interaction in optical resonators. *Phys. Rev. Lett.*, *114*, 203003.

19. Feis J., Beutel G., K¨opfler, J., Garcia-Santiago, J., Rockstuhl C., Wegener M., & Fernandez-Corbaton I.(2020). Helicity-preserving optical cavity modes for enhanced sensing of chiral molecules. *Phys. Rev. Lett.*, *124*, 033201.

20. Smith C. L. C., Stenger N., Kristensen A., Mortensen N. A., & Bozhevolnyi S. I. (2014). Efficient excitation of channel plasmons in tailored, UV-lithography defined V-grooves. *Nano Lett.*, *14*, 1659–1664.

21. Chen, Z., Wu, Z., Ming, Y., Zhang, X., & Lu, Y. (2014). Hybrid plasmonic waveguide in a metal V-groove. *AIP Adv.*, *4*, 017103.

22. Bian, Y., & Gong, Q. (2013). Low-loss hybrid plasmonic modes guided by metal-coated dielectric wedges for subwavelength light confinement. *Appl. Opt.*, *52*(23), 5733–5741.

23. Bian, Y., & Gong, Q. (2014). Deep-subwavelength light confinement and transport in hybrid dielectric-loaded metal wedges. *Laser Photon. Rev.*, *8*(4), 549–561.

24. Fan Z., Huang X., Tan C., & Zhang H. (2015). Thin metal nanostructures: Synthesis, properties and applications. *Chem. Sci.*, *6*, 95–111.

25. Zhuiykov, S. (Ed.). (2014). *6-Nanostructured Semiconductor Composites for Solar Cells*. Woodhead Publishing, Cambridge, UK.

26. Yi, G.-C. (Ed.). (2012). *Semiconductor Nanostructures for Optoelectronic Devices*. Springer, Berlin.

27. Martín-Palma R. J., Manso M., & Torres-Costa V. (2009). Optical biosensors based on semiconductor nanostructures. *Sensors*, *9*, 5149–5172.

28. Li M., Li J. C., & Jiang Q. (2010). Size-dependent bad-gap and dielectric constant of Si nanocrystals. *Int. J. Modern Phys. B*, *24*, 2297–2301.

29. Jiang R., Li B., Fang C., & Wang J. (2014). Metal/semiconductor hybrid nanostructures for plasmon-enhanced applications. *Adv. Mater.*, *26*, 5274–5309.

30. Fu M., Wang K., Long H., Yang G., Lu P., Hetsch, F. Susha A. S., & Rogach A. L. (2012). Resonantly enhanced optical nonlinearity in hybrid semiconductor quantum dot-metal nanoparticle structures. *Appl. Phys. Lett.*, *100*, 063117.

31. Kamarudheen R, Kumari G., & Baldi A. (2020). Plasmon-driven synthesis of individual metal@semiconductor core@shell nanoparticles. *Nat. Commun.*, *11*(3957), 1–10.

32. Shaviv E., Schubert O., Alves-Santos M., Goldoni G., Di Felice R., Vallée F., Del Fatti N., Banin U., & Sönnichsen C. (2011). Absorption properties of metal–semiconductor hybrid nanoparticles. *ACS Nano*, *5*(6), 4712–4719.

33. Tang X., Kröger E., Nielsen A., Schneider S., Strelow C., Mews A., & Kipp T. (2019). Fluorescent metal–semiconductor hybrid structures by ultrasound-assisted in situ growth of gold nanoparticles on silica-coated CdSe-Dot/CdS-rod nanocrystals. *Chem. Mater.*, *31*(1), 224–232.

34. Zámbó D., Schlosser A., Graf R. T., Rusch P., Kißling P. A., Feldhoff A., &. Bigall N. C. (2021). One-step formation of hybrid nanocrystal gels: Deposition of metal domains on CdSe/CdS nanorod and nanoplatelet networks. *Adv. Opt. Mater.*, *9*, 2100291.

35. Achermann, M. (2010). Exciton–plasmon interactions in metal–semiconductor nanostructures. *J. Phys. Chem. Lett.*, *1*(19), 2837–2843.

36. Wu J.-L., Chen F.-C., Hsiao Y.-S., Chien F.-C., Chen P., Kuo C.-H., Huang M. H., & Hsu C.-S. (2011). Surface plasmonic effects of metallic nanoparticles on the performance of polymer bulk heterojunction solar cells. *ACS Nano*, *5*(2), 959–967.

37. Gan Q., Bartoli F. J., & Kafafi Z. H. (2013). Plasmonic-enhanced organic photovoltaics: Breaking the 10% efficiency barrier. *Adv. Mater.*, *25*, 2385–2396.

38. Chou, C.-H., & Chen, F.-C. (2014). Plasmonic nanostructures for light trapping in organic photovoltaic devices. *Nanoscale*, *6*, 8444.

39. Tyagi P., Kumar A., Rhee S., Lee H., Song J., Kim J., & Lee C. (2016). Plasmon-induced slow aging of exciton generation and dissociation for stable organic solar cells. *Optica*, *3*(10), 1115–1121.

40. Ray D., Kiselev A., & Martin O. J. F. (2021). Multipolar scattering analysis of hybrid metal-dielectric nanostructures. *Opt. Exp.*, *29*(15), 24056–24067.

41. Barreda A., Vitale F., Minovich A. E., Ronning C., & Staude I. (2022). Applications of hybrid metal-dielectric nanostructures: State of the art. *Adv. Photon. Res.*, *3*(4), 2100286, 1–48.

42. Renaut C., Lang L., Frizyuk K., Timofeeva M., Komissarenko F. E., Mukhin I. S., Smirnova D., Timpu F., Petrov M., Kivshar Y., & Grange. R. (2019). Reshaping the second-order polar response of hybrid metal–dielectric nanodimers. *Nano Lett.*, *19*(2), 877–884.

43. Gili F. V., Ghirardini L., Rocco D., Marino G., Favero I., Roland I., Pellegrini G., Duò L., Finazzi M., Carletti L., Locatelli A., Lemaître A., Neshev D., De Angelis C., Leo G., & Celebrano M. (2018). Metal–dielectric hybrid nanoantennas for efficient frequency conversion at the anapole mode. *Beilstein J. Nanotechnol.*, *9*, 2306–2314.

44. Kim, K.-H. (2020). Low-index dielectric metasurfaces supported by metallic substrates for efficient second-harmonic generation in the blue-ultraviolet range. *Phys. Chem. Chem. Phys.*, *22*, 7300–7305.

45. Shibanuma T., Grinblat G., Albella P., & Maier S. A. (2017). Efficient third harmonic generation from metal–dielectric hybrid nanoantennas. *Nano Lett.*, *17*(4), 2647–2651.

46. Yao J., Yin Y., Ye L., Cai G., & Liu Q. H. (2020). Enhancing third-harmonic generation by mirror-induced electric quadrupole resonance in a metal–dielectric nanostructure. *Nano Lett.*, *45*(20), 5864–5867.

47. Veselago, V. G. (1968). The electrodynamics of substances with simultaneously negative values of ε and μ. *Sov. Phys. Usp.*, *10*, 509.

48. Pendry, J. B. (2000). Negative refraction makes a perfect lens. *Phys. Rev. Lett.*, *85*, 3966.

49. Pendry J. B., Holden A. J., Stewart W. J., & Youngs I. (1996). Extremely low frequency plasmons in metallic mesostructures. *Phy. Rev. Lett.*, *76*(25), 4773.

50. Pendry J. B., Holden A. J., Robbins D. J., & Stewart W. J. (1999). Magnetism from conductors and enhanced nonlinear phenomena. *IEEE Trans. Microw. Theor. Techn.*, *47*(11), 2075–2084.

51. Fang N., Liu Z, Yen T.-J., & Zhang X. (2003). Regenerating evanescent waves from a silver superlens. *Opt. Exp.*, *11*(7), 682–687.

52. Liu Z., Fang N., Yen T.-J., & Zhang X. (2003). Rapid growth of evanescent wave by a silver superlens. *Appl. Phys. Lett.*, *83*, 5184.

53. Zhang, X., & Liu, Z. (2008). Superlenses to overcome the diffraction limit. *Nat. Mater.*, *7*, 435–441.

54. Fang N., Lee H., Sun C., & Zhang X. (2005). Sub-diffraction-limited optical imaging with a silver superlens. *Science*, *308*(5721), 534–537.

55. Jacob Z., Alekseyev L. V., & Narimanov E. (2006). Optical hyperlens: Far-field imaging beyond the diffraction limit. *Opt. Exp.*, *14*, 8247–8256.

56. Salandrino, A., & Engheta, N. (2006). Far-field subdiffraction optical microscopy using metamaterial crystals: Theory and simulations. *Phys. Rev. B*, *74*, 075103.

57. Lee H., Liu Z., Xiong Y., Sun C., & Zhang X. (2007). Development of optical hyperlens for imaging below the diffraction limit. *Opt. Exp.*, *15*(24), 15886–15891.

58. Liu Z., Lee H., Xiong Y., Sun C., & Zhang X. (2007). Far-field optical hyperlens magnifying sub-diffraction-limited objects. *Science*, *315*(5819), 1686.
59. Lu, D., & Liu, Z. (2012). Hyperlenses and metalenses for far-field super-resolution imaging. *Nat. Commun.*, *3*(1205), 1–9.
60. Vanneste, C., & Sebbah, P. (2009). Complexity of two-dimensional quasimodes at the transition from weak scattering to Anderson localization. *Phys. Rev. A*, *79*, 041802(R).
61. Yeh P., Yariv A., & Marom E. (1978). Theory of Bragg fiber. *J. Opt. Soc. Am.*, *68*(9), 1196–1201.
62. Temelkurana B., Hart S. D., Benoit G., Joannopoulos J. D., & Fink Y. (2002). Wavelength-scalable hollow optical fibres with large photonic bandgaps for CO_2 laser transmission. *Nature*, *420*, 650–653.
63. Knight J. C., Birks T. A., Russell P. St. J., & Atkin D. M. (1996). All-silica single-mode optical fiber with photonic crystal cladding. *Opt. Lett.*, *21*(19), 1547–1549.
64. Cregan R. F., Mangan B. J., Knight J. C., Birks T. A., Russell P. St. J., Roberts P. J., & Allan D. C. (1999). Single-mode photonic band gap guidance of light in air. *Science*, *285*(5433), 1537–1539.
65. Dudley J. M, Genty G., & Coen S. (2006). Supercontinuum generation in photonic crystal fiber. *Rev. Mod. Phys.*, *78*, 1135.
66. Markos C., Travers J. C, Abdolvand A., Eggleton B. J., & Bang O. (2017). Hybrid photonic-crystal fiber. *Rev. Mod. Phys.*, *89*, 045003.
67. Zheltikov, A. M. (2003). The physical limit for the waveguide enhancement of nonlinear-optical processes. *Opt. Spectrosc.*, *95*, 410–415.
68. Almeida V. R., Xu Q., Barrios C. A., & Lipson M. (2004). Guiding and confining light in void nanostructure. *Opt. Lett.*, *29*(11), 1209–1211.
69. Xua Q., Almeida V. R., Panepucci R. R., & Lipson M. (2004). Experimental demonstration of guiding and confining light in nanometer-size low-refractive-index material. *Opt. Lett.*, *29*(14), 1626–1628.
70. Wiederhecker G. S., Cordeiro C. M. B., Couny F., Benabid F., Maier S. A., Knight J. C., Cruz C. H. B., & Fragnito H. L. (2007). Field enhancement within an optical fibre with a subwavelength air core. *Nat. Photon.*, *1*, 115–118.
71. Ruan Y., Ebendorff-Heidepriem H., Afshar S., & Monro T. M. (2010). Light confinement within nanoholes in nanostructured optical fibers. *Opt. Exp.*, *18*(25), 26018–26026.
72. Chen D., Tse M.-L. V., & Tam H. Y. (2010). Optical properties of photonic crystal fibers with a fiber core of arrays of sub-wavelength circular air holes: Birefringence and dispersion. *Progr. Electromagn. Res.*, *105*, 193–212.
73. Zhao X., Cheng J., Liu X., Jianga G., Hua L., Liu Z., & Li S. (2016). Multi-hole core photonic crystal fiber guiding light in subwavelength air holes. *IEEE Photon. Technol. Lett.*, *28*(22), 2625–2628.
74. Chen H.-T., Taylor A., & Yu N. (2016). A review of metasurfaces: Physics and applications. *Rep. Progr. Phys.*, *79*(7), 076401.
75. Hu J., Bandyopadhyay S., Liu Y. H., & Shao L. Y. (2021). A review on metasurface: From principle to smart metadevices. *Front. Phys.*, *8*, 586087.
76. Enkrich C., Wegener M., Linden S., Burger S., Zschiedrich L., Schmidt F., Zhou J. F., Koschny Th., & Soukoulis C. M. (2005). Magnetic metamaterials at telecommunication and visible frequencies. *Phys. Rev. Lett.*, *95*, 203901.
77. Yuan H.-K., Chettiar U. K., Cai W., Kildishev A. V., Boltasseva A., Drachev V. P., & Shalaev V. M. (2007). A negative permeability material at red light. *Opt. Exp.*, *15*(3), 1076–1083.
78. Cai W., Chettiar U. K., Yuan H.-K., de Silva V. C., Kildishev A. V., Drachev V. P., & Shalaev V. M. (2007). Metamagnetics with rainbow colors. *Opt. Exp.*, *15*(6), 3333–3341.
79. Smith D. J., Pendry J. B., & Wiltshire M. C. K. (2004). Metamaterials and negative refractive index. *Science*, *305*(5685), 788–792.
80. Husu H., Siikanen R., Mäkitalo J., Lehtolahti J., Laukkanen J., Kuittinen M., & Kauranen M. (2012). Metamaterials with tailored nonlinear optical response. *Nano Lett.*, *12*(2), 673–677.
81. Valev V. K., Smisdom N., Silhanek A. V., De Clercq B., Gillijns W., Ameloot M., Moshchalkov V. V., & Verbiest T. (2009). Plasmonic ratchet wheels: Switching circular dichroism by arranging chiral nanostructures. *Nano Lett.*, *9*(11), 3945–3948.
82. Linden S., Niesler F. B. P., Förstner J., Grynko Y., Meier T., & Wegener M. (2012). Collective effects in second-harmonic generation from split-ring-resonator arrays. *Phys. Rev. Lett.*, *109*, 015502.
83. Shcherbakov M. R., Neshev D. N., Hopkins B., Shorokhov A. S., Staude I., Melik-Gaykazyan E. V., Decker M., Ezhov A. A., Miroshnichenko A. E., Brener I., Fedyanin A. A., & Kivshar Y. S. (2014). Enhanced third-harmonic generation in silicon nanoparticles driven by magnetic response. *Nano Lett.*, *14*(11), 6488–6492.

84. Konishi K., Higuchi T., Li J., Larsson J., Ishii S., & Kuwata-Gonokami M. (2014). Polarization-controlled circular second-harmonic generation from metal hole arrays with threefold rotational symmetry. *Phys. Rev. Lett.*, *112*, 135502.

85. Boyd, R. W. (2008). *Nonlinear Optics*. Academic Press.

86. Yang Y., Wang W., Boulesbaa A., Kravchenko I. I., Briggs D. P., Puretzky A., Geohegan D., & Valentine J. (2015). Nonlinear fano-resonant dielectric metasurfaces. *Nano Lett.*, *15*(11), 7388–7393.

87. Chen S., Li G., Zeuner F., Wong W. H., Pun E. Y. B., Zentgraf T, Cheah K. W., & Zhang S. (2014). Symmetry-selective third-harmonic generation from plasmonic metacrystals. *Phys. Rev. Lett.*, *113*, 033901.

88. Timpu F., Sergeyev A., Hendricks N. R., & Grange R. (2017). Second-harmonic enhancement with mie resonances in perovskite nanoparticles. *ACS Photon.*, *4*(1), 76–84.

89. Timpu F., Escalé M. R., Timofeeva M, Strkalj N., Trassin M., Fiebig M., & Grange R. E. (2019). Enhanced nonlinear yield from barium titanate metasurface down to the near ultraviolet. *Adv. Opt. Mater.*, *7*(22), 1900936.

90. Almeida E., Bitton O , & Prior Y. (2016). Nonlinear metamaterials for holography. *Nat. Commun.*, *7*(12533), 1–7.

91. Cai, W., & Shalaev, V. (2010). *Optical Metamaterials: Fundamentals and Applications*. Springer.

92. Urbas A. M., Jacob Z., Negro L. D., Engheta N., Boardman A. D., Egan P., Khanikaev A. B., Menon V, Ferrera M., Kinsey N., DeVault C., Kim J., Shalaev V., Boltasseva A., Valentine J., Pfeiffer C., Grbic A., Narimanov E., Zhu L., Fan S., Alù A., Poutrina E., Litchinitser N. M., Noginov M. A., MacDonald K. F., Plum E., Liu X., Nealey P. F., Kagan C. R., Murray C. B., Pawlak D. A., Smolyaninov I. I., Smolyaninova V. N., & Chanda D. (2016). Roadmap on optical metamaterials. *J. Opt.*, *18*, 093005.

93. Tanaka T., & Ishikawa A. (2017). Towards three-dimensional optical metamaterials. *Nano Converg.*, *4*, 34.

94. Brener, I., Liu, S., Staude, I., Valentine, J., & Holloway, C. (Eds.). (2020). *Dielectric Metamaterials: Fundamentals, Designs, and Applications*. Elsevier Woodhead Publishing.

95. Jang M., Horie Y., Shibukawa A., Brake J., Liu Y., Kamali S. M., Arbabi A., Ruan H., Faraon A., & Yang C. (2018). Wavefront shaping with disorder-engineered metasurfaces. *Nat. Photon.*, *12*(6), 84–90.

96. Divitt S., Zhu W., Zhang C., Lezec H. J., & Agrawal A. (2019). Ultrafast optical pulse shaping using dielectric metasurfaces. *Science*, *364*(6443), 890–894.

97. Gutruf F., Zou C., Withayachumnankul W., Bhaskaran M., Sriram S., & Fumeaux C. (2016). Mechanically tunable dielectric resonator metasurfaces at visible frequencies. *ACS Nano*, *10*(1), 133–141.

98. Wang X., Kwon D.-H., Werner D. H., Khoo I.-C., Kildishev A. V., & Shalaev V. M. (2007). Tunable optical negative-index metamaterials employing anisotropic liquid crystals. *Appl. Phys. Lett.*, *91*, 143122.

99. Khoo I. C., Diaz A., Liou J., Stinger M. V., Huang J., & Ma Y. (2014). Liquid crystals tunable optical metamaterials. *IEEE J. Sel. Top. Quant. Electron.*, *16*(2), 410–417.

100. Komar, A., Fang, Z., Bohn, J., Sautter, J., Decker, M., Miroshnichenko, A., Pertsch, T., Brener, I., Kivshar, Y. S., Staude, I , & Neshev, D. N. (2017). Electrically tunable all-dielectric optical metasurfaces based on liquid crystals. *Appl. Phys. Lett.*, *110*, 071109.

101. Valentine J. Li J., Zentgraf T., Bartal G., & Zhang X. (2009). An optical cloak made of dielectrics. *Nat. Mater.*, *8*, 568–571.

102. Gabrielli L. H., Cardenas J., Poitras C. B., & Lipson M. (2009). Silicon nanostructure cloak operating at optical frequencies. *Nat. Photon.*, *3*, 461–463.

103. Chen H., Chan C. T., & Sheng P. (2010). Transformation optics and metamaterials. *Nat. Mater.*, *9*, 387–396.

104. Krishnamoorthy H. N. S., Jacob Z., Narimanov E., Kretzschmar I., & Menon V. M. (2012). Topological transitions in metamaterials. *Science*, *336*, 205–209.

105. Nash L. M., Kleckner D., Read A., Vitelli V., Turner A. M., & Irvine W. T. M. (2015). Topological mechanics of gyroscopic metamaterials. *PNAS*, *112*(47), 14495–14500.

106. Slobozhanyuk A. P., Khanikaev A. B., Filonov D. S., Smirnova D. A., Miroshnichenko A. E., & Kivshar Y.S. (2016). Experimental demonstration of topological effects in bianisotropic metamaterials. *Sci. Rep.*, *6*(22270), 1–7.

107. Yu R., Alaee R., Boyd R. W., & de Abajo F. J. G. (2020). Ultrafast topological engineering in metamaterials. *Phys. Rev. Lett.*, *125*, 037403.

4 Optical Nanoscopy

Kishore K. Madapu

CONTENTS

4.1 INTRODUCTION

The diffraction limit restricts the spatial resolution of conventional optical microscopy. The maximum spatial resolution achieved by an optical microscope with the highest available numerical aperture (N.A.) (1.3–1.4) is $\geq \lambda/2$, where λ is the wavelength of light. In other words, the spatial resolution achieved by the typical green light is not more than 250 nm. However, with its spectacular applications and fundamental research interest, nanotechnology deals with features with sizes ≤ 100 nm. Consequently, conventional optical microscopy is handicapped from detecting or resolving these nanostructures [1]. Similar to optical microscopy, optical spectroscopy is also limited by the diffraction limit, which poses a constraint on the spatial resolution of microanalysis. Moreover, the inherent capabilities of optical microscopy in extracting spectroscopic information, temporal resolution, and polarization capabilities cannot be utilized in the case of nanomaterials because of the diffraction limit. In the case of nanotechnology, vast information obtained from optical microscopy and spectroscopy cannot be retrieved except for ensemble information [1].

The limitation of optical microscopy can be understood from Abbe's diffraction theory. The light spot formed by the objective lens limits the spatial resolution achieved by optical microscopy. At the focal point, the lens system forms symmetric concentric rings with central maxima, Airy disk pattern (Figure 4.1). According to Abbe, the distance between the center of the central maxima and the first node of the diffraction pattern is given by

$$d = \frac{0.61\lambda}{n\sin\theta} = \frac{0.61\lambda}{N.A.} \tag{4.1}$$

where λ is the free-space wavelength, n is the refractive index of the medium, and θ is the light-converging angle of the objective lens. The product of n and $\sin\theta$ is the so-called numerical aperture

DOI: 10.1201/9781003248323-4

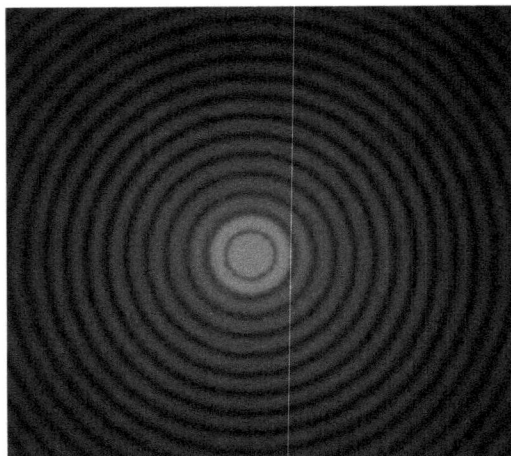

FIGURE 4.1 An Airy disk formed when a laser beam passed through a 90 µm pinhole aperture. Airy disk, https://en.wikipedia.org/w/index.php?title=Airy_disk&oldid=1064292651.

of an objective lens. According to the Rayleigh criterion, the minimum resolvable distance between the two objects by any optical system must be equal to the distance given by equation (4.1). Thus, the maximum achievable spatial resolution is $\sim\lambda/2$ with the highest N.A. of 1.3–1.4 [1,2]. The improvement of spatial resolution optical microscopy beyond the diffraction limit is inevitable to utilize the full potential of optical microscopy in nanotechnology.

4.2 NEAR-FIELD OPTICS AND EVANESCENT WAVE

As mentioned earlier, the spatial resolution of an optical system is limited by the diffraction limit. In 1928, Synge proposed an idea that could break the diffraction limit. The original concept of Synge was that the diffraction limit could be overcome by using sub-wavelength apertures. The original idea of Synge is shown in Figure 4.2. The high-intensity light is illuminated on the backside of the opaque metal screen with an aperture size less than the wavelength of the illuminated light. The aperture and the sample should be in close proximity. The illuminated light is confined in the aperture, which is used to image the sample surface point by point. In this case, the spatial resolution is dependent on the size of the aperture rather than the wavelength of the illuminated light. However, the idea of Synge was not materialized at that time because of difficulties in realizing the experimental setup. The experimental condition to get the high spatial resolution is that the sub-wavelength aperture must be scanned in a near-field regime.

The preceding discussion revealed the condition to be fulfilled for the high spatial resolution beyond the diffraction limit, i.e., near-field illumination through sub-wavelength apertures. The distance less than the optical wavelength (λ) from the sample surface is considered the near-field region (Figure 4.2). The optics that involves the excitation or collection of light in the near-field regions are defined as near-field optics. The importance of near-field optics in achieving high spatial resolution can be understood by the evanescent field. In a typical optical system, the spatial resolution depends on the extent of photon confinement. For example, in an optical microscope, the confinement of light is determined by the N.A. of an objective lens. Thus, the N.A. of the objective lens dictates the spatial resolution achieved for a given wavelength. The role of photon confinement in optical resolution can be understood from Heisenberg's uncertainty principle. According to Heisenberg's uncertainty principle, the product of confinement of photons (Δx) and momentum of the photons (ΔP_x) is always greater than or equal to $\hbar/2$ (equation 4.2).

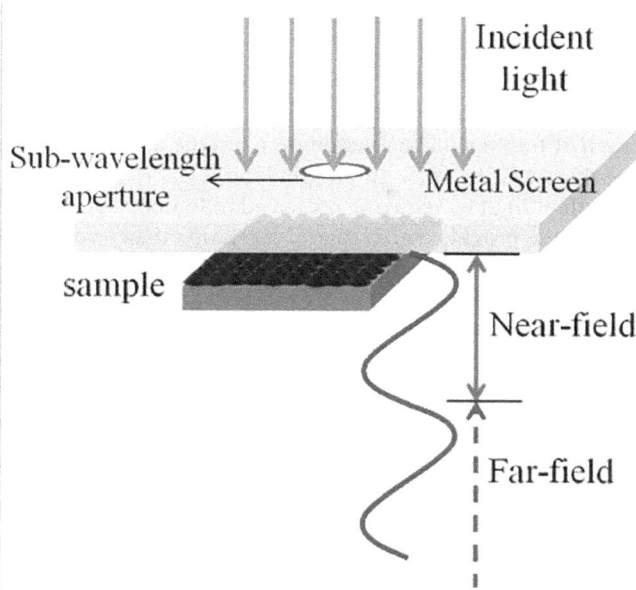

FIGURE 4.2 Schematic of Synge proposal to overcome the diffraction limit.

$$(\Delta x) \cdot (\Delta p_x) \geq \frac{\hbar}{2} \tag{4.2}$$

$$(\Delta x) \cdot (\Delta k_x) \geq \frac{1}{2} \tag{4.3}$$

$$(\Delta x) \cdot \geq \frac{\hbar}{2\Delta k_x} \tag{4.4}$$

According to equation (4.4), the confinement of photons is inversely proportional to the magnitude of the spread of the wave vector. In other words, a large spread of the wave vector magnitudes leads to high confinement of photons and consequently high spatial resolution. The total magnitude of the wave vector is given by

$$k_0 = \sqrt{k_x^2 + k_y^2 + k_z^2} \tag{4.5}$$

where $k_0 = 2\pi/\lambda$ is the total magnitude of the wave vector. In other words, equations (4.4) and (4.5) are reiterated that the free-space wavelength limits the maximum achievable resolution. However, the magnitude in-plane wave vector $\left(k_\parallel^2 = k_x^2 + k_y^2\right)$ can be increased more than the magnitude of the total wave vector (k_0^2) by making one of the components' wave vector imaginary (ik_z). Because of the imaginary component of the wave vector, the z component of the wave vector becomes negative, k_z is $-|k_z|$. As a result, the wave decays exponentially in the z-direction. The exponentially decaying wave is non-propagative, and it is called the "evanescent wave". Thus, the evanescent field possesses a large spread of in-plane wave vectors compared to free-space propagating waves. In other words, imaging with an evanescent field improves the spatial resolution beyond the diffraction limit. Equations (4.6) and (4.7) represent the wave solutions of the plane wave and evanescent wave, respectively. The E_0 in both equations is the electric field amplitude.

$$\vec{E} = E_0 e^{i\left(k_x + k_y \pm k_z\right)} \tag{4.6}$$

$$\vec{E} = E_0 e^{(ik_x + ik_y - |k_z|)} \tag{4.7}$$

Another way to look into improved resolution with the evanescent field is its slowness compared to free-space propagating waves. Equation (4.1) infers that the spatial resolution can be improved by reducing the wavelength of the light. The relation between the angular frequency and wavelength of light is $\omega = 2\pi * v/\lambda$, where v is the velocity of light, and it is c in the case of free space [3]. One can reduce the wavelength of light by reducing the speed of light by keeping the angular frequency constant. Thus, slow light can provide a high spatial resolution than free-space light. Reducing the speed of light is the idea behind oil-immersed optical microscopy. However, this technique cannot be applicable to all kinds of samples. The slow light, the evanescent wave can be available on the metal surface, which originates from collective oscillations of conduction band electrons with the interaction of electromagnetic (EM) waves [4]. The preceding discussion essential conveys that the evanescent wave is necessary for achieving high resolution. In this context, it is worth noting that the evanescent field can be realized by specialized techniques: (i) Total internal reflection by the prism, (ii) surface plasmon resonance (SPR) of metal nanoparticles, (iii) squeezing of light through sub-wavelength apertures, and (iv) activation of surface plasmon polaritons (SPPs) at the conductive and dielectric interface. The above-mentioned techniques are schematically depicted in Figure 4.3.

As the sub-wavelength apertures play an important role in achieving high resolution, the electric field distribution beyond the apertures and its dependence on the size of the aperture is inevitable to understand. The succeeding part discusses field distribution beyond sub-wavelength apertures. The near-field region of sub-wavelength aperture possesses a combination of the evanescent and propagative fields [5]. The evanescent wave is an imaginary or non-propagating EM wave that is confined to the surface [4]. The propagating EM wave is given by

$$\mathbf{E} = E_0 \exp[i(\mathbf{k}.r - \omega t)] \tag{4.8}$$

where $\mathbf{k}$ is the wave vector of the propagating EM wave and is given by,

$$\mathbf{k} = k_x i + k_y j + k_z k \tag{4.9}$$

The magnitude of the k is given by $|k| = 2\pi/\lambda$. Since the evanescent wave is imaginary, the evanescent wave is given by

$$\mathbf{E} = E_0 \exp[i(k_x i + k_y j - \omega t)] \exp[-k_z k] \tag{4.10}$$

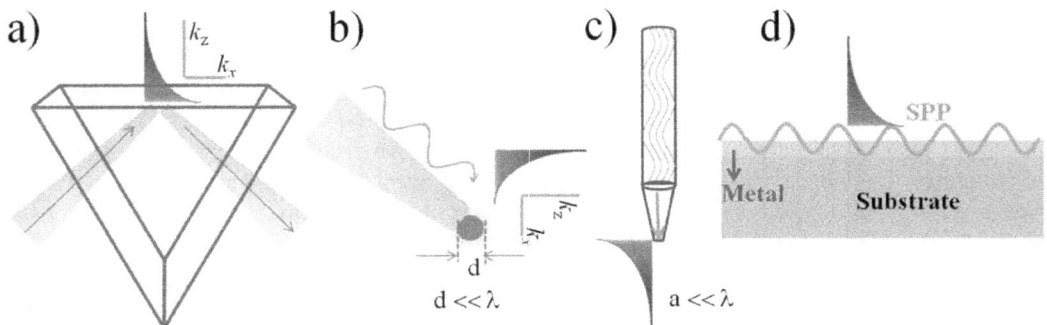

FIGURE 4.3 Different ways to produce the evanescent field: (a) total internal reflection at the prism, (b) surface plasmon resonance (SPR) at metal nanoparticles, (c) squeezing of light through sub-wavelength apertures, and (d) activation of SPPs on the metal surface.

Equation 4.10 indicates that the evanescent wave is exponentially decaying in the z-direction. As a consequence, the evanescent wave exists only in the near-field region. The magnitude of the **k** vector of the propagating EM wave is given by

$$k^2 = k_{\parallel}^2 + k_z^2, \text{ where } k_{\parallel}^2 = k_x^2 + k_y^2 \tag{4.11}$$

If light passes through an aperture of size 'a', $k_{\parallel}$ and k_z describe the EM wave properties of the field just after the aperture, which constitutes the plane and evanescent waves. The magnitude of the k_z has two possibilities depending on the magnitude of the $k_{\parallel}$. Thus, the k_z has the two following solutions:

$$k_z = \sqrt{k^2 - k_{\parallel}}, \ k_{\parallel} \leq k, \text{ and } k_z = i\sqrt{k^2 - k_{\parallel}} \ k_{\parallel} \leq k \tag{4.12}$$

The aperture of size 'a' possesses in-plane wave vector of $k_{\parallel} = 2\pi/a$. For the $k_{\parallel} = 2\pi/a$, the amplitude of the k_z is given by

$$k_z = \sqrt{\left(\frac{2\pi}{\lambda}\right)^2 - \left(\frac{2\pi}{a}\right)^2} \tag{4.13}$$

For $a < \lambda$, the propagating vector in the z-direction is imaginary. In other words, the field transmitted through this aperture is an evanescent field. As a result, the EM field is strongly confined to the aperture and has a non-propagative nature. Because of the strong confinement of light, the optical image constructed by the sub-wavelength aperture provides improved resolution [4]. In the case of $a > \lambda$, the $k_{\parallel} \sim 0$ and $k_z \sim k$, the transmitted wave is a propagating wave.

4.2.1 Transmission and Diffraction of Light through a Sub-Wavelength Aperture

Sub-wavelength apertures play a significant role in nanoscopy. For centuries, sub-wavelength apertures attracted researchers because of intriguing optical phenomena surrounding them and fascinating optical properties. This section of the chapter describes the transmission of light through sub-wavelength apertures. In addition, diffraction also takes place when the light transmits through sub-wavelength apertures. In 1940, Bethe gave the theoretical description for the transmission of light through sub-wavelength apertures. The circular hole in the opaque metal film was considered the sub-wavelength aperture. The transmission coefficient was derived based on two assumptions: the metal film was infinitely thin and perfectly conducting. The transmission coefficient of a sub-wavelength aperture of radius "r" impinged by a plane EM wave of wave wavelength λ is given by

$$T(\lambda) = \frac{64(kr)^4}{27\pi^2}, \tag{4.14}$$

where $k = 2\pi/\lambda$ is the wave vector of an incident plane wave. Equation (4.14) reveals that the transmission coefficient is directly proportional to the fourth power of ratio term of radius and wavelength (r/λ) [4]. In other words, it interprets that the transmission coefficient falls sharply with λ when the $\lambda \gg r$. The phenomenon is described schematically in Figure 4.4a. However, infinitely thin and perfectly conducting mental are practically not feasible. Thus, all apertures have a finite width. The finite width of the aperture enforces the waveguide nature on the sub-wavelength apertures when the light transmits through. As a result, the transmission of light through a finite-width aperture is different from the Bethe prediction as the waveguide nature of the aperture modifies the dispersion relation of light compared to the free space. The waveguide nature of the aperture imposes a cut-off wavelength (λ_c), where transmission is extremely feeble or seizes when $\lambda > \lambda_c$. In this case, transmitted intensity exponentially decays from the metal film and becomes non-propagative. However, in real

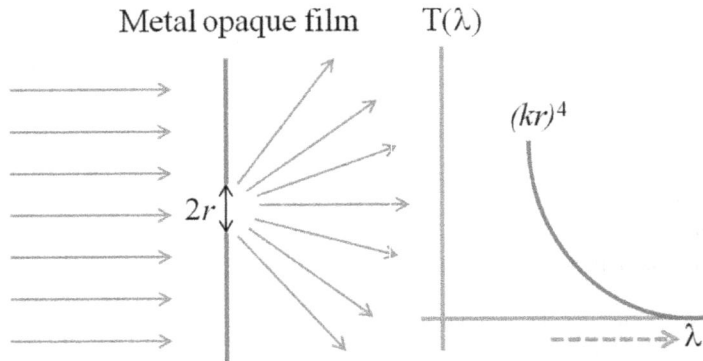

FIGURE 4.4 The typical transmission spectrum of light passing through the sub-wavelength aperture of radius 'r'.

metals, the λ_c is not sharp. Thus, when λ increases more than λ_c, transmission smoothly becomes an evanescent wave from a propagative wave. One can eliminate the λ_c by making the width of the film less than the cut-off wavelength. However, one cannot reduce the metal thickness indefinitely because of the skin depth of metals.

Bethe's theory predicts that the transmission through sub-wavelength apertures is extremely feeble if the $\lambda \gg r$. Because of the considerable width and finite conductivity, the actual sub-wavelength apertures transmission behavior deviates from Bethe's predictions. Degiron et al. studied the light transmission behavior of the sub-wavelength apertures milled in Ag metal films [6]. Resonance behavior was observed in the transmission spectrum sub-wavelength aperture (Figure 4.5a). In this case, the width and diameter of the aperture are 270 and 200 nm, respectively. This observation contradicts Bethe's theory. The observed resonance behavior in the transmission spectrum has been attributed to the excitation of localized surface plasmon resonance (LSPR). The LSPR modes were confined to the ridges of the circular aperture. The activation of the LSPR was further confirmed by the sub-wavelength rectangular apertures (Figure 4.5c). The transmission spectrum of rectangular apertures displays the two resonance peaks, and the spectrum can be tuned by incident light polarization (Figure 4.5d). The observed two peaks in the rectangular aperture are similar to the metal nanorods absorption spectrum, which possesses the two resonance peaks such as longitudinal and transverse modes. In the case of rectangular apertures, either of the modes is selectively excited by changing the polarization of incident light with respect to the aperture axis. In this case, the transmission of light is dictated by the LSPR dipoles, and the hollow space only plays a role in the tunneling of light and the coupling of evanescent waves on both sides [7]. The resonance behavior enhances the transmission intensity at specific wavelengths and sometimes beyond the incident light intensity. The subsequent section will show the possibility of enormous enhancement of transmission intensity and less diffraction for certain conditions.

The foregoing discussion revealed that light transmission is improved because of the activation of the surface plasmons (SPs). Interestingly, the transmission of light is further enhanced if the circular aperture is surrounded by the circular periodic groves such as Bull's eye structure (Figure 4.6) [8]. The transmission efficiency of light is enhanced by the factor of 10 if the circular aperture is surrounded by the groves [8]. However, the transmission is enhanced for only a specific wavelength and which is dictated by the periodicity of the groves. The activation of SPPs is the reason for the enhanced intensity. The SPP is coupled EM wave of free-space light and conduction electrons of metals. The SPPs are evanescently confined to the metal surface with enhanced field intensity. The complete description of SPPs and their importance in achieving the superlens will be discussed in the next section. However, one cannot be activated the SPPs with free-space EM waves because of the momentum mismatch. The circular groves surrounding the sub-wavelength aperture provide the

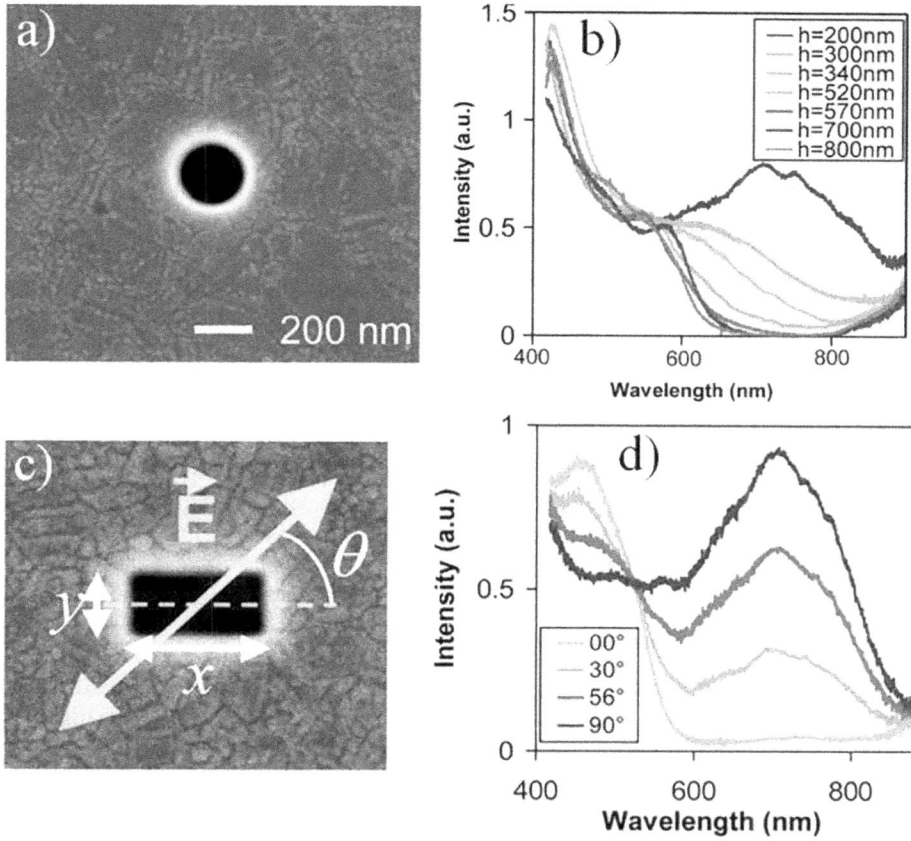

FIGURE 4.5 Light transmission through cylindrical and rectangular sub-wavelength apertures milled in Ag film. (a) SEM image of sub-wavelength aperture of radius 270 nm. (b) Transmission spectra of circular sub-wavelength aperture with varying thickness, h. (c) SEM image of the rectangular aperture and θ represents the angle between the electric field and longitudinal axis of the rectangular aperture. (d) Transmission spectra of rectangular aperture with different θ values. (Reprinted with permission from Ref. [6], © 2004 Elsevier B.V.)

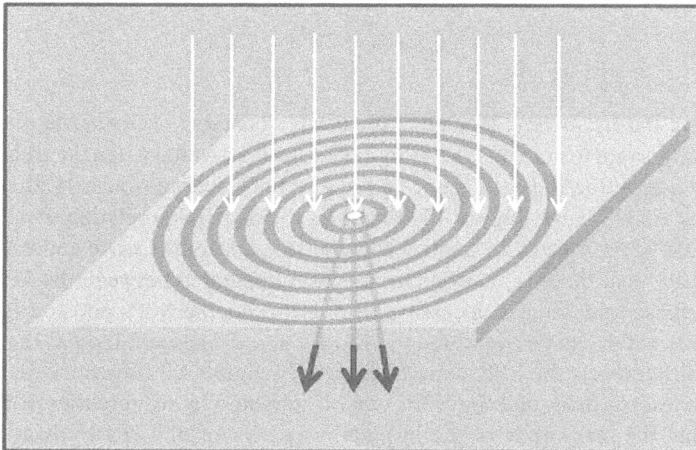

FIGURE 4.6 Schematic illustration of Bull's eye pattern milled around the circular aperture.

additional wave vector to overcome the momentum mismatch. Thus, the activation of SPPs further improved the transmission efficiency of the sub-wavelength aperture.

Even though the transmission improves because of SPs, the light transmitting from the aperture still suffers from diffraction. The diffraction from the sub-wavelength aperture was substantially reduced if the same corrugated circular grooves pattern was made on the exit side also (Figure 4.16). H. J. Lezec and co-workers [8] analyzed the transmission spectra of the sub-wavelength aperture, which possesses circular grooved patterns on both sides. In their study, Bull's eye structures are made on both sides of Ag film. The normal incident transmission spectra were collected at different angles from the exit side. It was found that transmission spectra possess a considerably sharp peak centered at 660 nm, which is slightly higher than the groove periodicity. In addition, transmission spectra are strongly dependent on the collection angles where transmitted light intensity reduces with increasing collection angle. These observations indicate that the emerging light beam is well defined, and the deduced divergence angle is ±3°. The confinement of light to the central area around the hole with lateral dimensions exceeding not more the 1 μm was clearly observed in the optical image. The observation of a low divergence emergent beam is attributed to the involvement of the SPPs wave. The activated SPPs form the front side tunnel through the circular aperture and reach the backside of the aperture. The Bull's pattern on the exit side helps to reradiate the SPPs to free-space propagating light. The emitted light normal to the aperture has a wavelength slightly higher than the periodicity of the grove due to the slowness of the SPPs waves [8].

4.2.2 SURFACE PLASMON POLARITONS: SUPER-LENSING

As discussed previously, the resolution of the optical system is limited by the diffraction limit. The inability to detect the evanescent field by the objective lens, which carries the high-resolution information, limits the resolution. The superlens or perfect lens can be realized by retrieving spatial frequencies associated with the evanescent field. In his pioneer paper, Pendry proposed an unconventional lens with a negative refractive index material that can act as a perfect or superlens [9]. In this case, high resolution is achieved by the combination of propagating and evanescent EM waves. In addition, the superlens can also be realized using the SPPs [3]. The SPPs are one of the fundamental excitations of SPs. The SPPs are EM waves that evanescently confine to the metal and dielectric interface [3,10,11]. In other words, the SPPs are collective oscillations of conduction band electrons with the interaction of EM waves. Many authors discussed the theoretical description of SPPs in the previous [2]. The dispersion relation of SPPs differs from the free-space propagative EM waves. The dispersion of relation of the SPP is written as,

$$k_{\mathrm{sp}} = \frac{\omega}{c} \sqrt{\frac{\varepsilon_d \varepsilon_m}{\varepsilon_d + \varepsilon_m}}$$

where ε_d and ε_m are the frequency-dependent dielectric constants of dielectric medium and metal, respectively. It is important to note that the SPPs can only be activated when the frequency-dependent dielectric constants have the opposite sign, i.e., $\varepsilon_d \varepsilon_m < 1$. This condition is readily fulfilled by the noble metals in the visible range. Figure 4.7 illustrates the dispersion relation of a free-space propagative EM wave and SPPs. One can visualize that the free-space EM wave curve and SPP curve do not intersect. It infers that there is always a momentum mismatch between the SPP and free-space EM wave. Thus, phase matching is necessary in order to couple SPPs and free-space EM waves. The momentum mismatch can be overcome by slow light or evanescent light as the dispersion curve of evanescent light intersects the SPP curve (Figure 4.7). Figure 4.3 illustrates the possible ways of producing the evanescent field, and the SPPs can be activated by using either of the techniques. It is worth noting that the magnitude of the in-plane wave vector of SPPs is always higher than the propagative EM wave. In other words, SPP is a slow light that is confined to the metal surface, and SPPs possess a large wave vector compared to the free-space EM wave. Thus, high resolution is

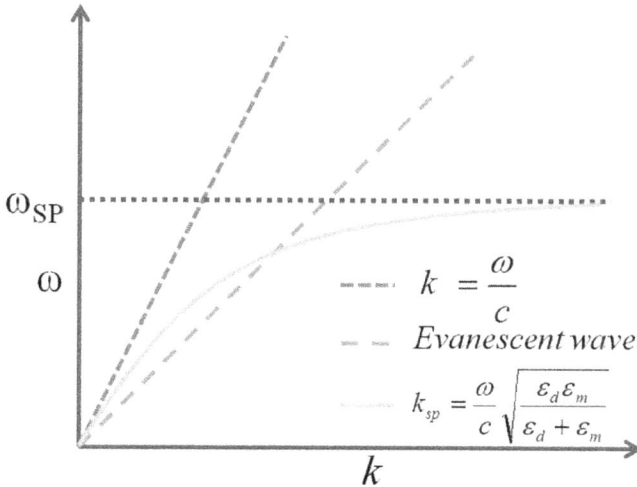

FIGURE 4.7 Dispersion curves of free-space, evanescent, and SPP EM waves.

possible if the imaging is carried out with SPPs. In principle, superlens can be realized at SPP resonance frequency (ω_{SP}) where k becomes infinity (Figure 4.7).

Smolyaninov et al. carried out the high-resolution imaging with the SPPs produced on the metal surface [12]. The typical experimental setup is shown in Figure 4.8. A parabolic glycerine dielectric droplet was used as the 2D optical element, which provides the large effective dielectric constant (Figure 4.8a). The boundary of parabolic glycerine droplets acts as the SPPs mirror. With the 502 nm laser line illumination, the expected SPPs of wavelength and effective dielectric constant are 69.8 and 7.14 nm, respectively. The high-resolution SPP imaging was carried out on the sample with nanohole arrays. The nanohole array test sample possesses a hole diameter of 100 nm, and the distance between holes is 40 nm. The SPPs produced 2D image that was visualized by the optical microscope. The far-field collection is possible as the SPPs scatter into free-space EM waves because of sample roughness and Rayleigh scattering. Figure 4.8b shows the SEM image of the triplet nanoholes array, whereas Figure 4.8c shows the optical image acquired with the 514 nm laser illumination. In addition, the zoom-in optical image clearly shows the triplet nanohole arrays with a clear of 40 nm nanohole gaps. These results indicate the possibility of high spatial resolution with SPP imaging.

The daunting problem of SPP imaging is to collect the optical signal in the near-field rather than far-field. Thus, SPP imaging poses technical challenges in getting high-resolution images. In this context, Fang et al. carried out sub-diffraction limit imaging with SPPs activated on Ag film indirectly [13]. A negative photoresist was used for the SPP imaging. In this case, sub-wavelength features are imaged onto the photoresist, and subsequently, AFM imaging is carried out on the photoresist after developing. The superlens is fabricated by depositing the different layers. The nanoscale objects are inscribed on the chrome film, which was deposited on the quartz substrate using the focused ion beam (FIB) technique. These nano-objects were flatted with a 40 nm thick polymethyl methacrylate (PMMA). Subsequently, a 30 nm thick Ag film was deposited, followed by a 120 nm thick negative photoresist. The substrate is illuminated with UV light of 365 nm wavelength. After developing the photoresist, the optical images of nano features are imprinted on the negative photoresist as topographic modulations. The AFM was employed to evaluate topographic modulations. In the AFM topographic analysis, the negative photoresist had shown the nanowire array with 60 nm wide and 120 nm pitch. For the comparison, the photoresist was illuminated without Ag film, where the Ag film was replaced with a 35 nm PMMA layer. These nano features failed to observe in atomic force microscopy (AFM) images without metal film. Thus, the results indicate

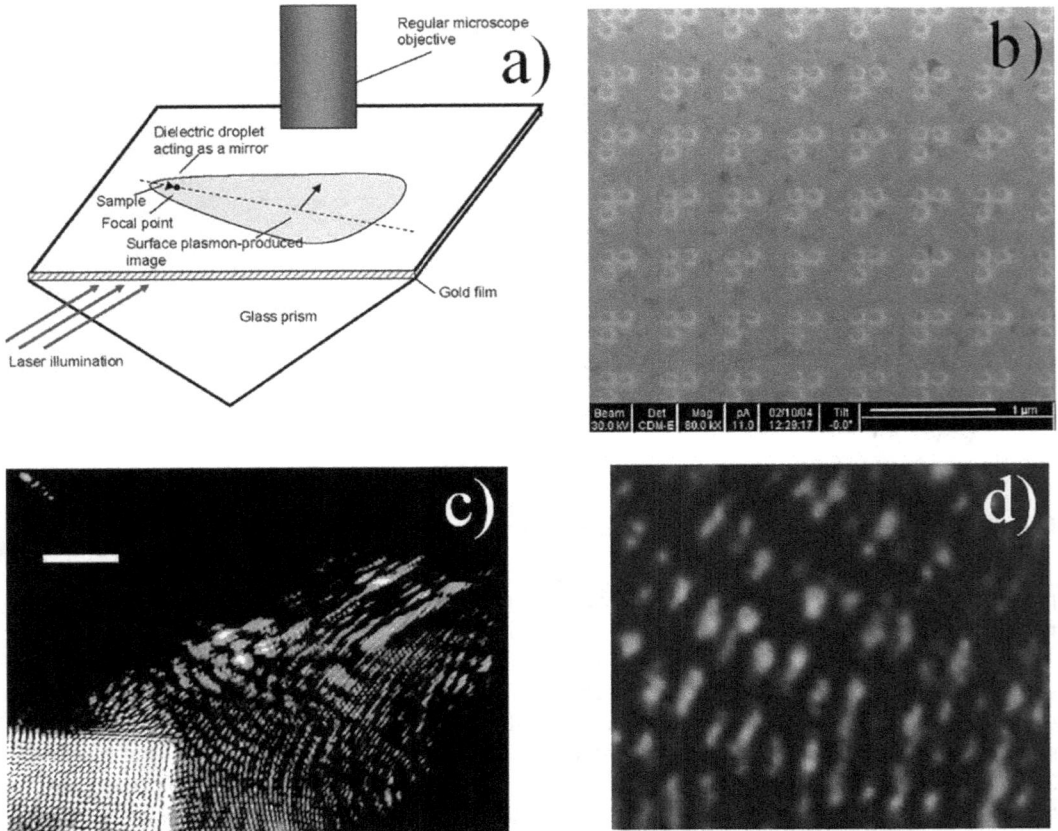

FIGURE 4.8 (a) Schematic illustration of typical experimental setup. (b) Electron microscope image of triplet nanohole arrays of test sample. (c) Optical image of triplet nanohole array obtained with 514 nm laser illumination. (d) Zoom-in optical image showing triplet nanohole array. (Reprinted with permission from Ref. [12], © 2005 Optical Society of America.)

that resolution is achieved down to $\lambda/6$ with the silver superlens. Furthermore, arbitrary nano-object such as 'NANO' inscribed on the chromium film is imaged with and without an Ag lens. The image recorded with Ag films resembles the SEM image in all aspects. The Ag superlens has shown to resolve the line width of 89 nm, which is far below the diffraction limit.

4.3 INSTRUMENTATION INVOLVED IN NANOSCOPY

4.3.1 NEAR-FIELD SCANNING OPTICAL MICROSCOPY: SCATTERING AND APERTURED TYPE PROBES

In order to achieve high-resolution imaging, one has to collect the evanescent field from the sample surface or imaging is carried out with the evanescent field. However, the evanescent field decays exponentially, and the typical decay length of the evanescent field is around 100 nm. The early technique to collect the evanescent field is scanning tunneling optical microscopy (STOM), also called photon scanning tunneling optical microscopy (PSTM). The frustrated total internal reflection (FTIR) is the principle behind the STOM or PSTM. However, the PSTM is only applicable for transparent samples. Present-day nanoscopy is carried out using near-field scanning optical microscopy (NSOM), and it is also called scanning near-field optical microscopy (SNOM). The NSOM is

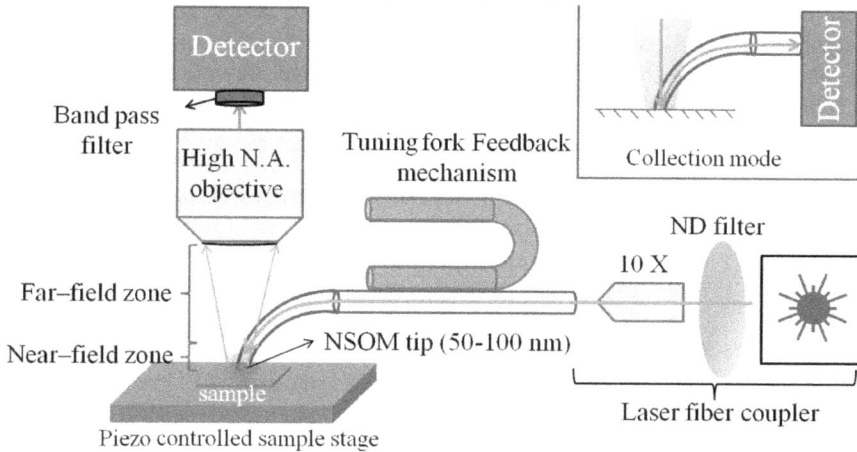

FIGURE 4.9 Schematic experimental setup of reflection mode NSOM configuration and inset figure shows the collection mode configuration.

a combination of AFM and optical microscopy. The AFM feedback mechanism accurately keeps and controls the NSOM tip in the near-field region of the sample surface. The typical experimental setup of the NSOM is shown in Figure 4.9. The depicted figure shows the tuning-fork-based feedback mechanism setup with an upright optical microscope. The NSOM tip is attached at the end of multimode or single-mode optical fibers. The NSOM measurements can be carried out in two configurations: reflection mode (near-field illumination-far-field collection) and collection mode (far-field illumination–near-field collection). Both modes are depicted schematically in the inset of Figure 4.9. The NSOM tip is raster-scanned over the sample surface in the reflection mode, and scattered light is collected in the far-field with a high N.A. objective lens. A single photon optical detector detects the scattered light, and the optical image is constructed with the help of associated electronics. The scanning is carried out using the AFM feedback mechanism, and the topography image is also acquired along with the near-field optical image. The sample is illumined from the far-field in collection mode, and the signal is collected in the near-field through the NSOM tip (inset of Figure 4.9). Similarly, near-field fluorescence imaging is carried out by replacing the bandpass filter with a notch or edge filter. The resolution of the NSOM image is dictated by the size of the NSOM tip rather than the laser wavelength. Moreover, in the case of transparent substrates, similar measurements can be carried out using the inverted optical microscope.

The NSOM measurements are usually carried out using two probes, such as aperture and scattering probes. Presently, the fabrication process of apertured probes is well established, and these probes are readily available commercially. The aperture probes are usually derived from dielectric optical fibers. The tapered optical fibers are fabricated by two methods such as heating and pulling [14], and chemical etching, i.e., Turner's etching technique [15]. These methods have their advantages and disadvantages. The heating and pulling method produce a very smooth tapered region. However, the tapered region has a small cone angle. It is worth noting that the transmission coefficient of tapered fibers depends on the cone angle. It was proven that the high cone angle improves light transmission [16]. In contrast, the cone angle can be tuned in the chemical etchant method. The hydrofluoric acid (HF) is used as the etchant which is covered by the organic solvent over the layer. The etching takes place in the meniscus region. Moreover, with the chemical etchant method, large-scale production of the tapered optical fibers is possible. However, the chemically etched tapered region possesses high toughness. The increased roughness of the taper region largely influences the smoothness of the metal coating. The size of the tapered region is smaller than the cut-off diameter waveguide. Thus, the light is prone to escape from the tapered region. To arrest the light leakage from the tapered region, metals are coated over it. The metal coating is carried out using thermal evaporation, and Al is the most used metal. Figure 4.10a and b shows the SEM image of a fiber probe fabricated using the heating and pulling method and the optical image of a fiber probe fabricated using the chemical etchant method, respectively. Compared to the chemically etched method (Figure 4.10b), the smaller cone angle and smoothness are visible for the tapered region manufactured by the heating and pulling method (Figure 4.10a). Figure 4.10c and d shows the SEM images of the Al-coated apex region for the pulled and etched probes, respectively. A typical NSOM study is depicted in Figure 4.11, and measurements are performed in reflection mode with tapping mode AFM feedback mechanism. The well-dispersed GaN nanowires are imaged with an aperture probe of size 100 nm (Figure 4.11) [17]. In this case, the even 10 nm nanowire is optically detected (Figure 4.11c).

The resolution of NSOM depends on the size of the aperture. Unfortunately, the size of the aperture cannot be reduced indefinitely. This limitation arises because of the poor throughput of the aperture probes. Moreover, the size of the aperture cannot be less than the skin depth of the coated metal. Thus, one has to compromise between the throughput and resolution. As a result, standard NSOM probes possess an aperture size in the range of 50–100 nm. The scattering-type probes can overcome the limitation posed by the aperture probes. The scattering-type probes exploit the SPR of the sub-wavelength metals nanoparticles to get a high resolution. In other words, high resolution is achieved by light confinement near the metal nanoparticles. In the case of metal nanoparticles, the SPR phenomenon prevails when the excitation frequency matches the resonance frequency

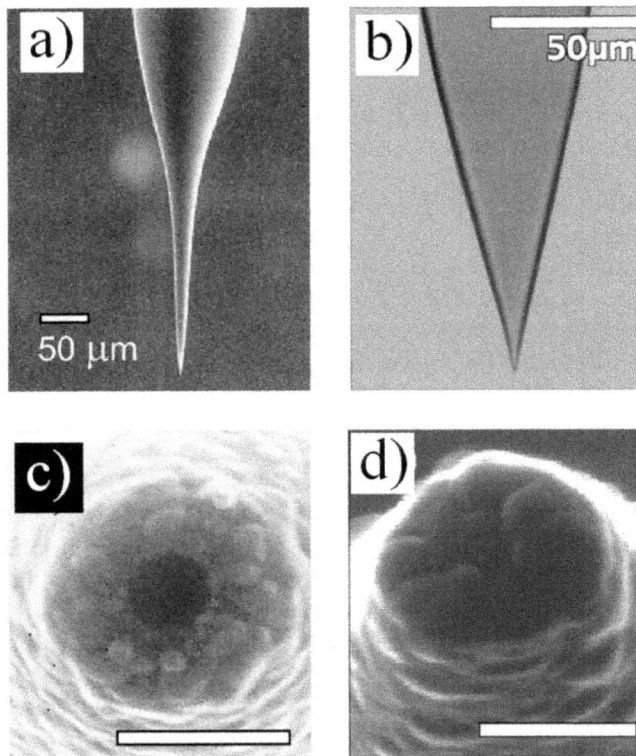

FIGURE 4.10 (a) SEM image of tapered fiber optic fabricated by heating and pulling method. (b) Optical image of tapered fiber optic fabricated by chemical etching (c and d) SEM images of apex regions fabricated by heating and pulling, and chemical etching method, respectively. The scale bar is 300 nm. (Reprinted with permission from Ref. [5], © 2020 American Institute of Physics.)

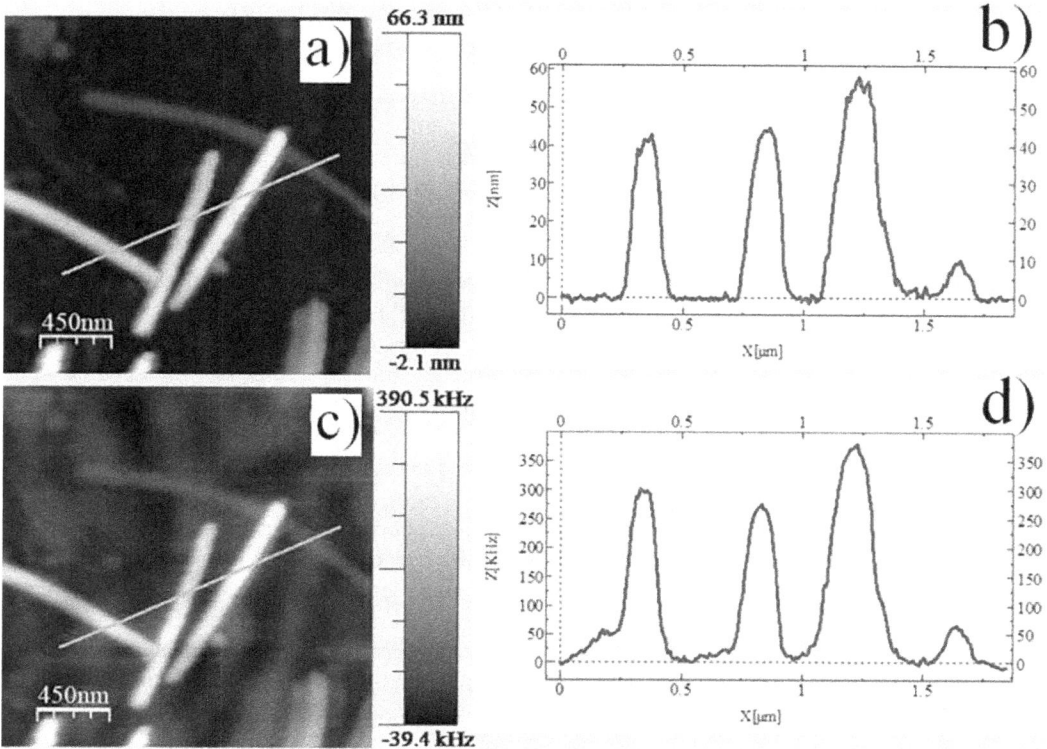

FIGURE 4.11 NSOM imaging of GaN nanowires was carried out using an aperture probe of size 100 nm. Topography (a) and corresponding line profile of (b) GaN nanowires. NSOM image (c) and corresponding line profile (d). (Reprinted with permission from Ref. [17], © 2000 American Institute of Physics.)

of the metal nanoparticles. As a consequence of SPR, light is confined and enhanced near metal nanoparticle surface. The confinement region is in the order of the metal nanoparticle size. Thus, the resolution achieved by the scattering-type probe is determined by the size of metal nanoparticles. Sharp tips can also achieve light confinement as a result of the lightning rod effect. When the sharp tip illuminates with monochromatic light, the electron oscillates with the frequency of exciting light. The charge accumulates near the tip apex, leading to the charge crowding. As a result, the enhancement and confinement of fields occur at the tip apex. The field enhancement depends on the tip material and metal tips most suitable for the high field enhancement. The polarization of light plays a crucial role in enhancing the field in the case of the scattering-type probes. The confinement takes place only when the polarization of excitation light is parallel to the tip axis. In contrast, polarization normal to the tip axis cannot enhance the field as charge crowding does not occur at the tip apex. The scattering-type probes provide a better resolution compared to the aperture probes. Unfortunately, in the case of scattering-type probes, information from the confined and enhanced field is superimposed over the far-field background. However, one can eliminate the far-field background using modulation techniques [18]. Usually, the tapping mode of AFM is employed to eliminate the far-field background. The distance between the tip and samples is periodically varied at the tip resonance frequency in the tapping mode. Thus, the near-field interaction is modulated over the oscillation frequency. The near-field optical signal is detected at the modulation frequency using the lock-in amplifier. However, to avoid the contribution from other tip parts, the optical signal is demodulated at the higher harmonics of fundamental oscillation, which is carried out using interferometry.

4.3.2 SUPER-RESOLUTION IMAGING: STOCHASTIC OPTICAL RECONSTRUCTION MICROSCOPY AND PHOTOACTIVATED LOCALIZATION MICROSCOPY

Fluorescence microscopy plays a major role in molecular and cell biology. Similar to optical microscopy, fluorescence microscopy is also limited by the diffraction limit. Several techniques have been developed to overcome the diffraction limit in fluorescence microscopy, and these techniques fall under the category of super-resolution fluorescence microscopy. Stochastic optical reconstruction microscopy (STORM) belongs to super-resolution fluorescence microscopy, which relies on the high-accuracy localization of single fluorescent emitters [19]. Using the STROM, spatial resolution was achieved down to 20 nm, tenfold higher than the conventional fluorescence microscopy. The STORM constructs an image by precisely locating the photo-switchable molecules. The photo-switchable molecules can exist in two states, such as fluorescent and non-fluorescent, and these states are switchable by exposing the light. For example, cyanine dye, Cy5, is switchable between fluorescent and non-florescent states by using two different wavelength lasers, such as red and green. The exposure of red laser leads to fluorescent emission from the Cy5 and subsequently converts into non-florescent. However, exposure to green light leads to the reconversion of Cy5 as the fluorescent molecule [19]. STORM imaging involves several numbers of cycles. In each cycle, a few photo-switchable molecules are selectively activated such that the separation between these molecules is more than diffraction resolvable distance. The fluorescent molecule's point spread function is broad enough, but the fluorescent molecule's position is accurately determined. A stochastically different set of fluorescent molecules is activated in each cycle, and their position is accurately determined. Subsequently, the overall image is constructed by the accurate positions of fluorescent molecules. In the STORM, the samples of specific interest are labeled with switchable fluorescent molecules. The position of each labeled fluorescent molecule is determined precisely. In each cycle, a stochastically different subset is activated and measured their position accurately. The high-resolution image is constructed from the position of fluorescent molecules after a sufficient number of cycles. In this case, the resolution depends on the precision of localization of the fluorescent molecule rather than the diffraction-limited spot. Photoactivated Localization Microscopy (PALM) is also based on a similar principle of STORM with different kinds of fluorescent molecules. In the case of PALM, the third state of fluorescent molecules exists, and it is not reversible like in STORM [20]. Figure 4.12 shows the comparative images of diffracted-limited and PALM optical techniques of a bacterial protein ParB tagged with the mNeon-Green photo-switchable molecule [21]. Bottom panel shows the zoomed-in images of the region of interest from the PALM image (Figure 4.12b). In this work, PALM imaging is enriched with illumination power and optimizing the sample preparation [21,22].

4.4 APPLICATIONS OF NANOSCOPY

4.4.1 NEAR-FIELD OPTICAL IMAGING OF COUPLED PLASMON–PHONON POLARITONS

SPPs are propagating EM waves at the metal and dielectric interface, which are confined along the interface region. The SPPs are evanescent in nature, which means that the field strength decreases exponentially from the surface. The most spectacular application of SPP is its ability to confine and guide the light in the sub-wavelength structures by overcoming the Abbe's diffraction limit [22]. The propagation of light in sub-wavelength nanostructures is reported to be utilized for the realization of nano-photonic devices as well as optical nano-connectors [22,23]. As the SPP field strength is confined to the interface, the near-field techniques play an important role in imaging the SP fields of SPPs. Several reports are available in the literature for the imaging of SPP propagation; however, the present section reviews the two early reports. Weeber et al. carried out the SPP imaging using the PSTM technique [24]. The SPPs were activated on the thin Ag film and Ag strips using the attenuated total internal reflection (ATR) method. The Ag film is micro-fabricated on indium-tin-oxide (ITO) doped glass using the e-beam lithography technique. The sample contains 60 nm thick

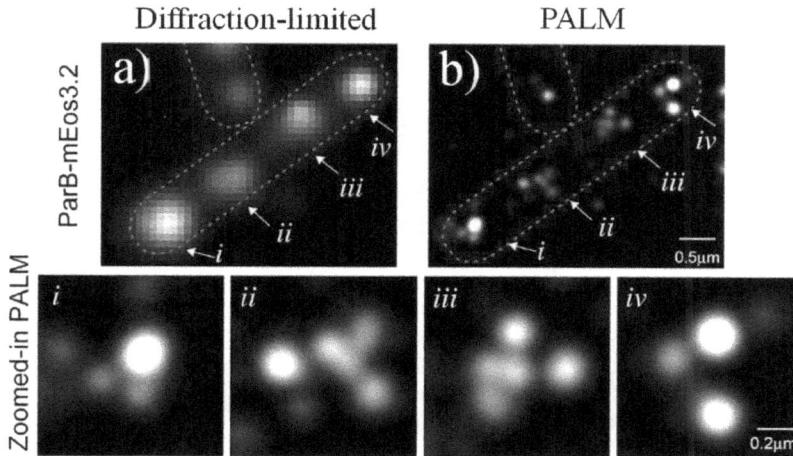

FIGURE 4.12 (a) Diffraction-limited image of bacterial protein ParB tagged with the mNeon-Green and corresponding PALM image is shown in (b). Bottom panel: Zoomed-in images of regions of specific interest in PALM image (b) such as *i–iv*. (Reprinted with permission from Ref. [21].)

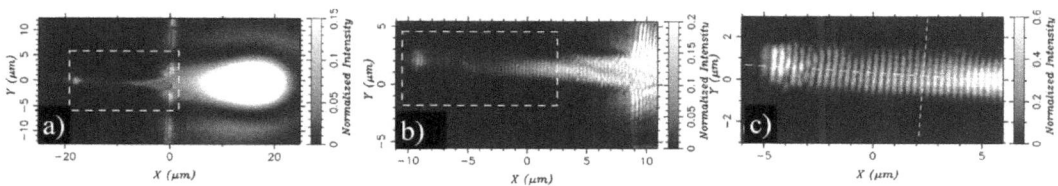

FIGURE 4.13 Near-field image collected at different proximities of Ag metal strip: (a) 350 nm, (b) 100 nm, and (c) 50 nm. (Reprinted with permission from Ref. [24], © 2001 The American Physical Society.)

Ag large area film, which includes micrometer width strips at edge regions. The SPPs are generated with a focused laser beam of 633 nm excited in the ATR configuration. The PSTM measurements were carried out using the standard aperture-type probes manufactured by the heating and pulling method. Figure 4.13 shows near-field images recorded at different proximities from the Ag strip: 350, 100, and 50 nm. In this case, the laser beam is focused on the large area region (Figure 4.13a). The fringes are clearly observed in the near-field images of the strip region. These observed fringes are intuitively attributed to the SPPs, which form standing waves after reflecting at the end of the strip. The fringe pattern is more clearly visible as the tip-sample distance reduces (Figure 4.13b and c). However, SPPs generation is also observed in large area regions but not only in the strip region.

Eloise Devaux and co-workers imaged SPP propagation on the thick film of Au [25]. Near-field imaging was carried out with PSTM technique similar to the above discussion. The micro-gratings were fabricated with arrays of sub-wavelength holes. These sub-wavelength hole patterns were utilized for the launching and decoupling of SPPs. The small area array is utilized as the "source array", which activates the SPPs after focusing the laser beam. The SPPs have propagated along with the Au film and decoupled at the large area array, which is termed as "probe array". The launching and decoupling are achieved by the additional wavevector provided by the sub-wavelength nanohole array. Figure 4.14a shows the near-field image recorded near the "source array" region. This image depicts bright intensity at the source region and a streak of light along the X-direction. Interestingly, the probe array region near-field image shows the fringes and bright intensity over the array (Figure 4.14b). The fringes observed in the image are attributed to the SPPs

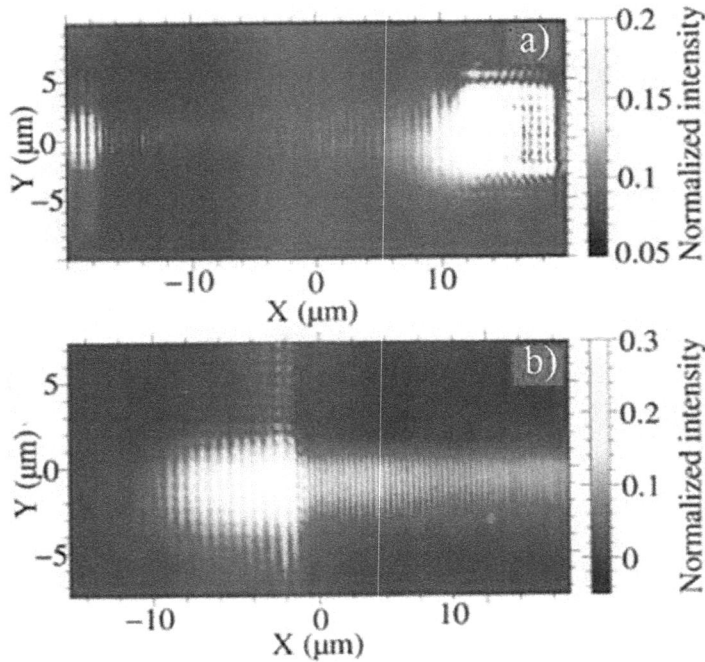

FIGURE 4.14   Near-field image collected near the (a) source array and (b) probe array. (Reprinted with permission from Ref. [25], © 2003 American Institute of Physics.)

which are generated by the source array. These SPPs are decupled at the probe array region and generate high-intensity propagative light.

4.4.2   Imaging of SPPs Generated by the 2D Plasmon

In recent times, 2D plasmons in graphene generated much curiosity because of their terahertz (THz) resonance frequency [26]. In the case of 3D plasmonic material, the *plasmon frequency* depends on the system carrier density. On the other hand, the *plasmon frequency* of 2D plasmons depends on the in-plane wave vector and the areal carrier density of two-dimensional electron gas (2DEG) [27,28]. A 2D plasmons system also exists in the semiconductor inversion layers of Si and GaAs [27–29]. In addition, InAs and InN possess the inherent surface electron accumulation near the surface region, which can act as a 2DEG [30]. This section discusses the imaging of SPPs corresponding to 2D plasmons. Z. Fei and co-workers imaged the propagation of the SPPs in single-layer graphene using the scattering-type SNOM [31]. The AFM sharp tip with the curvature of the radius of 25 nm is used as the scattering-type SNOM probe. The SNOM measurements are carried out by illuminating an infrared laser of wavelength 11.2 µm. The scattering amplitude, $s(\omega)$, represents the electric field strength between the tip apex and sample surface. The $s(\omega)$ corresponds normalized near-field amplitude: $s(\omega) = s_3(\omega) / s_3(\omega)^{SiO_2}$ where $s_3(\omega)$ and $s_3(\omega)^{SiO_2}$ are third-order the demodulated harmonics of near-field amplitude measured for the graphene and SiO_2 substrate. In this study, near-field SNOM images were acquired at arbitrary regions such as graphene and SiO_2 interface regions and near the defect regions. For all the cases, the periodic fringes are observed, which is attributed interference pattern of the SPPs. In this case, the AFM tip launches the SPPs, and the SPPs propagate radially outward from the tip apex. These propagated SPPs reflected at the edges and defect and produced the complex SPPs interference pattern. The wavelength of the SPPs calculated from the fringe width is around 100 nm. The short wavelength

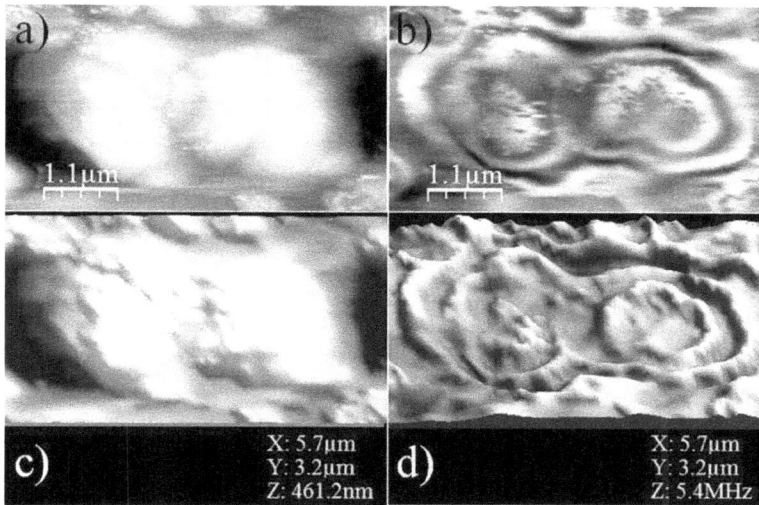

FIGURE 4.15 Topography (a) and corresponding NSOM (b) image of InN nanostructures with surface electron accumulation. Imaging is carried out in clustered regions. 3D images of topography (c) and NSOM (d). (Reprinted with permission from Ref. [33], © 2018 IOP Publishing Ltd Printed in the UK.)

of SPPs is attributed to the extraordinary confinement of infrared light normal to the graphene sheet [32].

Furthermore, Madapu and co-workers carried out the NSOM imaging on the InN nanostructures [33]. In this case, the InN nanostructures possess surface electron accumulation. The electrons in the surface electron accumulation behave like 2DEG. Aperture-type probe with the size of 100 nm is employed to study the near-field light-matter interaction. The NSOM measurements were carried out in the reflection mode configuration with a 532 nm laser excitation. The AFM topography and corresponding NSOM images are shown in Figure 4.15a and b. The NSOM image of each nanoparticle contains the periodic fringes (Figure 4.15b). These periodic fringes originated because of a standing wave formed by the generated and reflected SPPs at the edges. The origin of SPPs is solely attributed to 2D plasmons of surface electron accumulation, as the SPPs generation cannot be observed in InN nanostructures without surface electron accumulation. The measured SPP wavelength is in the range of 274–500 nm. The 3D images of topography and NSOM are depicted in Figure 4.15c and d, respectively.

4.4.3 Near-Field Optical Lithography

Optical lithography is also limited by the diffraction limit as it uses diffraction-limited optics. Near-field lithography can be employed to fabricate the sub-wavelength nanostructures [4]. Robert Riehn and co-workers fabricated the poly(p-phenylene vinylene) (PPV) sub-wavelength nanostructures using direct SNOM [34]. The PPV nanostructures are fabricated by inhibiting precursor solubility with the exposure of the UV laser light (325 nm) through the aperture probe of 40 and 80 nm size. The PPV soluble precursor, namely poly[p-phenylene[1-(tetrahydrothiophene-1-io ethylene chloride)]], was spin-coated from the methanol solution on the silica substrate. After UV writing with an aperture probe, the non-exposed material dissolved in methanol. The tuning-fork-based AFM feedback mechanism controls the probe-sample distance of SNOM. The sample with UV-exposed precursor is annealed at 220°C for 5 h at less than 10^{-5} mbar pressure to convert precursor into fully conjugated PPV. The topography of the patterned sample is studied by AFM. The top panel of Figure 4.16 shows an AFM image of patterned structures (before heat treatment) with different

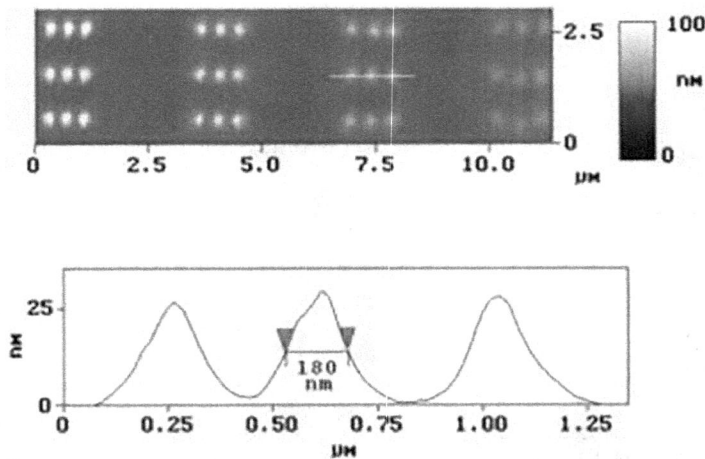

FIGURE 4.16 Top panel: AFM image of the patterned nanostructures. From right to left column represents the different exposure times, namely, 0.4, 0.3, 0.2, and 0.1 ms. Bottom Panel: Topographic profile of line indicated in AFM image, which corresponds to the 0.2 ms exposure time. (Reprinted with permission from Ref. [34], © 2003 American Institute of Physics.)

exposure times of 0.4, 0.3, 0.2, and 0.1 ms from right to left columns, respectively. The full-width half-maximum (FWHM) of the typical nanostructure with an exposure time of 0.2 ms is 180 nm (bottom panel of Figure 4.16). Interestingly, the FWHM is further reduced to 160 nm after conversion (after heat treatment). Thus, these results demonstrate the possibility of fabricating the sub-wavelength nanostructure with scanning near-field optical lithography.

In this context, Credgington et al. [35], fabricated the high-resolution PPV nanoparticles using the SNOM optical lithography. In this particular work, sub-wavelength PPV nanostructures were manufactured by the site-selective insolubilization of precursor polymer with UV light of 325 nm exposure through an NSOM aperture probe of 50 nm. The PPV nanostructures are prepared by insolubilizing the precursor polymer poly(*p*-xylene tetrahydrothiophenium chloride) (PXT) via a leaving-group type reaction. The nanoparticle's size is well below the diffraction limit, $\lambda/2$. The topography image of the dot particles patterned on a quartz substrate is shown in Figure 4.17a. The left to right columns shown in the image corresponds to the increased UV light exposure times such as 50, 100, 200, 500, and 1,000 ms. The 100 ms exposure time was found to produce the smallest feature size. Figure 4.17b shows the high-resolution topographic image of the PPV particle produced using the 100 ms exposure time. The corresponding particle is indicated by the square in Figure 4.17a. The FWHM of this dot particle is around 50 nm. Thus, the achieved feature size is in the order of $\lambda/6$, which is well below the diffraction limit. However, the theoretically predicted minimum feature size would be 90–100 nm with the above experimental conditions. The small size of features compared to the theoretically predicted size is attributed to the volume shrinkage of particles during the vacuum backing, which removes the solvent and reaction products.

4.4.4 BIO-IMAGING

Section 4.3.2 discusses super-resolution microscopy techniques such as STORM and PALM. In addition to these, several other techniques fall under the category of super-resolution techniques, namely stimulated emission depletion microscopy (STED) and structured illumination microscopy (SIM) [36]. These techniques play the most important role in cell biology. In this context, NSOM based fluorescence microscopy technique is one of the early high-resolution techniques [4,37,38].

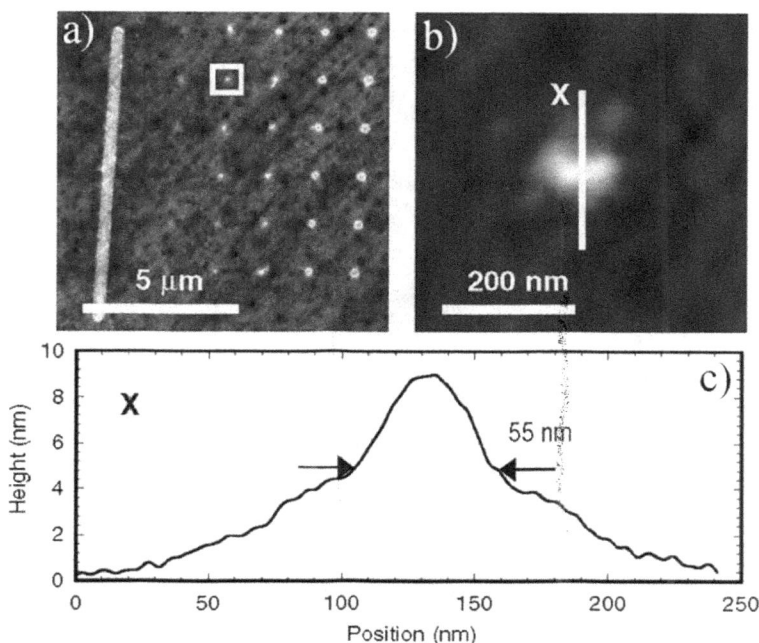

FIGURE 4.17 (a) AFM topographic image of PPV dot particles patterned on quartz substrate by SNOM UV lithography. Left to right columns correspond to the different UV exposure times such as 50, 100, 200, 500, and 1,000 ms. (b) High-resolution topographic image of PPV dot particle with exposure time 100 ms. The particular feature is indicated by the square in (a). (c) Line profile acquired across the dot particles (b) which delineates the FWHM of dot particle is around 50 nm. (Reprinted with permission from Ref. [35], © 2010 Wiley-VCH Verlag GmbH & Co. KGaA, Weinheim.)

This particular section discusses single-molecule detection using the NSOM. Veerman and co-workers [39] demonstrated the distribution and orientation of the single fluorescent molecules using NSOM fluorescence imaging. The NSOM imaging was carried out by FIB-modified high-quality aperture probes with 1.5 nm rms roughness at the end face. The shear-force feedback mechanism regulates the distance between the surface and the probe. Figure 4.18 shows the NSOM fluorescence image of $DiIC_8$ molecules embedded in the 10 nm thin PMMA film. The NSOM image is formed with a false color scale where green represents the optical field of molecules, dipoles, oriented along the vertical direction of the sample. In contrast, red represents the molecules oriented in the horizontal image direction. The near-field fluorescence images contain various intensity patterns: circular, elliptical to double arc, and ring structures. In other words, single-molecule detection is possible due to the small aperture size probe of 70 nm. The presence of different intensity patterns is attributed to the selective excitation molecule as a function of its position with respect to the probe. The present NSOM image has shown unprecedented optical resolution with the peak FWHM of 45 nm with a 70 nm size aperture probe.

In another independent work, Bärbel I. de Bakker and co-workers [40] investigated the fluorescently labeled DC-specific intercellular adhesion molecule (ICAM) grabbing nonintegrin (DC-SIGN; CD209) on intact isolated dendritic cells (DCs) employing NSOM fluorescence imaging. DC-SIGN, exclusively expressed in DCs, is a type II transmembrane protein containing a mannose-binding domain that forms the ligand-binding site [40]. Typical confocal fluorescence image of immature DC (imDC) expressing DC-SIGN at the cell surface is shown in Figure 4.19a. The confocal image has shown the typical imDC with the smallest visible feature size of 350 nm. However, from the confocal image, it is hard to distinguish the isolated components of the cell surface.

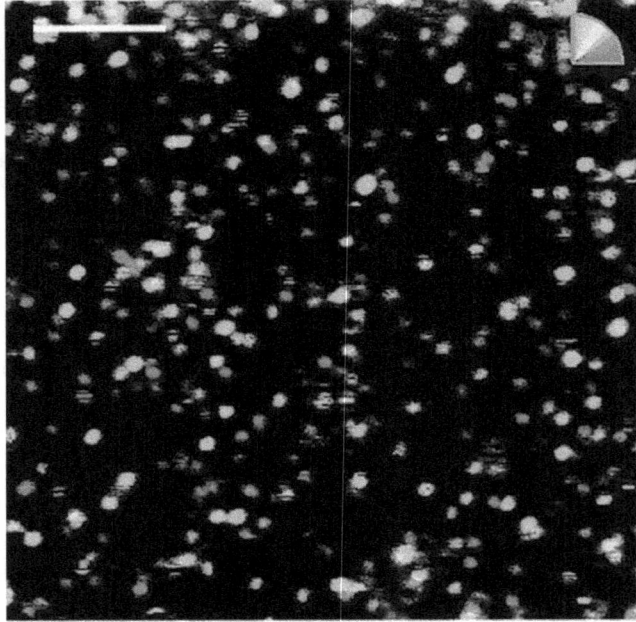

FIGURE 4.18 NSOM fluorescence image of DiIC$_8$ molecules embedded in PMMA films. Colors from green to red via yellow represent the different orientations of molecular dipole with respect sample (Right inset). Scale bar: 1 μm. (Reprinted with permission from Ref. [39], © 1999 The Royal Microscopical Society.)

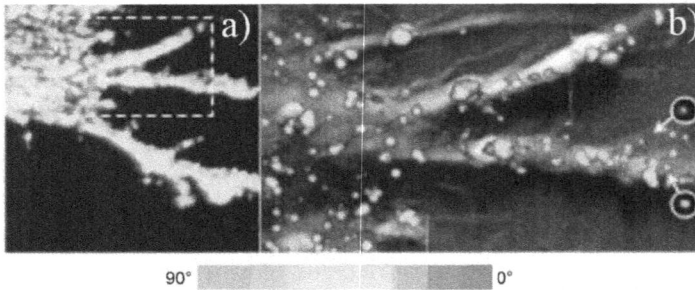

FIGURE 4.19 (a) Confocal image of a DC stretched on fibronectin-coated glass expressing DC-SIGN on the membrane (20×20 μm^2). (b) NSOM fluorescence image overlaid on AFM topographic image (12×7 μm^2). The dashed rectangle in the confocal image (a) represents where the NSOM image was carried out. Color coding in both images corresponds to the detected polarization. (Reprinted with permission from Ref. [40], © 2007 Wiley-VCH Verlag GmbH & Co. KGaA, Weinheim.)

Figure 4.19b shows the NSOM fluorescence image overlaid on the AFM topographic image. In contrast, the NSOM fluorescence image (Figure 4.19b) shows the distribution of DC-SIGN in greater detail. The location of NSOM imaging carried out is indicated by a dashed rectangle in a confocal microscopic image (Figure 4.19a). It is apparent from the NSOM image the fluorescent spots are decorated randomly over the dendrite region and cell body. The smallest fluorescent spot detected is 100 nm, which is the order of size of the aperture probe. In addition, the NSOM fluorescence image clearly shows the fluorescent spots with different intensities, sizes, and emission polarizations. These findings demonstrate that NSOM can resolve the static heterogeneity of the cell membrane with high resolution.

REFERENCES

1. R. C. Dunn, *Chemical Reviews* **99**, 2891 (1999).
2. L. Novotny and B. Hecht, *Principles of Nano-Optics*, Cambridge University Press (2012).
3. S. Kawata, Y. Inouye, and P. Verma, *Nature Photonics* **3**, 388 (2009).
4. E. Betzig and J. K. Trautman, *Science* **257**, 189 (1992).
5. B. Hecht, B. Sick, U. P. Wild, V. Deckert, R. Zenobi, O. J. Martin, and D. W. Pohl, *The Journal of Chemical Physics* **112**, 7761 (2000).
6. A. Degiron, H. Lezec, N. Yamamoto, and T. Ebbesen, *Optics Communications* **239**, 61 (2004).
7. C. Genet, and Ebbesen, T. W., *Nanoscience And Technology: A Collection of Reviews from Nature Journals* (2010), 205
8. H. J. Lezec, A. Degiron, E. Devaux, R. Linke, L. Martin-Moreno, F. Garcia-Vidal, and T. Ebbesen, *Science* **297**, 820 (2002).
9. J. B. Pendry, *Physical Review Letters* **85**, 3966 (2000).
10. S. A. Maier, *Plasmonics: Fundamentals and Applications*, Springer Science & Business Media (2007).
11. D. K. Gramotnev and S. I. Bozhevolnyi, *Nature Photonics* **4**, 83 (2010).
12. I. I. Smolyaninov, C. C. Davis, J. Elliott, and A. V. Zayats, *Optics Letters* **30**, 382 (2005).
13. N. Fang, H. Lee, C. Sun, and X. Zhang, *Science* **308**, 534 (2005).
14. E. Betzig, J. K. Trautman, T. Harris, J. Weiner, and R. Kostelak, *Science* **251**, 1468 (1991).
15. T. Yatsui, M. Kourogi, and M. Ohtsu, *Applied Physics Letters* **73**, 2090 (1998).
16. L. Novotny, D. Pohl, and B. Hecht, *Optics Letters* **20**, 970 (1995).
17. S. Parida, A. Patsha, K. K. Madapu, and S. Dhara, *Journal of Applied Physics* **127**, 173103 (2020).
18. F. Keilmann and R. Hillenbrand, *Philosophical Transactions of the Royal Society of London. Series A: Mathematical, Physical and Engineering Sciences* **362**, 787 (2004).
19. M. J. Rust, M. Bates, and X. Zhuang, *Nature Methods* **3**, 793 (2006).
20. S. W. Hell, *Science* **316**, 1153 (2007).
21. I. Stockmar, H. Feddersen, K. Cramer, S. Gruber, K. Jung, M. Bramkamp, and J. Y. Shin, *Scientific Reports* **8**, 1 (2018).
22. D. K. Gramotnev and S. I. Bozhevolnyi, *Nature Photonics* **8**, 13 (2014).
23. J. A. Schuller, E. S. Barnard, W. Cai, Y. C. Jun, J. S. White, and M. L. Brongersma, *Nature Material* **9**, 193 (2010).
24. J.-C. Weeber, J. R. Krenn, A. Dereux, B. Lamprecht, Y. Lacroute, and J.-P. Goudonnet, *Physical Review B* **64**, 045411 (2001).
25. E. Devaux, T. W. Ebbesen, J.-C. Weeber, and A. Dereux, *Applied Physics Letters* **83**, 4936 (2003).
26. A. Grigorenko, M. Polini, and K. Novoselov, *Nature Photonics* **6**, 749 (2012).
27. S. Allen Jr, D. Tsui, and R. Logan, *Physical Review Letters* **38**, 980 (1977).
28. D. Olego, A. Pinczuk, A. Gossard, and W. Wiegmann, *Physical Review B* **25**, 7867 (1982).
29. G. Fasol, N. Mestres, H. Hughes, A. Fischer, and K. Ploog, *Physical Review Letters* **56**, 2517 (1986).
30. K. K. Madapu, S. Polaki, and S. Dhara, *Physical Chemistry Chemical Physics* **18**, 18584 (2016).
31. Z. Fei, A. Rodin, G. Andreev, W. Bao, A. McLeod, M. Wagner, L. Zhang, Z. Zhao, M. Thiemens, and G. Dominguez, *Nature* **487**, 82 (2012).
32. J. Chen, M. Badioli, P. Alonso-González, S. Thongrattanasiri, F. Huth, J. Osmond, M. Spasenović, A. Centeno, A. Pesquera, and P. Godignon, *Nature* **487**, 77 (2012).
33. K. K. Madapu, A. Sivadasan, M. Baral, and S. Dhara, *Nanotechnology* **29**, 275707 (2018).
34. R. Riehn, A. Charas, J. Morgado, and F. Cacialli, *Applied Physics Letters* **82**, 526 (2003).
35. D. Credgington, O. Fenwick, A. Charas, J. Morgado, K. Suhling, and F. Cacialli, *Advanced Functional Materials* **20**, 2842 (2010).
36. J. Xu, H. Ma, and Y. Liu, *Current Protocols in Cytometry* **81**, 12.46.1 (2017).
37. E. Betzig and R. J. Chichester, *Science* **262**, 1422 (1993).
38. F. De Lange, A. Cambi, R. Huijbens, B. R. de Bakker, W. Rensen, M. Garcia-Parajo, N. van Hulst, and C. G. Figdor, *Journal of Cell Science* **114**, 4153 (2001).
39. J. Veerman, M. Garcia-Parajo, L. Kuipers, and N. Van Hulst, *Journal of Microscopy* **194**, 477 (1999).
40. B. I. de Bakker, F. de Lange, A. Cambi, J. P. Korterik, E. M. van Dijk, N. F. van Hulst, C. G. Figdor, and M. F. Garcia-Parajo, *ChemPhysChem* **8**, 1473 (2007).

5 Far-Field Spectroscopy and Surface-Enhanced Raman Spectroscopy (SERS)

Debanjan Bhowmik and Chandrabhas Narayana

CONTENTS

DOI: 10.1201/9781003248323-5

5.1 INTRODUCTION

Optical spectroscopic techniques, e.g., Raman, fluorescence, and infrared (IR) spectroscopy, probe either transmission or emission or scattering of light from molecules under investigation. Most of these techniques, inherently, can provide quantitative and specific information about the matter.

Scattering of light by any specimen, for example, can be either elastic or inelastic in nature. Inelastically scattered photons result from changes in the vibrational energy states of the molecules during their interaction with the incident light. The shift in the frequencies in the inelastically scattered photons, therefore, carries vibrational signatures of the specimen. Raman spectroscopists probe these frequency shifts to identify different types of vibrations inherent to the specimen under investigation [1,2].

Since its discovery in 1928 by C. V. Raman and K. S. Krishnan, Raman spectroscopy has been proven to be one of the most powerful spectroscopic techniques [3]. Both IR and Raman spectroscopies provide details—often complimentary to one another—about the vibrational energy states of the molecules [4–6]. These vibrational signatures typically are very sensitive toward even small structural/chemical changes, which open possibilities for a variety of applications. The strong IR-vibrational signal of water, the universal solvent, however, makes the application of IR spectroscopy very challenging in the presence of water and/or in an aqueous solution. The Raman scattering cross section of water is significantly weaker as compared to those of many of its typical solutes. This feeble Raman signature of water makes it possible to apply Raman spectroscopy in probing various specimens in aqueous solution and biological fluids.

Fluorescence spectroscopy is another versatile technique with a wide range of applications. However, fluorescence spectroscopic methods do not typically resolve individual vibrational states in a molecule. Therefore, unlike Raman spectroscopy, fluorescence spectroscopy fails to generate "fingerprint" like information about the specimen under investigation. Also, fluorescence-based techniques can only probe fluorescent molecules (fluorophores). Only a tiny subset of specimens has the required electronic energy states to exhibit fluorescence. On the other hand, every molecule has its own set of Raman signatures, which is why Raman spectroscopy is utilized as a tag-free and universal technique.

The only major drawback of Raman spectroscopy is its inherently low signal strength. A large number (10^5–10^{10}) of incident photons are required to generate one inelastically scattered photon from a molecule. However, with SERS, high enhancement of Raman signals (10^6–10^{10} times) is possible in molecules, on or near the surface of metal nanostructures. This enormous increase in Raman intensities in SERS can often generate signal strengths suitable for application even at the single molecular level.

The study of SERS started around 1974, when Fleischmann and co-workers unexpectedly reported a strong Raman signal from pyridine on roughened silver electrode [7]. They, however, could not conceive the real reason behind this increased Raman intensities. The first reports of the enhancement factors were made in 1977 by Jeanmaire and Van Duyne and Albrecht and Creighton [8,9]. Albrecht and Creighton spoke about resonance Raman effects originating from the plasmon excitations as the reason behind the observed signal enhancement in their studies. At that time, the study of oscillation of surface electrons in the presence of electromagnetic waves (with specific wavelengths) of incident light on the surface of metal nanoparticles (NPs) and thin films were already under extensive investigation in the field of plasmonics [10]. The relation between the remarkable strength of SERS signals and electromagnetic field enhancement, originating from the surface plasmons on metal nanostructured surfaces, was first described by Moskovits [11]. Later, alongside this electronic effect, additional "chemical enhancements"—up to a couple of orders of magnitude—due to modifications in the polarizability of the molecule from their interaction with SERS substrates were also noticed by several groups [12]. Since then, it has been realized that both electronic and chemical enhancement processes contribute to the SERS signal intensities, with the electronic factors being the major contributor [13].

Surface-enhanced Raman spectroscopy provided an excellent opportunity to extend the application of the robust and highly informative Raman spectroscopy to a level that was unimaginable before. Scientists have made enormous progress while applying SERS in analytical quantification, chemosensing, single-molecule probing, remote applications, material science, medical science, chemical warfare, and the field of biology. The parallel advancement in instrumentation and detection also contributed immensely to the studies of SERS. However, there are some issues with SERS, e.g., problems of reproducibility, high sensitivities toward impurities interfering with signal form sample showing changes in the Raman peak shape, relative peak intensities, and shifts in Raman peak positions on SERS substrates among others, which have not fully been resolved yet and are currently under extensive research.

This chapter aims to provide the readers with (i) a basic understanding of the theory behind the surface enhancement, (ii) the knowledge of different types of SERS substrates and instruments, and (iii) an account of different types of applications of SERS.

5.2 THE MECHANISM BEHIND SURFACE-ENHANCED RAMAN SPECTROSCOPY

From the discussion in Chapter 1, we learn how the strength of a Raman signal is related to the field strength of the incident laser light and the polarizability of the specimen under investigation. If the electromagnetic field experienced by the molecules can be enhanced and/or the polarizability of the molecule can be increased, the overall Raman scattering resulting from the specimen can also be strengthened accordingly. SERS uses the large local electromagnetic field enhancements due to plasmonic effects and alteration of the polarizability of the molecules that can occur at certain metallic surfaces to drastically enhance the strength of Raman scattering. The related concepts behind plasmonics and optical confinements are the topics of Chapter 2. In Section 5.2.1 of this chapter, we intend to discuss why certain materials are best suited to generate an adequate amount of surface plasmon at desired wavelengths for SERS applications.

5.2.1 Suitable Materials for the Application in SERS

Understanding the optical response of a metal nanostructure is a prerequisite for its application as a SERS nanoconstruct. Au, Ag, Cu, Al, etc. have unique optical properties; for example, these metals are the excellent reflector of visible light, making them very different from regular dielectrics [14]. Such optical responses result from the abundance of free conduction electrons in these metals. These free mobile electrons are balanced out by the presence of an equivalent number of relatively more stationary positive charges in the metals resulting in the formation of "free-electron plasma". The "free-electron plasma" at and near to its characteristic resonance frequency (which is within the visible range for metals like Au and Ag) defines the optical responses of the corresponding metal. One can understand the effect from the knowledge of dielectric function in these materials. The relative dielectric function of a metal $\epsilon(\omega)$ can be obtained from the Drude model, as follows:

$$\epsilon(\omega) = \epsilon_\infty - \frac{\epsilon_\infty \omega_P^2}{\omega(\omega + i\gamma)} \tag{5.1}$$

where ϵ_∞ is the high-frequency permittivity. ϵ_∞ adds a constant background dielectric function that originates from the optical response of the positive ions in the material. γ is the damping term related to the collision rate of free electrons, which also causes electrical resistivity. γ is significantly smaller compared to the frequency ω. ω_P is the plasma frequency (oscillation frequency of the free-electron-plasma), which is given by the following equation:

$$\omega_P = \sqrt{\frac{ne^2}{m\epsilon_0\epsilon_\infty}} \tag{5.2}$$

where m, e, and n are the mass, the charge, and the number of carrier electrons, respectively. ϵ_0 is the permittivity of free space. The real part of the expression (5.1) can be written as follows:

$$\text{Re}\big[\epsilon(\omega)\big] = \epsilon_\infty - \frac{\epsilon_\infty \omega_P^2}{\omega^2 + \gamma^2} \tag{5.3}$$

From equation (5.3), we learn that $\text{Re}\big[\epsilon(\omega)\big] < 0$ for $\omega < \omega_P$. Similarly, the imaginary part of the expression (5.1) can be written as follows:

$$\text{Im}\big[\epsilon(\omega)\big] = \frac{\epsilon_\infty \omega_P^2 \gamma}{\omega\big(\omega^2 + \gamma^2\big)} \tag{5.4}$$

The absorption of light by the material is characterized by the $\text{Im}\big[\epsilon(\omega)\big]$. Therefore, under the condition $\omega < \omega_P$ (where $\text{Re}\big[\epsilon(\omega)\big] < 0$) and for $\omega \approx \omega_P$ the absorption of light is small. Having a small value of $\text{Im}\big[\epsilon(\omega)\big]$ and a negative value for $\text{Re}\big[\epsilon(\omega)\big]$ at certain wavelengths is important for the generation of plasmon resonances. Let us understand this by assuming a metallic sphere, which is kept inside a medium of dielectric constant (ϵ_M). The sphere is allowed to interact with an external electromagnetic field (E_0) of an incident laser beam. The diameter of the sphere is distinctly smaller than the wavelength of the laser and E_0 is uniform in space. The electric field generated inside the sphere (E) in such a scenario can be expressed as follows:

$$E = \frac{3\epsilon_M E_0}{\epsilon(\omega) + 2\epsilon_M} \tag{5.5}$$

From equation (5.5), we can infer that an enormously large electromagnetic field can be generated inside the small sphere for $\epsilon(\omega) \approx -2\epsilon_M$. For almost every non-metal, $\epsilon(\omega)$ is typically between 1 and 10, and therefore, these materials generally do not fulfill the aforementioned conditions to support plasmon resonances. For metals, however, $\text{Re}\big[\epsilon(\omega)\big]$ can be negative for incident light (preferably in the UV-visible range) frequencies $\omega < \omega_P$. Therefore, for metals, this condition can be closely met if at a frequency where $\text{Re}\big[\epsilon(\omega)\big] \approx -2\epsilon_M$, the absorption is small (i.e., $\text{Im}\big[\epsilon(\omega)\big] \approx 0$). At this point, we should note that the above expression applies only to a small spherical geometry. For different geometries, other expressions can be derived with different denominators. Therefore, the condition for achieving plasmon resonance is governed by both the material and the geometry of the nanoconstruct.

The condition for supporting a plasmonic resonance by material for a given wavelength of light is determined by the value of $\text{Re}\big[\epsilon(\omega)\big]$. A considerable strength of the localized surface plasmon resonance (LSPR), however, is only possible for a small value of $\text{Im}\big[\epsilon(\omega)\big]$. Larger values of $\text{Im}\big[\epsilon(\omega)\big]$ would result in stronger absorption incident light, resulting in weaker plasmon. We can define a "quality factor (Q)" to quantify 'loss' due to such absorption in nanoconstructs, as follows:

$$Q = \frac{\omega \dfrac{d\big[\text{Re}(\epsilon)\big]}{d\omega}}{2\big[\text{Im}(\epsilon)\big]^2} \tag{5.6}$$

The higher the value of Q, the better it is for supporting strong LSPR. In fact, only the metal-nanoconstructs those have (i) $\text{Re}(\epsilon) < 0$, in the visible and IR region (LSPR application requires $-20 \leq \text{Re}(\epsilon) \leq -1$) and (ii) $Q > 2$ (preferably >10) support strong LSPR and consequently are suitable for SERS application. The Q is very low for metals like Pd and Pt due to too much absorption. For Al, however, sufficiently strong LSPR can be obtained at the UV range. In this regard, Ag and Au are the most promising material for LSPR application. Cu can also support strong LSPR but at longer wavelengths ($\geq 600\,\text{nm}$). Li supports large Q values for a wide range of wavelengths, but it is too reactive for LSPR applications. Alongside the value of Q, one should also consider the cost, reactivity, toxicity, etc., of the nanoconstructs before using them for SERS applications. Au, for example, is best suited for applications that require incident light in the red and near infrared region and/or for performing SERS in the biological systems. Ag finds it use when significant field enhancement is required with wavelengths shorter than 600 nm.

5.2.2 SERS ENHANCEMENT FACTORS

Enhancement in SERS has two different origins: *electromagnetic enhancement* and *chemical enhancement*. The electromagnetic enhancement is purely related to the plasmonic enhancement of the electromagnetic fields near the metal nanostructures, and chemical enhancements result from changes in the molecular polarizability originating from possible chemical bonding interactions that can occur between the specimen and the probe (metal nanostructures). However, the mechanism or even the mere existence of chemical enhancement is still under debate in the field of SERS. Even if chemical enhancement does exist in some instances, its contribution is small, and electromagnetic enhancement has been accepted to be the main contributing enhancement factor in SERS.

From the classical description of the Raman effect, we learn that a Raman dipole, $P = \alpha E$. The power of the radiation of this Raman dipole that we detect as Raman signal in the far-field is proportional to the $|P|^2$. The same description of the strength of the Raman signal can be adopted for SERS with the consideration that the presence of the metal nanostructure in the vicinity can modify both α and E. We have discussed how the massive enhancement of the electromagnetic field is possible at the position of the molecule if it is kept near plasmonic nanostructures. This results in *local field enhancement* in the Raman signal. Additionally, modification in the P also results in a *radiation enhancement*. Both local field and radiation enhancement contribute to the electromagnetic enhancement factors.

5.2.2.1 The Local Field Intensity Enhancement

By now, we know that the magnitude and orientation of the local electric field (E_L) at the position of a molecule near a SERS substrate are strongly dependent upon the wavelength (with respect to the plasmon resonance) and the polarization of the incident light beam. The E_L experienced by the molecule can also strongly vary based on the relative position and distance between the specimen and the SERS substrate. For SERS-based applications, the design of the experiment should ensure many fold increment of the incident $|E_L|$. The E_L induces a Raman dipole in the molecule, $P_{E_L} = \alpha E_L(\omega_L)$, at frequency ω_L. Therefore, the radiated energy of the Raman dipole would be enhanced due to the enhancement associated with the excitation of the Raman dipole under E_L. This enhancement factor (EF_L) is known as *the local field intensity enhancement factor*.

$$EF_L(\omega_L) = \frac{|P_{E_L}|^2}{|P|^2} = \frac{|E_L(\omega_L)|^2}{|E_I|^2} \tag{5.7}$$

Under certain experimental designs (e.g., in a "hot spot" generated within a nanoscopic gap between two metallic nanostructures), EF_L of even ~10^5 can be achieved.

5.2.2.2 The Radiation Enhancement

The local field intensity enhancement factor is associated with the excitation of the Raman dipole. However, this Raman dipole is not emitting in free space but near a metallic structure, which would result in an additional modification to the overall enhancement factor. At this stage, we would like to emphasize that it would be incorrect to assume that the *radiation enhancement* of the emitted field happens post-radiation. The presence of the metal would directly affect the emission process. The electromagnetic environment near the metal nanostructure surface would extract many orders of magnitude more power (i.e., stronger emission) from the oscillating dipole than what the same oscillating dipole would radiate in the free space. Enhancement factor associated with radiation enhancement (EF_R) thus can be defined as follows:

$$EF_R(\omega_R) = \frac{|E_L(\omega_R)|^2}{|E_I|^2} \tag{5.8}$$

where, $E_L(\omega_R)$ is the local electric field at the frequency of the Raman radiation. Both EF_L and EF_R are related to the local electric field strength. Therefore, just like EF_L of ~10^5, "hot-spots" can generate EF_R of ~10^5 as well.

The total *electromagnetic enhancement* factor (EF_{EM}), which is the combination of the *local field enhancement* factor and the *radiation enhancement* factor, can be expressed as follows:

$$EF_{EM} = EF_L(\omega_L)EF_R(\omega_R) = \frac{|E_L(\omega_L)|^2}{|E_I|^2}\frac{|E_L(\omega_R)|^2}{|E_I|^2} \tag{5.9}$$

Now for Raman signature at low wavenumbers $\omega_L \approx \omega_R$ and we can approximate the EF_{EM} as follows:

$$EF_{EM} = \frac{|E_L(\omega_R)|^4}{|E_I|^4} \tag{5.10}$$

Equation (5.10) is arguably the simplest and the most useful relation in SERS. In fact, in many applications, the SERS enhancement factors can be derived very close to the accuracy using this approximation. However, one should always keep in mind that this is an approximation and in certain cases (for example, in probing the polarization effects) EF_R can be very different than the more simply obtainable EF_L.

The magnitude of EF_{EM} also strongly depends upon the distance between the analyte and the SERS substrate. For a spherical NP of radius "r", the ratio of the EF_{EM} at a distance, "d" from its surface vs EF_{EM} at zero distance is given by the following relation:

$$\frac{EF_{EM}(d)}{EF_{EM}(0)} = \left(\frac{r}{r+d}\right)^{12} \tag{5.11}$$

5.2.2.3   The Chemical Enhancement

Adsorption of analyte molecules on the SERS substrates (metal nanostructures) may lead to the modification of the Raman polarizability, which consequently may alter the Raman intensities. Such "chemical" effects can enhance or quench Raman signal strengths. Enhancement can happen if the modified polarizability ensures a better resonance with the local field than the unaltered polarizability. Let us understand the chemical enhancement (CE) considering a formation of a charge-transfer state between the metal and the analyte. Although *the charge-transfer mechanism* is the most examined mechanism for explaining CE, we must note that there are other possible processes that can result in the chemical enhancement as well. In practice, the metal may perturb the electronic arrangement in the molecules in the vicinity even without a metal-ligand bond formation, resulting in a minor alteration in the polarizability affecting Raman efficiency of the corresponding mode(s) of vibration. However, a stronger effect is expected if the analyte is involved in chemical interaction with the metal surface. Such interaction can create electronic states with energy gap matching (or close to matching) the frequency of the local field, resulting in resonance Raman-type enhancements of the SERS signal. In special cases, if the energy gap between the highest occupied molecular orbital (HOMO) and/or lowest occupied molecular orbital (LUMO) of the molecule and the Fermi level of the metal is matched by the local field, a photo-induced transfer of electron can occur between the electron-occupied states (close to the Fermi level) of the metal and the LUMO of the analyte or between the HOMO of the molecule and unoccupied states above the Fermi level of the metal. This charge-transfer process can substantially contribute to the total enhancement factor of SERS.

5.2.3 SURFACE SELECTION RULES

In practice, the SERS spectrum cannot simply be derived or explained from the knowledge of the Raman spectrum of the free molecule and the estimation of the electromagnetic and chemical enhancement factors. The electromagnetic field near the SERS substrate has both perpendicular and parallel components, whose magnitudes depend upon both the geometry and material of the nano-construct, as well as their alignment with respect to the polarity of the incident light beam. Also, on the SERS substrates, analyte molecules often are not freely rotating but are oriented in certain ways due to their interaction with the NPs. Consequently, vibrational modes with polarizability changing parallel to the direction of the local field show stronger surface enhancements. The surface selection rules address this situation. A full description of the surface selection rules is beyond the scope of this chapter. In brief, the surface selection rules take into account the polarization of the enhanced electromagnetic fields and modifications of the polarizability tensors of the analytes. Surface selection rules are often applied to comprehend the observation of various extents of SERS enhancements between different vibrational signatures of the same analyte on nanoconstruct surfaces.

From the theoretical description of the enhancement factors of SERS, we can observe that the enhancement process depends on many factors. Any theoretical treatment that aims to predict enhancement factors should take into account the details of the incident light (wavelength, power, polarization, solid angle of incidence, etc.), configuration of the detection part of the instrument (e.g., whether the same optics are used for excitation and collection or not; whether it is the detection of polarized or unpolarized photons, etc.), properties of the SERS substrates (e.g., its material, geometry, arrangement in space, and orientation with respect to the excitation, etc.), and properties of the analytes (e.g., their intrinsic Raman cross section, their adsorption process on the metal, concentration and orientation of the analytes on the SERS substrates, their distance from the SERS substrate, and any modification of the polarizability tensor, etc.).

5.2.4 EXPERIMENTAL DETERMINATION OF THE SERS ENHANCEMENTS

To determine the practical sensitivity (detection limit) of a SERS experiment and to find out the effect of geometry/morphology of a newly developed SERS substrate on its performance, it is important to experimentally determine the SERS enhancement factor and comprehend it with theory. The estimation of the average enhancement of Raman signal (EF_{avg}) in a SERS experiment can be done by knowing all the terms on the right side of the following relation:

$$EF_{avg} = \frac{I_{SERS}}{I_R} \frac{N_R}{N_{SERS}} \tag{5.12}$$

where I_{SERS} and I_R are the intensity of the Raman signals measured in the presence and in the absence of the SERS substrate, and N_R and N_{SERS} are the number of molecules from which the Raman and the SERS signals were detected, respectively. Equation (5.12) is straightforward and easy to comprehend, but caution should be practiced while determining the number of molecules involved in the respective processes. For any liquid sample or solution, it is easy to find N_R by knowing the probe volume of the instrument and the density/concentration of the analyte but determination of N_{SERS} from a solution is tricky. The user must know the packing density of the analyte on the nanoconstruct, the average surface area of the nanoconstruct, and the concentration of the nanoconstructs in the solution in order to accurately estimate N_{SERS} in those cases. Instead, it is easier to estimate N_{SERS} for an assembly of NPs deposited on the surface or if the SERS substrate is prepared via lithography, as it is uncomplicated to determine the total SERS-active surface area under the probe volume in those cases. However, one must make sure that both the Raman and the SERS experiments are performed with a monolayer of the analyte, as enhancement factors are strongly dependent upon the distance from the surface of the SERS substrate. The monolayer of an analyte, in practice, may not give enough signal for regular Raman experiments. Stronger Raman mode of a different reference molecule can be used in those cases on the condition that the ratio of the Raman scattering cross section of the reference molecule and the analyte is available. Photo-induced degradation of the analyte also needs to be avoided in any of these experiments.

In some experiments (e.g., determination of the secondary structure of peptides or active site of proteins at low concentration), however, estimation of the enhancement factors may not be necessary. It would still be important to apply surface selection rules for comprehensive interpretation of the SERS signals in those cases.

5.3 COMMONLY USED AND PROMISING SURFACE-ENHANCED RAMAN SPECTROSCOPY SUBSTRATES

Both the enhancement factors and the relative intensities of different Raman lines in a SERS experiment are highly susceptible to even small changes in the metallic nanostructures and their arrangements [15,16]. A good SERS substrate should generate a uniform and reliable optical response, and its manufacturing (and/or synthesis) should be highly reproducible. Preferably, the substrate should be chemically less reactive, cost-effective, durable, yet easily recyclable such that it can be reliably reused. However, not every SERS substrate needs to fulfill all of the above-said qualities. The choice of the SERS substrate should be made based on how closely its properties suit the type of application it is chosen for. Therefore, while designing a SERS substrate, it is very important to have a clear idea about the questions at hand and the hurdles associated with them. For example, for single-molecule detection, it is highly beneficial to work with reproducible "hot-spots" having well-defined and strong electromagnetic fields. On the other hand, the SERS substrate needs to show selectivity and sensitivity toward the analyte molecules to be applied as a sensing platform. For chemical analysis, however, the preferred SERS substrate should support homogeneous and predictable electromagnetic field enhancements.

5.3.1 COLLOIDAL METAL NANOPARTICLES

The benefits of using a dispersion of colloidal Au or AgNPs as substrates are many. (i) Their synthesis is cost-effective and can provide reproducible NP geometries. (ii) NPs with a variety of optical properties and plasmon resonances can be synthesized relatively easily. (iii) Their surfaces can be deftly modified with different functional groups and coatings for a variety of applications. In fact, spherical metal NPs are some of the most used SERS substrates [17–19]. The signal strength of such systems can also be drastically enhanced by trapping the analyte molecules in "hot-spots" generated within aggregating NPs. However, it is challenging but very important to precisely control the aggregation condition and replicate the time window of data accusation to obtain reproducibility. High enhancement factors can also be achieved in a non-aggregating colloidal dispersion of anisotropic metal NPs (e.g., nanorods [20,21], nanobipyramids [22,23], and nanostars [24–26]), as they can support plasmonic "hot-spots" at their sharp surface features (Figure 5.1).

5.3.2 SELF-ASSEMBLED METAL NANOPARTICLES

Achieving high enhancement factors with an aggregating metal NP solution is tricky. Once spontaneous aggregation is triggered, the dynamic process of the formation of larger aggregates will continue with time. The SERS signal strengths are subject to change drastically with aggregation and

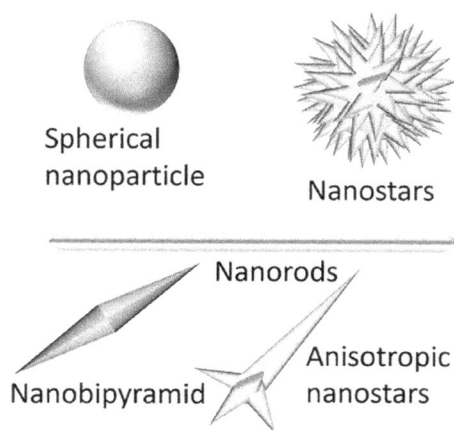

FIGURE 5.1 Shapes of different types of metal nanoparticles.

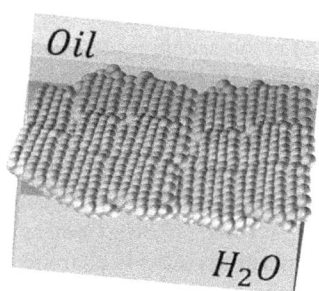

FIGURE 5.2 Graphical representation of self-assembled monolayer of metal nanoparticles. A layer of spherical NPs is assembled spontaneously at the interface of water and oil.

would only peak within a certain time window for which the maximum number of "hot-spots" with trapped analyte molecules would be available for detection. Also, the orientation of the analyte molecules with respect to the local electric field directions may not always be controllable, especially if the initial NP solution had a heterogeneous distribution of size and shape. This can result in variability in the relative peak intensities in the SERS spectra. Ordered arrays of self-assembled metal NPs with organized hot-spots technically can address these issues [27–30]. Controlled arrangements of densely packed hot-spots with precise size and geometry over large areas can also be obtained with lithography. Still, such top-down approaches are associated with significantly higher costs. However, an assembly of self-assembled monolayer (SAM) of monodisperse metal NPs is a cost-effective way of obtaining strong and relatively uniform SERS signals over a large surface area.

The repulsive charge interaction of the capping ligands leads to stable dispersions of Au or Ag NPs in solution. One needs to overcome this repulsive interaction to form self-assembled layers of NPs. The common practice, in this regard, is to replace the charged ligands with neutral but strongly adsorbing molecules (e.g., organic thiols) [29,31]. With this approach, monolayers of densely packed metal NPs can be assembled (supporting the dense distribution of stable hot-spots) at the interface of immiscible liquids (Figure 5.2). These SAMs, unlike aggregating colloidal NPs, provide strong and uniform optical enhancements and, at the same time, remain stable for several days if handled properly. These SAMs are typically formed at the organic solvent/oil-water interface, which can be used to detect and/or probe both hydrophilic as well as hydrophobic solutes from either of the solvents at very low concentrations. The ability to probe analytes from both phases opens up the possibility of applying these monolayers (and deposits) in various SERS-based applications, such as in situ probing of organic reaction [32], detecting trace amounts of pollutants and explosive materials in solution, etc. [33]. Also, these monolayers can be easily transferred to various substrates by simple dip-coating [27,34]. The monolayer-coated substrates can be used as portable SERS-active surfaces. The dip-coated systems are typically stabilized by weak Van der Waals' interaction between the NP layer and the underneath solid surface and, therefore, can suffer damages upon drying. Polymer-based thin films are sometimes deposited from the organic solvent side on the SAM of NPs before the dip-coating step to overcome this issue [35]. These polymer layers only cover a small fraction of the NP surface—keeping a large area available for the generation of SERS—while, all together, providing necessary stability and flexibility to the NP layers from underneath. This type of SERS platform can withstand the physical pressing of solid analytes into the NP layers for solvent-free studies [36]. Uniform enhancement of the electromagnetic field over a large enough area on this polymer stabilized NP arrays even allows probing of distribution of immobile analytes with SERS-based imaging.

5.3.3 COATED METAL NANOPARTICLES

A coating of either organic molecules [e.g., poly(ethylene glycol) PEG] or biomolecules (e.g., lipid bilayers, proteins, nucleic acids, etc.) or inorganic molecules (e.g., silica-coated NPs) has been in use to stabilize NPs in solution (Figure 5.3) [19,37–45]. These coatings, in general, create barriers that keep different molecules away from interacting with the NP surface. At the same time, some of these coatings allow only a select class of molecules to get embedded close to the NP surface or even penetrate through to have direct interaction with the NP [19,38,39]. For example, a lipid bilayer coating on AgNP or AuNP would create a 4–5 nm wide barrier that selectively allows lipophilic molecules to remain close to the NP surface [37,38]. Coating of negatively charged mesoporous silica, on the other hand, would only permit positively charged and neutral molecules with a diameter smaller than their pore sizes to approach the NP underneath [19]. At this point, we should recall that the magnitude of electromagnetic enhancements strongly depends upon the distance from the nanoparticle surface (equation 5.12). Therefore, these coated SERS substrates can be used as excellent platforms for sensing applications [19], used as nanotags [46], and handy in determining the structure of membrane-bound peptides and proteins [38]. Some of these aspects would be further elaborated in Section 5.4, where we have discussed various applications of SERS.

5.3.4 SINGLE HOT-SPOTS

A comprehensive understanding of the theory behind SERS requires experimental testing of different factors (e.g., the geometry of the hot spot, material, dielectric environment, etc.) linked to its mechanism [47–49]. The experimentation is particularly challenging as it requires manufacturing of different SERS substrates with high precision and accuracy. Also, ideally, such testing needs to be

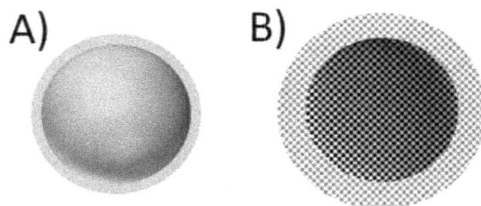

FIGURE 5.3 Depiction of coated metal NPs. (a) Metal nanoparticle coated with a relatively thin layer of PEG or biomolecules (e.g., lipid bilayer). (b) Metal nanoparticle coated with a porous inorganic layer of mesoporous silica. Thickness of these coatings can be controlled.

done preferably at the single hot-spot label to avoid ensemble effects [47–49]. In this regard, SERS substrates with precise geometries, carrying a single or only a few "hot-spots" have been successfully manufactured by lithography [47,48,50]. Both far-field and near-field optical responses of these plasmonic substrates have been successfully correlated with their structure (as probed by electron microscopy) [49,51,52]. Some of these studies, however, have shown greater than expected differences between the far-field and near-field resonances, the reason for which has not fully been understood yet [50,53]. In many cases, it has been challenging to determine the number of molecules at the "hot-spots", their positions and orientations [53–55]. Measured SERS enhancements often turned out to be significantly larger than predicted from simulation. Thus, these experiments point toward the possibility that the theoretical understanding of SERS may be incomplete at the moment. Therefore, these "hot-spots" with discrepancies between their theoretically predicted and experimental enhancement factors provide excellent platforms to fully comprehend the SERS effects.

5.3.5 Nanorattles

These are modified versions of core-shell plasmonic nanostructure that can be used as extremely bright SERS reporters [56,57]. The synthesis process typically begins with a AuNP core that is coated with Ag. These core-shell NPs are then subjected to *galvanic replacement* to make the Ag-shell porous with the aim to produce a nanocage structure with AuNP inside. A large enough concentration of Raman reporter molecules can then be packed through the porous Ag-shell into the nanocages. The final step involves closing the nanocages with a layer of Au-coating to produce the *nanorattles* (Figure 5.4). These nanostructures can demonstrate up to three orders of magnitude stronger SERS signal per particle compared to a simple metal nanoparticle with Raman reporter molecules on their surface. Initially, the spherical nanorattles were made (Figure 5.4a) [56]. In later studies, their structure has been upgraded to a cubical nanostructure (Figure 5.4b) [57]. The nanorattles have been successfully used to detect *Plasmodium falciparum* deoxyribonucleic acid (DNA) from a blood sample at a very low concentration [57]. The details of this study are discussed in Section 5.4.

5.3.6 Leaning Nanopillars

Lithography finds another use in fabricating substrates with dense vertical nanopillars [58,59]. These substrates are successfully made with block copolymer lithography [60], nanosphere lithography [61], phase shift lithography [62], and recently by non-lithographic techniques like reactive ion etching (maskless) [59]. Strong and uniform electric field enhancements over large areas have

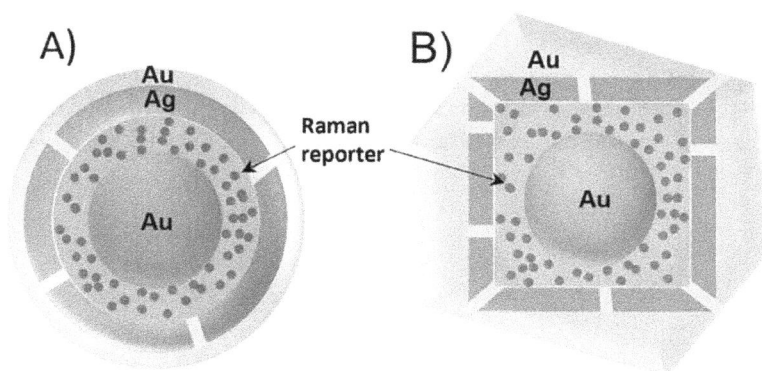

FIGURE 5.4 Graphical representation of different types of nanorattles. (a) A spherical nanorattle. (b) A nanorattle with a cubical Ag-cage.

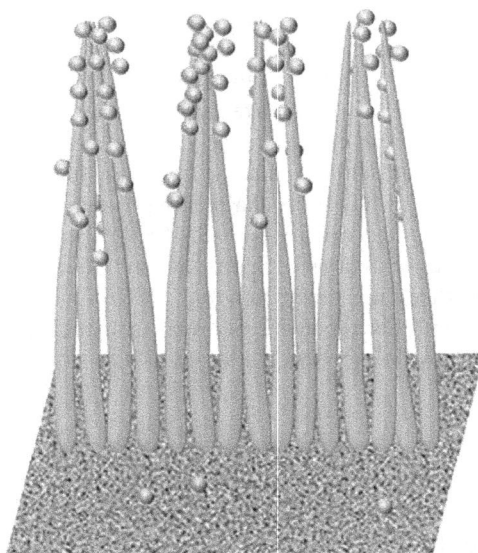

FIGURE 5.5 Depiction of a SERS substrate containing dense nanopillars. The gold nanoparticles attached to the surface of the nanopillars form clusters as the nanopillars lean on each other during the solvent evaporation step and thereby, create several hot-spots.

been achieved with some of these substrates. Often while applying photo-lithographic techniques, followed by solvent exposure and evaporation, the formed nanopillars can lean onto each other in a cluster due to the capillary effect and thereby trapping analytes close to the hot-spots. For example, J. Chen, H. Xu, and co-workers have probed SERS induced by AuNPs attached to InP nanow-ires [58]. The analyte molecules (rhodamine 6G) were adsorbed onto the AuNPs, and the clusters formed due to the collective leaning of the nanowires resulted in bringing the AuNPs close enough to create hot-spots (Figure 5.5). Raman mapping was performed to prove that a significantly stron-ger SERS signal originated only from the AuNPs trapped in the clusters of the leaning nanowires compared to the AuNPs located on the wafer surface.

5.4   IMPORTANT APPLICATIONS OF SURFACE-ENHANCED RAMAN SPECTROSCOPY

Surface-enhanced Raman spectroscopy has been brilliantly applied in several different ways, including the monitoring of special distribution and chemical recognition of the targeted specimen with SERS tags, analytical identification and/or quantification of the molecule of interest, identifi-cation of molecules based on their chiralities, medical applications, in the prevention of chemical warfare, application in the material science and biological application, etc. Some of the use of SERS has been listed below.

5.4.1   SERS-NANOTAGS

Fluorescence tags (fluorescent molecules) have long been used to localize and track various objects and molecules, especially in biology [63–68]. Molecular recognition is the key for such probing, often achieved via highly specific non-covalent interaction between the targeting moiety (e.g., anti-bodies, oligonucleotide strands) and the molecule/specimen of interest (e.g., antigens, complemen-tary oligonucleotide sequence, etc.). The targeting moieties, in this case, are typically tagged with a fluorophore to probe the position of the analytes. SERS-nanotags are increasingly finding their

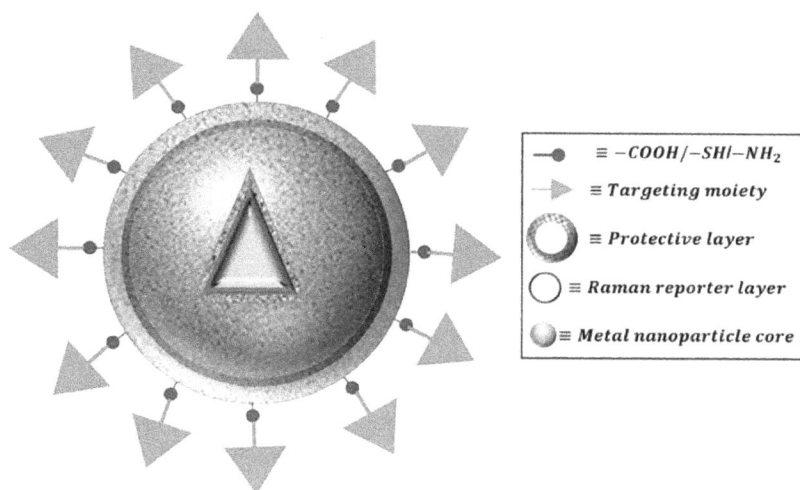

FIGURE 5.6 Depiction of classical SERS-nanotags. The plasmonic metal nanoparticle (golden-colored sphere) is coated with a layer of Raman reporter molecules (red color ring) and the whole structure is stabilized by a protective layer (textured, gray color, and ring). The surface functionalities (blue-colored discs) of the protective layer are attached to the targeting molecules (green-colored triangles). A cross section (triangular) of the different layers is shown in the middle.

use in such applications. A SERS-nanotag typically consists of a plasmonic metal NP core (e.g., spherical, star-shaped, rod-shaped NPs of Au and Ag) (Figure 5.6), with a surrounding layer of Raman reporter molecules (e.g., organic dyes or a SAM of thiolated aromatic molecules with high Raman cross section) adhered or attached on the NP surface (Figure 5.6, shown as a thin, red-colored ring around the core), which is then stabilized by an additional protective layer from the outside (e.g., a layer of thiol terminated polyethylene glycol molecules or an adsorbed layer of bovine serum albumin protein or a silica layer, etc.) (Figure 5.6, shown as a thick, gray-colored ring on top of the Raman reporter layer). The functional groups on the outer surface of the protective layers (e.g., $-COOH$, $-NH_2$, $-SH$, etc.) (Figure 5.6, blue discs) are then used to attach (bio-conjugate) antibodies or oligonucleotides, etc. (Figure 5.6, green triangles) to specifically target the analytes [69–71].

SERS-nanotags carry several advantages. Unlike fluorophores, which are susceptible to fast photobleaching, SERS-nanotags are photostable. The broad nature of fluorescence spectra often results in the spectral overlap between different fluorophores and limits the possibilities of parallel detection in fluorescence-based applications. Raman features, however, are considerably sharper and thereby provide excellent multiplexing in SERS-based imaging. The fluorophores at close spatial proximity to one another often result in self-quenching. Still, SERS signal strengths from nanotags are additive and thus can be used to quantify the number of SERS-nanotags. Also, simultaneous probing of different fluorophores—with differentiable absorption and emission profiles—typically requires excitation via a variety of different incident wavelengths. However, SERS-nanotags with different types of Raman signatures can be excited with the same laser source, allowing easy simultaneous detection of the probes. Certain biological samples emit autofluorescence if excited with UV-Vis light, creating an unwanted background during imaging. Alternatively, SERS-nanotags with NP-core resonating at red or NIR frequency (e.g., Au nanorods, Au nanostars, etc.) can be used to avoid complications of autofluorescence [70,71].

One of the major disadvantages of SERS-nanotags is their large size (several 10s of nm in diameter), which limits their in vivo applications. Also, fluorescence tags with various spectral features e.g., green, yellow, red fluorescent proteins (GFP, YFP, and RFP) can be co-expressed in conjugation with other proteins with the help of genetic engineering [64]. SERS-nanotags, however, always need to be administered from outside, which often leads to unwanted adsorption of serum proteins on the nanotag surfaces. The absorption on the nanotag-surface ultimately may result in the burial of their designed functionalities and reduction in their ability to exert specific interactions [26,72,73]. However, successful targeting with functionalized metal NPs has been achieved with proper nanoconstruct designing strategies [26,73], which bring forth promising possibilities of even more versatile application of SERS-nanotags in the future.

Conventional SERS-based experiments often require extremely strong enhancements of the local electric field, which is typically achieved via aggregation of metal NPs. This practice may not be easily adapted for biological applications of SERS-nanotags. Often there can be spatial constraints to accommodate more than one NP at proximity and/or the distribution of the targeted biomolecules can be highly dispersed. Therefore, ideally, a SERS-nanotag should be bright enough to enable single-particle detection and should provide reproducible SERS intensities at the single-particle level [74,75]. The specific condition would require a very conscious approach for producing and purifying the SERS-nanotags to ensure a highly uniform distribution of self-similar and bright particles.

5.4.2 CHEMOSENSORS

A chemosensor works by identifying or sensing the presence of a molecule of interest. SERS-based chemosensors consist of two main components: (i) a plasmonic nanostructure capable of producing strong enhancements of the Raman signals and (ii) a mechanism that can selectively allow only the analyte-molecule(s) to be present at the proximity of the plasmonic nanostructure. The most common practice for making SERS-based chemosensors involves functionalizing metal NPs or nanostructures with ligands that selectively interact with molecules of interest. These interactions would thereby hold the analyte molecules at the proximity of the NPS and result in modification of the SERS signals [76–79]. Similar concepts can be easily extended in sensing biomolecules in a platform where the ligands can be antibodies (or receptors) or oligonucleotides based on the sensing requirements [80–82].

Alternatively, a coating can be used to "hide" the NP-core from non-specific interactions with approaching molecules. Instead, the coating would possess selective permeability for analyte molecules based on their molecular charge and/or size. Such an application has been demonstrated by H. Fathima, K.G. Thomas, and co-workers with mesoporous silica-coated AgNPs [19]. The negatively charged mesoporous silica coat generates a sieving effect based on its pore size and a charge-based selection to only allow neutral or positively charged small molecules to approach close to the AgNP surface for chemosensing.

A typical Raman spectrum of a molecule consisting of several lines (or bands) with relatively small bandwidths and molecules with even very small differences in their respective molecular structure often have easily distinguishable Raman signatures. This is an added advantage of SERS-based chemosensors, as the analytes are identifiable from their respective SERS spectrum.

5.4.3 QUANTITATIVE ANALYSIS WITH SERS

Fluorescence techniques are regularly employed for investigating analytes of small quantities. However, their usefulness is limited to only probing fluorophores. It is, therefore, imperative to extend the use of SERS to identify and quantify analytes from their vibrational signatures at low concentrations. We have discussed many factors that control the SERS signal strengths. By now, we understand that the SERS intensities can greatly vary even with ~1 nm differences in the relative position of the analytes from plasmonic NP and slight changes in their orientations with respect to the NP surface. This variability originates from the strong distance and polarization dependence in the enhancement factors. Any quantitative analysis using SERS is, therefore, quite challenging.

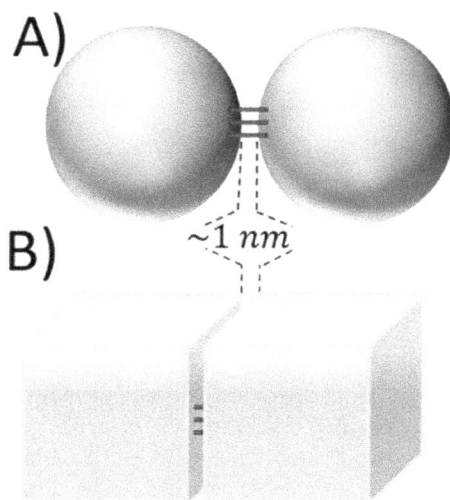

FIGURE 5.7   Cartoon of nano-dumbbells with ~1 nm interparticle gaps. (a) Nano-dumbbells formed between two spherical nanoparticles. (b) Nano-dumbbell formed with metal nanocubes. Nanocube-nano-dumbbells provide more uniform enhance factors across the nanogap.

It is obvious that this would require extremely high but mostly homogeneous electric field enhancements originating from the plasmonic structures. Plasmonic nanogaps that can generate strong but uniform electric field enhancements within the hot-spots, are suitable for such applications. Theoretically, the ideal width of a nanogap that can generate high enhancement factors in SERS experiments is ~1 nm. Further lowering of gap distance (~0.7 nm or lower) may result in quantum tunneling effects [50], which may decrease the magnitude of the electromagnetic field. Nano-dumbbells formed between Au and/or AgNPs can match this requirement. For example, J-E, Park, J-M Nam, and co-workers have demonstrated such uniform, and strong enhancement factors in the nanogap between two Au-nanocubes (Figure 5.7b) [83,84]. Schematics of such nano-dumbbells are given in Figure 5.7.

5.4.4   SERS with Remote Excitation

The dimensions of SERS hot-spots are typically a couple of orders of magnitude smaller than the excitation laser spots, which can result in unwanted background signals coming from regions outside the hot-spots. Also, intense laser sources often result in photodamage for biological samples. Therefore, it would be beneficial to have a system where the excitation laser spot can be placed away from the sample sites. Such a system was materialized by remotely exciting a hot-spot created at the junction of a nanowire and a nanoparticle (Figure 5.8) [85,86]. The surface plasmon polaritons created by a laser source parked at one of the ends of the nanowire [87], resulted in electric field enhancements at the nanogap that was located several microns away from the primary excitation site. Such a device has been successfully used to generate SERS signals with high signal-to-noise ratios from live cells while keeping the photo damages at a minimum [88].

5.4.5   SERS of Biofluids

Biofluids like blood, urine, cervical fluid, cerebrospinal fluid, saliva, etc. contain many different constituents in a variety of different amounts (sub $-nM$ to $> mM$). Concentrations of these constituents also have inherent variations even in normal conditions. Therefore, it is somewhat challenging to properly investigate the effect of a disease and/or treatment that can only be tracked by

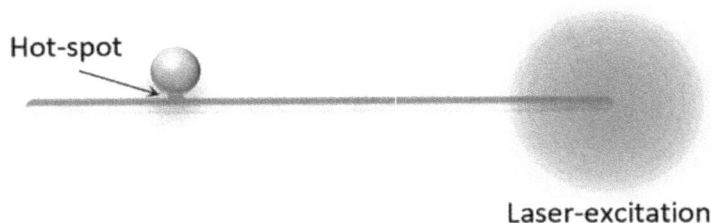

Laser-excitation

FIGURE 5.8 Remote SERS. Cartoon showing creation of a hot-spot between a plasmonic nanoparticle and a plasmonic nanowire, by remote excitation.

investigating relative changes in the concentration of biofluid constituents. It is, however, easier to check biofluids for the presence of an analyte that originates only in a given disease condition. In fact, SERS has been successfully employed to detect the presence of a virus [89], bacteria [90,91], fungi [92], etc. Also, by utilizing strong interaction between AuNPs and the amino acid stretch "-Met-Lys-His-Met-" present in the sequence of Creutzfeldt-Jakob prions, the presence of the prionic protein in the blood sample has been detected by SERS [93]. In this study by R. A. Alvarez-Puebla, L. M. Liz-Marzán, and co-workers, SERS-based quantification of this prion protein up to a few pM concentrations was achieved. There is, in fact, an added advantage in doing SERS if the analyte of interest has a sufficiently stronger affinity toward the SERS substrate or its surface functionalities (e.g., antibodies, complimentary DNA sequence, etc.), compared to the other constituents of the biofluid. An alternative approach would be to purify biofluid to separate the analyte of interest before subjecting it for SERS-based quantification. In this way, for example, nucleic acid fragments have been probed to look for the presence of single-point mutations.

5.4.6 APPLICATION OF SERS IN DETECTING GENETIC MATERIAL

The presence and concentration of certain genetic materials, e.g., DNA, messenger ribonucleic acid (mRNA), micro-RNA (miRNA), etc., in biofluids can be indicative of the progression of different diseases. There are several ways by which various SERS substrates are used to detect DNA and other genetic materials. One of the examples involves SERS-based *molecular beacons*. A traditional molecular beacon consists of a fragment of nucleic acid with a fluorophore and a quencher labeled at its two ends. This nucleic acid fragment would initially have a collapsed structure with the fluorophore in the "off" state (quenched). In the presence of a complementary nucleotide (analyte) in solution, the fluorescence would get turned "on", as the hybridization between the two nucleotides would increase the distance between the fluorophore and the quencher [94,95]. In a SERS-based *molecular beacon*, the targeting nucleotide fragment is attached with a plasmonic surface or NP [96]. The other end of this nucleotide would typically contain a fluorophore. Therefore, the hybridization events are detected from the changes in the SERS intensities in this case.

A related method for detecting specific nucleotides in solution is called *inverse molecular sentinel* [97–99]. This method uses stem-loop DNAs (Figure 5.9, black lines). The two ends of these DNAs can form intramolecular base pairs to manifest a noose shape. The stem-loop DNAs used here are end-tethered to a plasmonic surface (e.g., Au nanostars), and their free ends are labeled with Raman reporter molecules. Initially, with a 2nd DNA strand (Figure 5.9, gray lines), part of which contains complementary bases to the stem-loop DNA; the two ends of the stem-loop DNA are held apart (Figure 5.9a). Therefore, the Raman reporter does not show high SERS signals initially. The 2nd DNA, at this point, has a significantly long stretch of unhybridized bases as well. This system, as depicted in Figure 5.9a, can efficiently detect the DNA molecule (target DNA) that is complementary to the full length of the 2nd DNA mentioned here. In the presence of the target DNA (Figure 5.9, blue dotted line), the unhybridized part of the 2nd DNA would first form intermolecular base pairs with the target DNA (Figure 5.9b). Then with branch migration (Figure 5.9c),

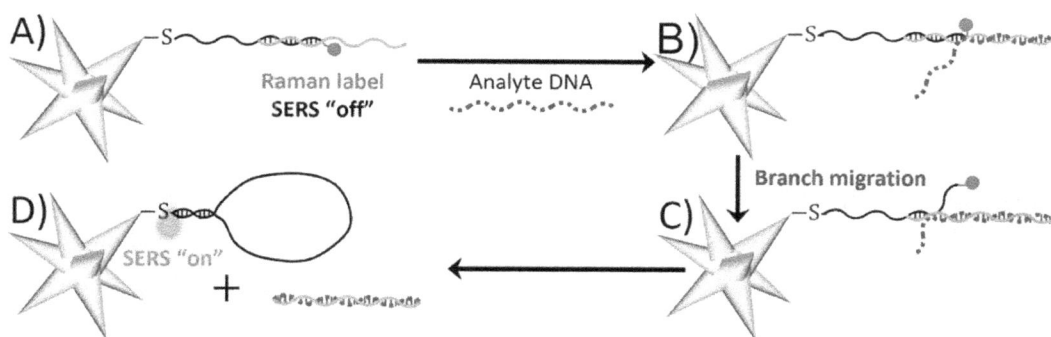

FIGURE 5.9 Schematic showing the steps of detection of target DNA by inverse molecular sentinel method. (a) Initially, the SERS signal is "off" due to the hybridization between the NP-tethered stem-loop DNA (black line) and the 2nd DNA (gray line) with a sequence that is complimentary to the targeted DNA (blue dotted line). (b) In the presence of the sample, the unhybridized part of the 2nd DNA strand starts hybridizing with the analyte DNA. (c) The analyte DNA keeps on replacing the stem-loop DNA from its complimentary DNA via branch migration. (d) The analyte DNA fully hybridizes with its complimentary DNA. This allows the stem-loop DNA to hybridize intramolecularly and that brings the Raman reporter close to the NP surface.

the stem-loop DNA would ultimately be fully detached by the target DNA from the 2nd DNA with time (Figure 5.9d). This final step would allow the untethered end of the stem-loop DNA to fold back to its NP-tethered end (Figure 5.9d). Consequently, there would be a jump in the SERS signal intensities from the Raman reporter molecules, reporting the presence of the target DNA molecules. The inverse molecular sentinel method is inherently hassle-free without much need for purification and isolation of the target DNA strands. The technique can be used to detect, for example, cancer-specific miRNA and genetic biomarkers of various viral infections.

Another interesting method of DNA detection involves *nanorattles*. This technique has been successfully employed to detect viral DNAs of *Plasmodium falciparum* from blood samples [57]. The structure of *nanorattles* has already been discussed in Section 5.3. The outer surface of the *nanorattles* has Au-coating, which can be easily functionalized with DNA molecules (Figure 5.10, blue dotted line), having a sequence complementary to one of the end part of the viral DNA (Figure 5.10, black line) present in the infected blood. The functionalized *nanorattles* are then mixed with magnetic beads containing another DNA molecule (Figure 5.10, gray line), which is complementary to the other end of the target DNA molecules. Therefore, only in the presence of the targeted viral DNA molecules, both the magnetic bead and the functionalized *nanorattles* would bind to the viral DNA strands, forming a sandwiched structure (Figure 5.10). Then with the help of an external magnet, all the magnetic beads (alongside the bound *nanorattles*) can be pulled together for SERS detection. This method was demonstrated to be powerful enough to differentiate between a mutant DNA that results in drug resistance and the wild-type (WT) DNA at low concentrations.

5.4.7 Application of SERS in Detecting Biomarkers

The application of SERS is not only limited to detecting genetic materials [100–102]. Many different types of biomarkers starting from small molecules (e.g., neurotransmitters, toxins, etc.), to large biomolecules (e.g., antigens, growth factors, bacterial proteins, nucleic acids, etc.), to even detection of pathogens (e.g., bacteria, virus, fungi, etc.) have been demonstrated with SERS [99,102–106]. By now, we understand that detecting biomarkers by SERS does not typically involve Raman tags. Rather, the spectrum of the biomarker itself (intrinsic SERS) is directly probed for its detection, which makes this type of application more challenging to execute. In the absence of any Raman reporters (or SERS-nanotags), the sensitivity of these experiments highly depends upon the optical properties

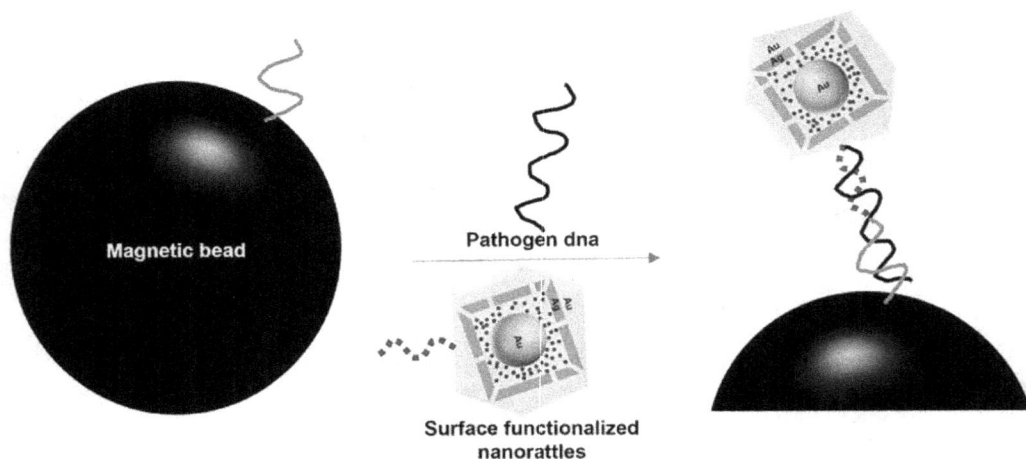

FIGURE 5.10 Schematic showing viral DNA detection using nanorattles. The DNA molecule used for the functionalization of the magnetic beads (green) and the DNA strand attach to the nanorattles are complimentary to different parts of the viral DNA (red). The hybridization between these three DNA strands results in the formation of the sandwiched structure involving both the nanorattle particle and the magnetic bead (right side).

of the plasmonic NPs, their distribution, and their aggregation in the biological system. It is, therefore, important to standardize every new nanoconstruct design and strategy by mass spectrometry, along with fast optical mapping of the cellular distribution of the NPs with techniques like differential interference contrast (DIC) microscopy (for in situ detection in live cells) and imaging of NP-distribution and aggregation by transmission electron microscopy (in fixed cells). Also, the effect of the formation of "protein corona" from the non-specific absorption of various biomolecules (mainly serum proteins) on the NP surface needs to be elucidated for almost every use of metal-nanoconstructs in biology, including their application in detecting biomarkers. As a matter of fact, by its very nature, SERS can be used to understand the composition and dynamics of the composition of such protein coronas [107].

It is important to have means with which the biomarker(s) of interest can be selectively held close to the NP surface for the surface enhancements to happen. In certain cases, the biomarker molecule itself might have an affinity toward the NP surface. The detection can be achieved by simple incubation of the sample with the plasmonic material. For example, the presence of *Bacillus anthracis* infection can be detected by SERS by utilizing the inherent affinity of dipicolinic acid molecules (a biomarker) toward plasmonic material, as demonstrated by X. Zhang, R. P. Van Duyne, and coworkers [108]. In the absence of such inherent affinity between the biomarker and the plasmonic surface, smart capture mechanisms are required.

Small assemblies of NPs with controllable gap sizes have been shown to reach sensitive detection limits for biomarkers. For example, L. Xu, H. Kuang, and co-workers developed a DNA-mediated pyramidal assembly of three AgNPs. The so-formed pyramids were encoded with three different Raman tags. These DNA aptamers have specificity toward certain biomarkers [106]. The interaction between the aptamers and their targets resulted in reconfiguration in the pyramidal structure to produce smaller gaps, causing stronger SERS. With this technique, several biomarkers, e.g., thrombin, prostate-specific antigens, and mucin-1, were detected with near attomolar sensitivity.

5.4.8 SERS OF GLOBULAR PROTEINS

The spectral positions of the vibrational modes involving the backbone of a protein (especially amide I and amide III modes) are sensitive to the secondary structure of the biomolecule. Therefore, the determination of the different secondary structural elements (e.g., α-helix, β-sheet, random coil, β-turn, etc.) is possible by SERS [17,38,109–114].

Before discussing the application of SERS in this field, we should recall a few things about the SERS mechanism. (i) The magnitude of the surface enhancement drops sharply as a function of the distance (d) from the plasmonic surface with the relation $\left(\dfrac{r}{r+d}\right)^{12}$ (equation 5.11) and SERS can only be obtained from the analytes located within a few nm from the surface. (ii) The orientation of the analyte-molecule with respect to the SERS substrate is extremely important too. Vibrational modes with polarizability changing in the direction of the enhanced local field show a stronger presence in the SERS spectra.

A small peptide with a hydrodynamic radius in the order of 1–2 nm can be accommodated close to the plasmonic surface. Therefore, the determination of the overall secondary structure of a peptide is possible by SERS, provided that its orientation allows enhancements of Amide I and/or Amide III modes from its different structural elements. For a large protein, however, predominantly the part of the molecule that is close to the plasmonic surface with the "right" orientation would contribute to the overall SERS spectrum [115]. Therefore, it is extremely important to predict the mode of nanoconstruct–protein interaction (electrostatic vs Van der Waals vs covalent) and identify the amino acids directly involved in those interactions before interpreting SERS spectra from protein molecules. The identification is typically achieved by inspecting the electrostatic potential maps of the protein surface and by identifying the arrangements of the protruding side chains of the amino acids at the outer surface of the protein.

Vibrational modes of side chains of certain amino acids like phenylalanine, tryptophan, tyrosine, and histidine—from close vicinities of the plasmonic surface—often exhibit strong SERS signal within the spectral range of 1,200–1,300 cm^{-1}, which overlaps with the amide III mode. However, the amide I region (i.e., 1,600–1,710 cm^{-1}) is background free and, therefore, should be interpreted for obtaining secondary structural information about the protein samples.

It is extremely challenging to perform single-crystal X-ray diffraction (XRD), nuclear magnetic resonance (NMR), and circular dichroism (CD) experiments while maintaining physiological conditions, for proteins with inherently low solubility, for membrane proteins, for proteins that are meant to undergo drastic structural changes upon small perturbation (e.g., at certain ionic strength and/or pH and in the presence of cofactor molecules or ions, etc.), and for transient protein structures those are kinetically formed (e.g., aggregation intermediate of amyloid proteins). With its ability to probe secondary structural changes at low concentrations of proteins, SERS finds many uses to answer unknowns about these kinds of proteins and peptides. For example, S. Aggarwal, C. Narayana, and co-workers reported structural changes associated with drastic functional modification of restriction endonuclease Kpnl by SERS [116]. KpnI has fascinating properties. It can engage in complexation with the GGTACC sequence in DNAs even in the absence of its cofactor ions Mg^{2+} and Ca^{2+} but cannot function as an endonuclease. Kpnl, in response to Mg^{2+} binding, turns into a promiscuous endonuclease, a functional response that is very different compared to most other RNA interference (RNAi) nucleases from the same family of proteins. However, in the presence of Ca^{2+}, it undergoes a functional transition to turn into a high-fidelity enzyme. Any corresponding structural changes those are associated with this functional shift were unknown in the absence of crystallographic structures. The group performed SERS and figured out that Ca^{2+} binding leads to the conversion of some of the anti-parallel β-sheet features (Figure 5.11a, peak ~ 1,620 cm^{-1} in the black spectrum) to random coils (Figure 5.11a, the region highlighted in the gray spectrum) in the enzyme, which could be the allosteric link to its observed functional transition. Their SERS results were reaffirmed by molecular dynamics (MD) simulations (Figure 5.11b). It is in fact, essential to confirm the retention of activity of enzymes on the plasmonic surface and ensure good agreement between the SERS and the simulation and/or complementary experimental results to interpret SERS data with full conviction.

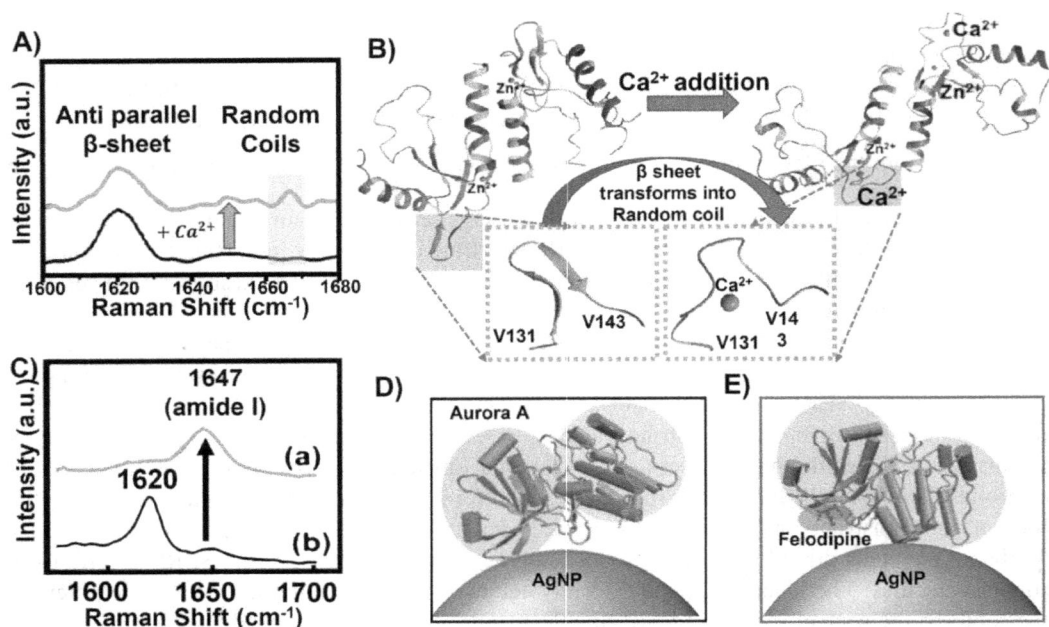

FIGURE 5.11 Key examples of SERS of globular proteins. (a) Changes in the SERS spectra (Amide I region) of KpnI upon Ca^{2+} addition. Ca^{2+} binding resulted in increase in the SERS intensity at $1,667\,cm^{-1}$ (gray spectrum), indicating formation of random coils. (b) Molecular dynamics simulation showing structural transformation from β-sheet to random coil near the Ca^{2+} binding site (enlarged in the inset). (c) Amide I region of SERS spectrum of Aurora A (black) and Aurora A complexed with felodipine (gray). (d) Mode of attachment of Aurora A to the AgNP surface and (e) change in orientation of Aurora A on the AgNP surface upon complexation with felodipine. (Figures 5.11a and b are adapted with permission from Aggarwal, S. et al., *The Journal of Physical Chemistry B* 2021, 125(9), 2241–2250. Copyright 2021 American Chemical Society; Figure 5.11c–e are adapted with permission from Karthigeyan, D. et al., *Proceedings of the National Academy of Sciences* 2014, 111(29), 10416–10421. Copyright 2014 National Academy of Sciences.)

D. Karthigeyan, C. Narayana, and co-workers have demonstrated an innovative drug screening approach that involves probing protein–drug interaction by SERS [115]. They showed that the amide I SERS signal from Aurora A (an oncogenic kinase protein) shifts from 1,620 to $1,647\,cm^{-1}$ (Figure 5.11c) in the presence of a felodipine drug. This kind of spectral shift can result from a significant loss in the β-sheet content in the protein. They, however, proved that this spectral shift resulted from the reorientation of the Aurora A molecules on the AgNP surface upon their interaction with the drug. The SERS results indicated that the drug felodipine binds on the surface of the Aurora A and thereby intervenes in a particular region with significant β-sheet content of Aurora A to remain close to the AgNP surface (Figure 5.11d). Instead, in the felodipine bound state, an α-helical region of the protein moves toward the NP surface (Figure 5.11e), resulting in the spectral shift (Figure 5.11c). Such spectral shift was not observed for a homologous kinase named Aurora B. Indeed, a functional assay proved that the felodipine action of inhibiting phosphorylation only worked for Aurora A but not for Aurora B. Alongside the biochemical assays, they also backed their SERS-findings with molecular docking and point-mutation-based studies.

5.4.9 SERS OF MEMBRANE PROTEINS

Membrane proteins are one of the primary regulators of cell function. It is somewhat counter-intuitive to imagine that the limited two-dimensional space of the membrane houses the types of proteins those cover ~30% of the human genome. It is, therefore, imperative to decipher the structure of each membrane protein to have a comprehensive idea about their functionalities, a task which has remained extremely challenging to achieve. The apparently simpler task of just expression and purification of these proteins, in practice, is quite demanding due to the complexities associated with their folding and their low solubility. Moreover, it is tricky to prepare the samples containing membrane proteins for NMR and XRD-based structural studies while maintaining them close to their physiological environment.

Aggregation intermediates of disease-related amyloids (e.g., amyloid β, α-synuclein, amylin, etc.) are often membrane-active, and their interaction with membranes may lead to toxic effects [68]. Added to the inherent problems associated with studying membrane proteins (as discussed in the previous paragraph), structural studies of these aggregation intermediates have quite a few more challenges. These oligomers are transient and only can be stabilized for sufficient time at very low concentrations (~μM), often making techniques like CD and solution NMR redundant. Also, the membrane-active oligomers, being aggregation intermediates, always coexists with other membrane non-interacting species (including monomers) in solution. Therefore, most experimental approaches would suffer from convoluted signals arising from membrane-bound populations and membrane non-interacting free species in solution. A good experimental technique, in this regard, ideally should be able to selectively probe the membrane-bound population at very low (~μM) concentrations. SERS-based studies of these membrane proteins with lipid bilayer-coated metal NPs (Figure 5.12a, ii) fulfill these requirements [38,117,118]. These coatings can easily be made with biological lipids that can capture and hold the membrane-interacting proteins close to the plasmonic NP surface. Since the surface enhancement effects of the NP-core would drop sharply beyond the thin coating of the lipid bilayer, there would be no SERS possible from the membrane unbound proteins present in the solution. Additionally, the lipid coating makes the NPs stable at physiological pH and salt concentrations, providing an opportunity to probe membrane-attached protein conformation under near-physiological conditions. For example, D. Bhowmik, S. Maiti, and co-workers have probed the structure of membrane-interacting small oligomers of amyloid β by this method [38]. Their studies revealed signatures of β-sheet and a peak at 1,689 cm^{-1} (Figure 5.12e, top panel) from the presence of β-turn(s) in the membrane-bound structure. By monitoring SERS-spectral shifts (Figure 5.12e, the middle and the bottom panels) with oligomers having heavy atoms (^{13}C and ^{15}N labeling) at selected positions (Figure 5.12d, residues marked with gray circles and black circles), they could also narrow down the location of the β-turn at near the middle (near the 26th amino acid from the N-terminal) of the protein sequence with flanking β-sheets on either side of the β-turn (Figure 5.12f). These observed structural features of membrane-bound amyloid-β oligomers have similarities with pore-forming transmembrane proteins called "porins". Such membrane pores (Figure 5.12g, schematic), if formed, may disturb the ionic homeostasis across the neuronal membrane to cause neuronal death. We should keep in mind that in the absence of a full understanding of the position of the proteins with respect to the plasmonic surface, SERS-based findings need to be backed by parallel experiments. In this study, the researchers proved the membrane insertion of the oligomers independently with other techniques like small-angle X-ray scattering (Figure 5.12c), confocal imaging (Figure 5.12b), and fluorescence correlation spectroscopy. Complementary findings from solid-state NMR experiments also backed their SERS-based findings.

5.4.10 Detection of Circulating Tumor Cells with SERS

Detection of circulating tumor cells (CTC) in the blood can be consequential for understanding the progression of the disease. CTC can be present at very low concentrations (~one CTC/mL for stage IV cancer) in cancer patients' blood. CTCs, however, are unique in many ways. For example, their deformability, density, size, expression label of certain membrane proteins, etc. can be very different from a healthy cell. Several methods utilize one or a few of these properties to capture and isolate CTCs from blood samples [119]. For example, since circulating cells in healthy blood are not epithelial, antibodies that bind to epithelial cell adhesion molecules can isolate epithelial CTCs.

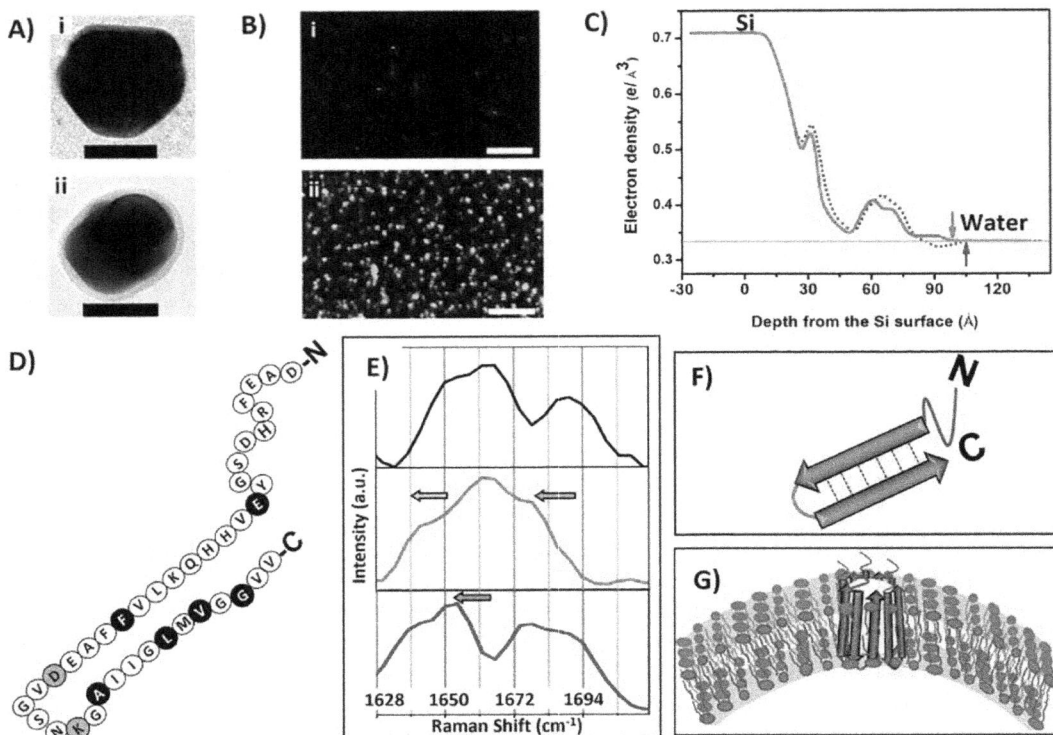

FIGURE 5.12 A method for obtaining secondary structural information about membrane proteins with SERS. (a) TEM image of silver nanoparticles: (i) before and (ii) after coating with lipid bilayer. Lipid coating is visible as an extra layer around the particles. Scale bar is 50 nm. (b) Binding of amyloid β oligomers was probed by the confocal imaging of bilayer-coated AgNPs (i) before and (ii) after incubation with 2 µM of the fluorescein labeled oligomers. Scale bar: 5 µm. (c) X-ray reflectivity measurements proved the insertion of the amyloid β oligomers deep into the lipid bilayer (supported on Si-surface), as the electron density (ED) profiles of the lipid bilayer were changed before (red-colored solid line) and after (blue-colored dotted line) oligomer incubation. (d) Amino acid sequence of amyloid β. A total of three sequences were used for SERS measurements: unlabeled sequence, sequence with two ^{13}C–^{15}N labeled amino acids (D23 and K28) near the suspected β-turn region (gray), and sequence with six ^{13}C–^{15}N labeled amino acid (E11, F19, A30, L34, V36 and G38) residues in the suspected β-sheet region (black). (e) Amide I region of SERS spectra of lipid membrane-bound amyloid β oligomers having no ^{13}C–^{15}N labeled amino acids (top panel), having ^{13}C–^{15}N labeled amino acids near the β-turn (middle panel), and having ^{13}C–^{15}N labeled amino acids in the β-sheet region (bottom panel). Isotopic shifts in the spectral features are emphasized by arrows. In the middle panel, the arrows are pointing toward the direction of the shifts in the spectral features originating from ^{13}C to ^{15}N labeling near the β-turn. In the bottom panel, the arrow is accentuating the isotopic shift in the β-sheet spectral feature. (f) Cartoon showing a probable structure of one monomeric unit of membrane-bound amyloid β oligomer. (g) Cartoon showing a plausible structural reasoning behind the amyloid β toxicity. The aforementioned structural findings support the hypothesis that the amyloid β oligomers can form transmembrane pores. (Figure 5.12 is adapted with permission from Bhowmik, D. et al., *ACS Nano* 2015, 9, 9070–9077. Copyright 2015 American Chemical Society.)

However, since the detection and/or isolation of CTCs must be done in the presence of large excess of healthy cells, extreme precautions should be taken to avoid false-positive results.

SERS-based detection of CTCs with plasmonic NPs provides added advantages of having a high-spectral resolution that can be analyzed to avoid false-positive results [120,121]. However, direct detection of one CTC from blood samples is still challenging using SERS (requiring several milliseconds of acquisition time) in the presence of nine orders of magnitude higher number of healthy circulating cells [120,121]. Thus, practically, SERS can be applied to ensure the presence of CTCs in a sample that has already been subjected to other methods [104] those aspire enrichment of CTCs.

5.4.11 SERS OF EXTRACELLULAR VESICLES

Extracellular vesicles (EVs) play a major role in intercellular communication [122,123]. EVs carry lipids, different types of nucleic acids, proteins, etc. in them [124]. The biogenesis, composition, size, and shape of EVs (namely microvesicles, exosomes, autophagic EVs, etc.) are quite diverse [123]. Their characterization can provide important physiological and pathophysiological clues about the processes involved. With its ability to report the complete or partial composition of complex aqueous mixtures, SERS can be used to study EVs. A. Merdalimova, A. Yashchenok, and co-workers have extensively reviewed different SERS-based strategies for the identification of cancer-related exosomes [125].

S. Stremersch, K. Braeckmans, and co-workers have shown the use of cationic AuNPs that can electrostatically bind to exosome-like vesicles (several hundred NPs per vesicle) and help identify them by SERS [126]. In a separate study, by drying exosomes (derived from cancer cells and normal cells) on a surface containing AuNPs, J. Park, Y. Choi, and co-workers were able to obtain SERS signals from exosomes [127]. With principal component analysis (PCA) of the so obtained SERS data, they were able to identify the cancer cell-derived exosomes with high specificity [127]. Characterization of EVs from cells under autophagy was achieved by D. Chalapathi, C. Narayana, and co-workers with AgNP-SERS [128]. They too employed PCA to obtain compositional variation in EVs [128]. These label-free approaches are simple and efficient; however, they often require an extra step to separate the exosomes. Antibody labeled SERS-platforms and SERS-nanotags can be used to probe exosomes directly from blood or serum samples. For example, by targeting CD63 and HER-2 proteins on tumor-derived exosome surface, with silica-coated anti-CD63-labeled Fe_3O_4 NPs (anti-CD63-labeled magnetic-NPs) and anti-HER-2-labeled SERS-nanotags, S. Zong, Y. Cui, and co-workers were able to form a sandwich structure (similar to the structure shown in the Figure 5.10) that could be pulled together for SERS-based detection without a need of any other purification steps [129].

5.4.12 DETECTION OF TOXINS IN FOOD BY SERS

With a proper mechanism for capturing and holding toxin molecules at proximity to the plasmonic surface, SERS can be engaged in detecting and identifying toxin molecules [130–132]. In some cases, the toxin itself might have a strong affinity toward the bare SERS substrate (e.g., the affinity of the Creutzfeldt-Jakob prions toward Au and the dipicolinic acid toward Ag surfaces, respectively) [93,108], allowing easy implementation of SERS. In all other cases, affinity agents of some kinds are used to capture specific toxins near the nanoconstruct surfaces. Affinity agents should be chosen, keeping in mind that the SERS spectrum of the agent itself should not have many similarities with that of the targeted toxin, and the size of the agent should accommodate the toxin molecule near to the plasmonic surface for maximum enhancements. For example, Porter and co-workers have demonstrated one excellent way of SERS-based detection of toxins (in their case, antigens) [133]. They functionalized a 200 nm thin layer of Au-coated platform with affinity molecules (using antibodies) (Figure 5.13a) and captured toxins on the platform (Figure 5.13b). They then chased it with antibody-coated SERS-nanotags to produce a sandwiched arrangement suitable for SERS-based detection (Figure 5.13c). In practice, functionally active fragments of antibodies, instead of the full-length antibodies should be used as affinity agents, whenever possible. Smaller size of the fragments would allow the analytes (toxins) to reside closer to the plasmonic surfaces for stronger SERS enhancements.

Molecularly imprinted polymers (MIPs) are another class of affinity agents that have been used to separate out targeted analytes (e.g., toxin) for SERS-based detection. Polymerizations that produce

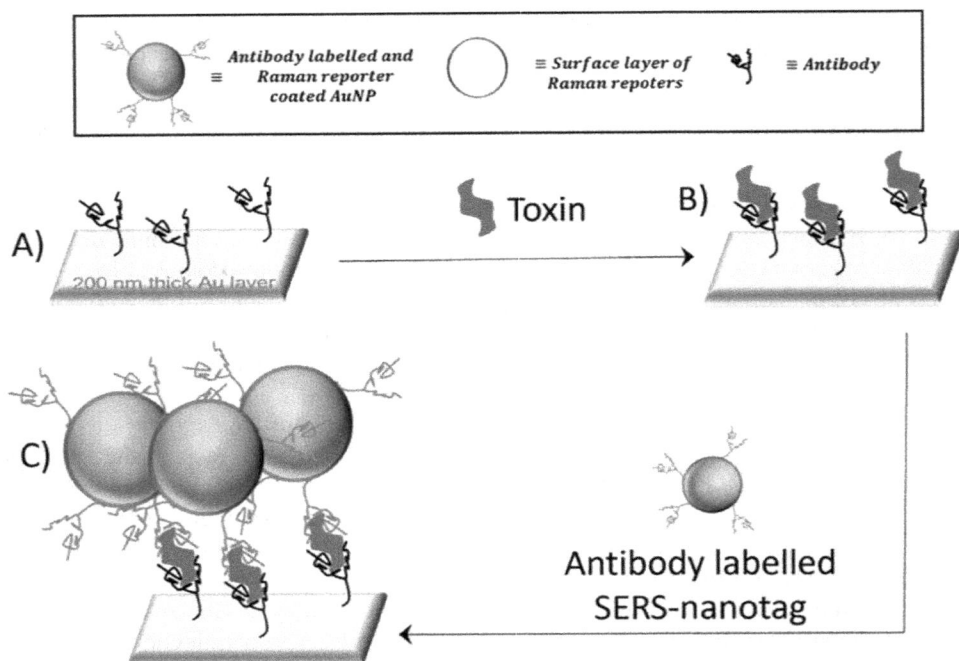

FIGURE 5.13 A scheme for detection of toxin (e.g., antigens) by SERS. (a) A platform containing a thin layer of gold functionalized with affinity agents (shown as black lines, e.g., antibodies). (b) Toxins are immobilized by the affinity agents on the platform. (c) The toxin molecules are sandwiched between thin layer of gold and SERS-nanotags (consisted of an AuNP core [sphere], followed by a layer of Raman reporter molecules, and a final layer of affinity agents [gray lines]).

MIPs are carried out in the presence of the targeted toxins. Some of the monomeric units used in the polymerization process are functionalized to interact with the toxin molecules. The so-formed polymer, therefore, carries the toxin molecules at several sites. Upon removal of these toxin molecules in the next step, the MIPs with embedded imprints of the toxin molecules are produced. MIPs, thus, have a strong affinity and specificity toward the toxin molecules and can be used to separate out and concentrate the toxins from the solution. Therefore, SERS of toxin-containing MIPs can report the presence of the toxins in food items. SERS of MIPs has been successfully demonstrated to detect the presence of chloramphenicol—a broad-spectrum antibiotic—in milk and honey [134], as well as the presence of Sudan I (a carcinogenic compound) in paprika [135].

5.4.13 PROBING NUTRIENTS IN FOOD BY SERS

We need a hassle-free way to choose food with the best nutritional balance in this ever-changing world with a fast-paced lifestyle. With its ability to probe different constituents of a mixture from their respective vibrational signatures, SERS has excellent potential in examining different nutrients in food. For example, A. I. Radu, J. Popp, and co-workers have estimated β-carotene, and lycopene contents in tomatoes plucked at various stages of ripening by SERS [136]. These carotenoids, with their radical scavenging activities, may play a significant role in balancing human health. The researchers used a lithographically produced SERS substrate to first generate a SERS database with known mixtures of the two carotenoids and established that with PCA, the correct ratios of the β-carotene and lycopene could be predicted. Then they used this database to interpret the SERS spectra of tomatoes to estimate the amounts of the two carotenoids and proved the accuracy of their technique by cross-checking these contents with high-performance liquid chromatography

(HPLC). A sophisticated technique like HPLC requires an adequate laboratory set-up, but SERS can be in principle, taken out from a laboratory for easier implementation in day-to-day life. A. I. Radu, J. Popp, and co-workers, in a separate study, also have demonstrated the detection of vitamins (riboflavin and cyanocobalamin) in cereals by SERS [137].

5.4.14 SERS-Based Investigation to Determine the Quality of Water and Drinks

The use of pesticides, antibiotics on a massive scale, and ever-increasing pollution levels have threatened to contaminate major sources of drinking water on our planet. The relative presence of these toxins and pollutants in water varies largely based on the source and treatment procedures of the water. It is, therefore, imperative to monitor water quality to ensure the effectiveness of a water-purification protocol/routine. The probing of certain vibrational fingerprints of the pollutants by SERS can be implemented to report the presence of pollutants and toxins (up to very low concentrations) in water. For example, S. Patze, J. Popp, and co-workers have developed a protocol to detect the amount of a common antibiotic, Sulfamethoxazole, in water using a SERS substrate containing microfluidic cartridge [138]. Their protocol is sensitive enough to detect the presence of the antibiotic molecules in water even at a level of concentrations well below the permissible level of the compound.

In the three studies of J. Popp and co-workers described so far, the detection of the target molecules was performed without any additional recognition sites other than the bare plasmonic surfaces. Many similar applications, however, may require higher specificity toward the target. One such example is discussed in Section 5.4.12, where MIPs were used to capture certain toxins from food. We have also discussed methods of functionalizing the plasmonic surface with recognition sites to immobilize the targeted molecules in the hot-spots. The SERS signal of the captured molecule itself is used for detection in those cases.

Sometimes, the structure and/or orientation of the capturing molecules—present on the plasmonic surface—can change upon their interaction with the targeting molecules. In some cases, these changes in the surface ligands can be monitored by SERS to indirectly detect the binding of the targeting pollutants and/or toxins. This method is especially useful where no detectable SERS signals can be obtained from the analytes. For example, N-[4-Methylthiobenzyl]-N, N-bis(2-pyridylmethyl)amine functionalized AuNPs have a strong and selective affinity toward Cu (II) ions. The ligand and the Cu (II) interaction happen via coordination from three nitrogen lone pairs. Two of these lone pairs come from two pyridine moieties. Upon Cu (II) binding, thus, electronic redistribution in the pyridine rings results in a decrease in the intensity of a SERS band at 1,016 cm^{-1} and the simultaneous rise of the band at 1,030 cm^{-1}. This concept was used in a study to determine Cu (II) contamination in white wine [139].

The presence of some of the organic pollutants (e.g., polycyclic aromatic hydrocarbons, pesticides, etc.) in air, water, and food, even at very low concentrations, can be extremely toxic for humans and many other life forms. SERS-based detection of such nonpolar pollutants is not straightforward due to their low affinities toward standard plasmonic surfaces. Some excellent examples of capturing and sensing pollutants like 1-naphthol and gas-phase detection of polycyclic aromatic hydrocarbons have been demonstrated by R. A. Á-Puebla, L. M. L-Marzán, and co-workers and by M. Mueller, A. Fery, and co-workers by using a core-shell nanoparticle [140,141]. These core-shell NPs consist of a metal NPs core (e.g., Au nanostars or Au nanosphere, etc.) and a poly-(N-isopropyl acrylamide) microgel shell. The poly-(N-isopropylacrylamide) shell used here can capture the nonpolar pollutants. This layer is also thermosensitive and shrinks at a temperature above 32°C. This shrinking effect brings the captured-pollutant molecules closer to the plasmonic core for SERS-based detection.

5.5 CONCLUSIONS AND OUTLOOK

This chapter began with a discussion of the mechanism behind surface enhancement effects. We understood the required optical properties in a material that allows the generation of strong surface plasmon and thereby electrical field enhancements at and near the surface of plasmonic nanostructures. We then comprehended how using these surface plasmons, we can obtain SERS.

SERS is a powerful technique that can provide details about molecular structure, identity, and their arrangement in space within a relatively fast time scale. The invention of SERS has greatly extended the application of weak but extremely informative Raman spectroscopy to a previously unimaginable level. However, during the first 20–25 years after the discovery of the SERS phenomena, most of the works in this field were performed with roughened metal surfaces and uncontrolled NP aggregates. Electric field enhancements in these poorly-designed substrates are somewhat haphazard, and that caused a bad reputation that the SERS experiments are inherently not reproducible. Since then, there has been reinvigoration in the field with a better theoretical understanding of the phenomena and fabrication of more ordered nanostructures. We now understand the importance of surface selection rules and the "right" methods for experimentally obtaining the enhancement factors. In this regard, we have discussed some of the most useful substrates (e.g., self-assembled or tethered monolayer of uniform metal NPs, core-shell NPs, coated NPs, uniform plasmonic substrates fabricated by lithography, etc.) those can generate reproducible SERS results. The use of SERS has now broadened to address several important chemicals, analytical, and biological queries in medical, environmental, and material sciences. In fact, accomplishments in the precise designing and fabrication of "hot-spots" in the nanogaps and on the sharp NP surface features are one of the main reasons behind such widening of the range of practical applications of SERS. These ingenious substrate developments and parallel advancements in instrumentation hold enormous potential for the field of SERS. In future, the fabrication and/or synthesis of the SERS substrates are expected to be even cheaper, faster, and reproducible for larger-scale of productions. Also, simultaneous identification of several types of molecules in complex analytes and the quantification of their relative amounts by SERS is now possible with the application of modern statistical data analysis tools.

The SERS substrates and devices—to become more cost-effective—need to be chemically robust and easily recyclable. Engineering of the SERS platform with even better optimization of optical properties and desirable chemical characteristics (e.g., ability to hold the analyte molecules within the hot-spots) is a need for the future. Also, many experimental results (especially those involving hot-spots) still cannot be rationalized entirely with the current understanding of the mechanism behind SERS. There is, therefore, a lot of scope and need for further theoretical developments in understanding the phenomena behind SERS. Better computational models that can incorporate fine details of the nanoconstruct structure—to completely rationalize the characteristics of the near-field—are also required to realize the experimental findings. We also need even brighter SERS-nanotags and better portable devices to apply SERS with its full potential in the field of medical diagnostics, environmental probing, and in chemical and biological warfare. Being an optical technique, SERS suffers from the opacity of biological tissues and is limited to probing only up to a certain depth if applied from the exterior. Therefore, the current efforts are trying to combine fiber-optics-based methods (e.g., upper endoscopy and colonoscopy, etc.) with SERS to identify and detect biomarkers in vivo.

We can expect SERS to be applied as a commercial analytical and imaging platform with this future insight. SERS is already in use to complement results from other tools (e.g., fluorescence spectroscopy and traditional pathology tests, etc.). In the future, it will be a portable and robust device-based technique with the ability for even faster and more reliable data collection and in situ data interpretation, such that it can be used in the remotest of places and under the harshest conditions.

REFERENCES

1. Ewen Smith, Geoffrey D., Introduction, basic theory and principles. In *Modern Raman Spectroscopy – A Practical Approach*. John Wiley & Sons, Ltd, West Sussex, England 2004; pp. 1–21.
2. Ferraro, J. R.; Nakamoto, K.; Brown, C. W., Chapter 1: Basic theory. In *Introductory Raman Spectroscopy* (2nd ed.), San Diego, USA, 2003; pp. 1–94.

3. Raman, C. V.; Krishnan, K. S., A new type of secondary radiation. *Nature* 1928, *121* (3048), 501–502.

4. Larkin, P. J., Chapter 6: IR and Raman spectra–structure correlations: Characteristic group frequencies. In *Infrared and Raman Spectroscopy* (2nd ed.), Larkin, P. J., Ed. Amsterdam: Elsevier, 2018; pp. 85–134.

5. Larkin, P. J., Chapter 7: General outline for IR and Raman spectral interpretation. In *Infrared and Raman Spectroscopy* (2nd ed.), Larkin, P. J., Ed. Amsterdam: Elsevier, 2018; pp. 135–151.

6. Chandra, B.; Bhowmik, D.; Maity, B. K.; Mote, K. R.; Dhara, D.; Venkatramani, R.; Maiti, S.; Madhu, P. K., Major reaction coordinates linking transient amyloid-β oligomers to fibrils measured at atomic level. *Biophys. J* 2017, *113* (4), 805–816.

7. Fleischmann, M.; Hendra, P. J.; McQuillan, A. J., Raman spectra of pyridine adsorbed at a silver electrode. *Chem. Phys. Lett.* 1974, *26* (2), 163–166.

8. Jeanmaire, D. L.; Van Duyne, R. P., Surface Raman spectro electrochemistry: Part I. Heterocyclic, aromatic, and aliphatic amines adsorbed on the anodized silver electrode. *J. Electroanal. Chem. Interf. Electrochem.* 1977, *84* (1), 1–20.

9. Albrecht, M. G.; Creighton, J. A., Anomalously intense Raman spectra of pyridine at a silver electrode. *J. Am. Chem. Soc.* 1977, *99* (15), 5215–5217.

10. Philpott, M. R., Effect of surface plasmons on transitions in molecules. *J. Chem. Phys.* 1975, *62* (5), 1812–1817.

11. Moskovits, M., Surface roughness and the enhanced intensity of Raman scattering by molecules adsorbed on metals. *J. Chem. Phys.* 1978, *69* (9), 4159–4161.

12. Campion, A.; Ivanecky, J. E.; Child, C. M.; Foster, M., On the mechanism of chemical enhancement in surface-enhanced Raman scattering. *J. Am. Chem. Soc.* 1995, *117* (47), 11807–11808.

13. Langer, J.; Jimenez de Aberasturi, D.; Aizpurua, J.; Alvarez-Puebla, R. A.; Auguié, B.; Baumberg, J. J.; Bazan, G. C.; Bell, S. E. J.; Boisen, A.; Brolo, A. G.; Choo, J.; Cialla-May, D.; Deckert, V.; Fabris, L.; Faulds, K.; García de Abajo, F. J.; Goodacre, R.; Graham, D.; Haes, A. J.; Haynes, C. L.; Huck, C.; Itoh, T.; Käll, M.; Kneipp, J.; Kotov, N. A.; Kuang, H.; Le Ru, E. C.; Lee, H. K.; Li, J.-F.; Ling, X. Y.; Maier, S. A.; Mayerhöfer, T.; Moskovits, M.; Murakoshi, K.; Nam, J.-M.; Nie, S.; Ozaki, Y.; Pastoriza-Santos, I.; Perez-Juste, J.; Popp, J.; Pucci, A.; Reich, S.; Ren, B.; Schatz, G. C.; Shegai, T.; Schlücker, S.; Tay, L.-L.; Thomas, K. G.; Tian, Z.-Q.; Van Duyne, R. P.; Vo-Dinh, T.; Wang, Y.; Willets, K. A.; Xu, C.; Xu, H.; Xu, Y.; Yamamoto, Y. S.; Zhao, B.; Liz-Marzán, L. M., Present and future of surface-enhanced Raman scattering. *ACS Nano* 2020, *14* (1), 28–117.

14. Sharma, B.; Frontiera, R. R.; Henry, A.-I.; Ringe, E.; Van Duyne, R. P., SERS: Materials, applications, and the future. *Mater. Today* 2012, *15* (1), 16–25.

15. Le Ru, E. C.; Etchegoin, P. G., Chapter 4: SERS enhancement factors and related topics. In *Principles of Surface-Enhanced Raman Spectroscopy*, Le Ru, E. C.; Etchegoin, P. G., Eds. Amsterdam: Elsevier, 2009; pp. 185–264.

16. Le Ru, E. C.; Etchegoin, P. G., Chapter 5: Calculations of electromagnetic enhancements. In *Principles of Surface-Enhanced Raman Spectroscopy*, Le Ru, E. C.; Etchegoin, P. G., Eds. Amsterdam: Elsevier, 2009; pp. 265–297.

17. Bhowmik, D.; MacLaughlin, C. M.; Chandrakesan, M.; Ramesh, P.; Venkatramani, R.; Walker, G. C.; Maiti, S., pH changes the aggregation propensity of amyloid-β without altering the monomer conformation. *Phys. Chem. Chem. Phys.* 2014, *16* (3), 885–889.

18. Kumari, G.; Kandula, J.; Narayana, C., How far can we probe by SERS? *J. Phys. Chem. C* 2015, *119*, 20057.

19. Fathima, H.; Paul, L.; Thirunavukkuarasu, S.; Thomas, K. G., Mesoporous silica-capped silver nanoparticles for sieving and surface-enhanced Raman scattering-based sensing. *ACS Appl. Nano Mater.* 2020, *3* (7), 6376–6384.

20. D'Elia, V.; Rubio-Retama, J.; Ortega-Ojeda, F. E.; García-Ruiz, C.; Montalvo, G., Gold nanorods as SERS-substrate for the ultratrace detection of cocaine in non-pretreated oral fluid samples. *Coll. Surf. A: Physicochem. Eng. Aspects* 2018, *557*, 43–50.

21. Rekha, C. R.; Nayar, V. U.; Gopchandran, K. G., Synthesis of highly stable silver nanorods and their application as SERS substrates. *J. Sci.: Adv. Mat. Dev.* 2018, *3* (2), 196–205.

22. Feng, J.; Chen, L.; Xia, Y.; Xing, J.; Li, Z.; Qian, Q.; Wang, Y.; Wu, A.; Zeng, L.; Zhou, Y., Bioconjugation of gold nanobipyramids for SERS detection and targeted photothermal therapy in breast cancer. *ACS Biomater. Sci. Eng.* 2017, *3* (4), 608–618.

23. Lin, B.; Wang, Y.; Yao, Y.; Chen, L.; Zeng, Y.; Li, L.; Lin, Z.; Guo, L., Oil-free gold nanobipyramid@Ag microgels as a functional SERS-substrate for direct detection of small molecules in a complex sample matrix. *Anal. Chem.* 2021, *93* (49), 16727–16733.

24. Lu, G.; Forbes, T. Z.; Haes, A. J., SERS detection of uranyl using functionalized gold nanostars promoted by nanoparticle shape and size. *Analyst* 2016, *141*, 5137.

25. Indrasekara, A. S. D. S.; Meyers, S.; Shubeita, S.; Feldman, L. C.; Gustafsson, T.; Fabris, L., Gold nanostar substrates for SERS-based chemical sensing in the femtomolar regime. *Nanoscale* 2014, *6* (15), 8891–8899.

26. Bhowmik, D.; Culver, K. S. B.; Liu, T.; Odom, T. W., Resolving single-nanoconstruct dynamics during targeting and nontargeting live-cell membrane interactions. *ACS Nano* 2019, *13* (12), 13637–13644.

27. Konrad, M. P.; Doherty, A. P.; Bell, S. E. J., Stable and uniform SERS signals from self-assembled two-dimensional interfacial arrays of optically coupled Ag nanoparticles. *Anal. Chem.* 2013, *85*, 6783.

28. Xu, Y.; Konrad, M. P.; Lee, W. W. Y.; Ye, Z.; Bell, S. E. J., A method for promoting assembly of metallic and nonmetallic nanoparticles into interfacial monolayer films. *Nano Lett.* 2016, *16*, 5255.

29. Duan, H.; Wang, D.; Kurth, D. G.; Möhwald, H., Directing self-assembly of nanoparticles at water/oil interfaces. *Angew. Chem., Int. Ed.* 2004, *43*, 5639.

30. Li, X.; Lin, X.; Zhao, X.; Wang, H.; Liu, Y.; Lin, S.; Wang, L.; Cong, S., Self-assembled monolayer film of concave star-shaped Au nanocrystals as highly efficient SERS-substrates. *Appl. Surf. Sci.* 2020, *518*, 146217.

31. Andryszewski, T.; Iwan, M.; Hołdyński, M.; Fiałkowski, M., Synthesis of a free-standing monolayer of covalently bonded gold nanoparticles. *Chem. Mater.* 2016, *28*, 5304.

32. Xie, W.; Walkenfort, B.; Schlücker, S., Label-free SERS monitoring of chemical reactions catalyzed by small gold nanoparticles using 3D plasmonic superstructures. *J. Am. Chem. Soc.* 2013, *135* (5), 1657–1660.

33. Ben-Jaber, S.; Peveler, W. J.; Quesada-Cabrera, R.; Cortés, E.; Sotelo-Vazquez, C.; Abdul-Karim, N.; Maier, S. A.; Parkin, I. P., Photo-induced enhanced Raman spectroscopy for universal ultra-trace detection of explosives, pollutants and biomolecules. *Nat. Commun.* 2016, *7*, 12189.

34. Kelly, J.; Patrick, R.; Patrick, S.; Bell, S. E. J., Surface-enhanced Raman spectroscopy for the detection of a metabolic product in the headspace above live bacterial cultures. *Angew. Chem., Int. Ed.* 2018, *57*, 15686.

35. Xu, Y.; Konrad, M. P.; Trotter, J. L.; McCoy, C. P.; Bell, S. E. J., Rapid one-pot preparation of large freestanding nanoparticle-polymer films. *Small* 2017, *13*, 1602163.

36. Xu, Y.; Ye, Z.; Li, C.; McCabe, H.; Kelly, J.; Bell, S. E. J., Pressing solids directly into sheets of plasmonic nanojunctions enables solvent-free surface-enhanced Raman spectroscopy. *Appl. Mater. Today* 2018, *13*, 352.

37. Ip, S.; MacLaughlin, C. M.; Gunari, N.; Walker, G. C., Phospholipid membrane encapsulation of nanoparticles for surface-enhanced Raman scattering. *Langmuir* 2011, *27* (11), 7024–7033.

38. Bhowmik, D.; Mote, K. R.; MacLaughlin, C. M.; Biswas, N.; Chandra, B.; Basu, J. K.; Walker, G. C.; Madhu, K.; Maiti, S., Cell-membrane-mimicking lipid-coated nanoparticles confer Raman enhancement to membrane proteins and reveal membrane-attached amyloid-β conformation. *ACS Nano* 2015, *9*, 9070.

39. Shanthil, M.; Thomas, R.; Swathi, R. S.; George Thomas, K., Ag@SiO2 core–shell nanostructures: Distance-dependent plasmon coupling and SERS investigation. *J. Phys. Chem. Lett.* 2012, *3*, 1459.

40. Liz-Marzán, L. M.; Giersig, M.; Mulvaney, P., Synthesis of nanosized gold-silica core–shell particles. *Langmuir* 1996, *12*, 4329.

41. Küstner, B.; Gellner, M.; Schütz, M.; Schöppler, F.; Marx, A.; Ströbel, P.; Adam, P.; Schmuck, C.; Schlücker, S., SERS labels for red laser excitation: Silica-encapsulated SAMs on tunable gold/silver nanoshells. *Angew. Chem., Int. Ed.* 2009, *48*, 1950.

42. Pierre, M. C. S.; Haes, A. J., Purification implications on SERS activity of silica coated gold nanospheres. *Anal. Chem.* 2012, *84*, 7906.

43. Schmit, V. L.; Martoglio, R.; Scott, B.; Strickland, A. D.; Carron, K. T., Lab-on-a-bubble: Synthesis, characterization, and evaluation of buoyant gold nanoparticle-coated silica spheres. *J. Am. Chem. Soc.* 2012, *134*, 59.

44. Schütz, M.; Schlücker, S., Towards quantitative multi-color nanodiagnostics: Spectral multiplexing with six silica-encapsulated SERS labels. *J. Raman Spectrosc.* 2016, *47*, 1012.

45. Zheng, X. S.; Hu, P.; Cui, Y.; Zong, C.; Feng, J. M.; Wang, X.; Ren, B., BSA-coated nanoparticles for improved SERS-based intracellular pH sensing. *Anal. Chem.* 2014, *86*, 12250.

46. Schütz, M.; Salehi, M.; Schlücker, S., Direct silica encapsulation of self-assembled-monolayer-based surface-enhanced Raman scattering labels with complete surface coverage of Raman reporters by non-covalently bound silane precursors. *Chem. Asian J.* 2014, *9*, 2219.

47. Nordlander, P.; Oubre, C.; Prodan, E.; Li, K.; Stockman, M., Plasmon hybridization in nanoparticle dimers. *Nano Lett.* 2004, *4*, 899.

48. Hentschel, M.; Saliba, M.; Vogelgesang, R.; Giessen, H.; Alivisatos, A. P.; Liu, N., Transition from isolated to collective modes in plasmonic oligomers. *Nano Lett.* 2010, *10*, 2721.

49. Yanai, A.; Grajower, M.; Lerman, G. M.; Hentschel, M.; Giessen, H.; Levy, U., Near- and far-field properties of plasmonic oligomers under radially and azimuthally polarized light excitation. *ACS Nano* 2014, *8*, 4969.

50. Zhu, W.; Crozier, K., Quantum mechanical limit to plasmonic enhancement as observed by surface-enhanced Raman scattering. *Nat. Commun.* 2014, *5*, 5228.

51. Luk'yanchuk, B.; Zheludev, N. I.; Maier, S. A.; Halas, N. J.; Nordlander, P.; Giessen, H.; Chong, C. T., The fano resonance in plasmonic nanostructures and metamaterials. *Nat. Mater.* 2010, *9*, 707.

52. Lombardi, A.; Grzelczak, M. P.; Crut, A.; Maioli, P.; Pastoriza-Santos, I.; Liz-Marzán, L. M.; Del Fatti, N.; Vallée, F., Optical response of individual Au−Ag@SiO$_2$ heterodimers. *ACS Nano* 2013, *7*, 2522.

53. Lim, D. K.; Jeon, K. S.; Hwang, J. H.; Kim, H.; Kwon, S.; Suh, Y. D.; Nam, J. M., Highly uniform and reproducible surface-enhanced Raman scattering from DNA tailorable nanoparticles with 1-nm interior gap. *Nat. Nanotechnol.* 2011, *6*, 452.

54. Wustholz, K. L.; Henry, A. I.; McMahon, J. M.; Freeman, R. G.; Valley, N.; Piotti, M. E.; Natan, M. J.; Schatz, G. C.; Van Duyne, R. P., Structure-activity relationships in gold nanoparticle dimers and trimers for surface-enhanced Raman spectroscopy. *J. Am. Chem. Soc.* 2010, *132*, 10903.

55. Talaga, D.; Comesaña-Hermo, M.; Ravaine, S.; Vallee, R. A. L.; Bonhommeau, S., Colocalized dark-field scattering, atomic force and surface-enhanced Raman scattering microscopic imaging of single gold nanoparticles. *J. Opt.* 2015, *17*, 114006.

56. Ngo, H. T.; Gandra, N.; Fales, A. M.; Taylor, S. M.; Vo-Dinh, T., Sensitive DNA detection and SNP discrimination using ultrabright SERS nanorattles and magnetic beads for malaria diagnostics. *Biosens. Bioelectron.* 2016, *81*, 8.

57. Ngo, H. T.; Freedman, E.; Odion, R. A.; Strobbia, P.; Indrasekara, A. S. D. S.; Vohra, P.; Taylor, S. M.; Vo-Dinh, T., Direct detection of unamplified pathogen RNA in blood lysate using an integrated lab-in-a-stick device and ultrabright SERS nanorattles. *Sci. Rep.* 2018, *8*, 4075.

58. Chen, J.; Mårtensson, T.; Dick, K. A.; Deppert, K.; Xu, H. Q.; Samuelson, L.; Xu, H., Surface-enhanced Raman scattering of rhodamine 6G on nanowire arrays decorated with gold nanoparticles. *Nanotechnology* 2008, *19*, 275712.

59. Schmidt, M. S.; Hübner, J.; Boisen, A., Large area fabrication of leaning silicon nanopillars for surface enhanced Raman spectroscopy. *Adv. Mater.* 2012, *24*, OP11.

60. Li, T.; Wu, K.; Rindzevicius, T.; Wang, Z.; Schulte, L.; Schmidt, M. S.; Boisen, A.; Ndoni, S., Wafer-scale nanopillars derived from block copolymer lithography for surface-enhanced Raman spectroscopy. *ACS Appl. Mater. Interf.* 2016, *8*, 15668.

61. Karadan, P.; Aggarwal, S.; Anappara, A. A.; Narayana, C.; Barshilia, H. C., Tailored Periodic Si nanopillar based architectures as highly sensitive universal SERS biosensing platform. *Sens. Actuators, B* 2018, *254*, 264.

62. Jeon, T. Y.; Park, S. G.; Kim, D. H.; Kim, S. H., Standing-wave-assisted creation of nanopillar arrays with vertically integrated nanogaps for SERS-active substrates. *Adv. Funct. Mater.* 2015, *25*, 4681.

63. Ruedas-Rama, M. J.; Walters, J. D.; Orte, A.; Hall, E. A. H., Fluorescent nanoparticles for intracellular sensing: A review. *Anal. Chim. Acta* 2012, *751*, 1.

64. Thorn, K., Genetically encoded fluorescent tags. *Mol. Biol. Cell* 2017, *28* (7), 848–857.

65. Jensen, E. C., Use of fluorescent probes: Their effect on cell biology and limitations. *Anatom. Rec.* 2012, *295* (12), 2031–2036.

66. Gauer, J. W.; LeBlanc, S.; Hao, P.; Qiu, R.; Case, B. C.; Sakato, M.; Hingorani, M. M.; Erie, D. A.; Weninger, K. R., Chapter 10: Single-molecule FRET to measure conformational dynamics of DNA mismatch repair proteins. In *Methods in Enzymology*, Spies, M.; Chemla, Y. R., Eds. Oxford: Academic Press, 2016; Vol. 581, pp. 285–315.

67. Herrick-Davis, K.; Mazurkiewicz, J. E., Chapter 10: Fluorescence correlation spectroscopy and photon-counting histogram analysis of receptor–receptor interactions. In *Methods in Cell Biology*, Conn, P. M., Ed. Oxford: Academic Press, 2013; Vol. 117, pp. 181–196.

68. Bhowmik, D.; Das, A. K.; Maiti, S., Rapid, cell-free assay for membrane-active forms of amyloid-β. *Langmuir* 2015, *31* (14), 4049–4053.

69. Lane, L. A.; Qian, X.; Nie, S., SERS nanoparticles in medicine: From label-free detection to spectroscopic tagging. *Chem. Rev.* 2015, *115*, 10489.

70. Schlücker, S., SERS microscopy: Nanoparticle probes and biomedical applications. *ChemPhysChem* 2009, *10*, 1344.

71. Wang, Y.; Yan, B.; Chen, L., SERS tags: Novel optical nanoprobes for bioanalysis. *Chem. Rev.* 2013, *113*, 1391.

72. Madathiparambil Visalakshan, R.; González García, L. E.; Benzigar, M. R.; Ghazaryan, A.; Simon, J.; Mierczynska-Vasilev, A.; Michl, T. D.; Vinu, A.; Mailänder, V.; Morsbach, S.; Landfester, K.; Vasilev, K., The influence of nanoparticle shape on protein corona formation. *Small* 2020, *16* (25), 2000285.

73. Choo, P.; Liu, T.; Odom, T. W., Nanoparticle shape determines dynamics of targeting nanoconstructs on cell membranes. *J. Am. Chem. Soc.* 2021, *143* (12), 4550–4555.

74. Pallaoro, A.; Braun, G. B.; Moskovits, M., Biotags based on surface-enhanced Raman can be as bright as fluorescence tags. *Nano Lett.* 2015, *15*, 6745.

75. Tran, V.; Thiel, C.; Svejda, J. T.; Jalali, M.; Walkenfort, B.; Erni, D.; Schlücker, S., Probing the SERS brightness of individual Au nanoparticles, hollow Au/Ag nanoshells, Au nanostars and Au core/Au satellite particles: Single-particle experiments and computer simulations. *Nanoscale* 2018, *10*, 21721.

76. Tsoutsi, D.; Montenegro, J. M.; Dommershausen, F.; Koert, U.; Liz-Marzán, L. M.; Parak, W. J.; Alvarez-Puebla, R. A., Quantitative SERS ultradetection of atomic inorganic ions: The case of chlorine. *ACS Nano* 2011, *5*, 7539.

77. Tsoutsi, D.; Guerrini, L.; Hermida-Ramon, J. M.; Giannini, V.; Liz-Marzán, L. M.; Wei, A.; Alvarez-Puebla, R. A., Simultaneous SERS detection of copper and cobalt at ultratrace levels. *Nanoscale* 2013, *5*, 4776.

78. Alvarez-Puebla, R. A.; Liz-Marzán, L. M., SERS detection of small inorganic molecules and ions. *Angew. Chem., Int. Ed.* 2012, *51*, 11214.

79. Cao, Y.; Li, D. W.; Zhao, L. J.; Liu, X. Y.; Cao, X. M.; Long, Y. T., Highly selective detection of carbon monoxide in living cells by palladacycle carbonylation-based surface enhanced Raman spectroscopy nanosensors. *Anal. Chem.* 2015, *87*, 9696.

80. Cao, W. Y. C.; Jin, R.; Mirkin, C. A., Nanoparticles with Raman spectroscopic fingerprints for DNA and RNA detection. *Science* 2002, *297*, 1536.

81. Vo-Dinh, T.; Allain, L. R.; Stokes, D. L., Cancer gene detection using surface-enhanced Raman scattering (SERS). *J. Raman Spectrosc.* 2002, *33*, 511.

82. Feliu, N.; Hassan, M.; Garcia Rico, E.; Cui, D. X.; Parak, W.; Alvarez-Puebla, R., SERS quantification and characterization of proteins and other biomolecules. *Langmuir* 2017, *33*, 9711.

83. Park, J. E.; Jung, Y.; Kim, M.; Nam, J. M., Quantitative nanoplasmonics. *ACS Cent. Sci.* 2018, *4*, 1303.

84. Park, J. E.; Lee, Y.; Nam, J. M., Precisely shaped, uniformly formed gold nanocubes with ultrahigh reproducibility in single-particle scattering and surface-enhanced Raman scattering. *Nano Lett.* 2018, *18*, 6475.

85. Fang, Y. R.; Wei, H.; Hao, F.; Nordlander, P.; Xu, H. X., Remote-excitation surface-enhanced Raman scattering using propagating Ag nanowire plasmons. *Nano Lett.* 2009, *9*, 2049.

86. Hutchison, J. A.; Centeno, S. P.; Odaka, H.; Fukumura, H.; Hofkens, J.; Uji-i, H., Subdiffraction limited, remote excitation of surface enhanced Raman scattering. *Nano Lett.* 2009, *9*, 995.

87. Agreda, A.; Sharma, D. K.; Colas des Francs, G.; Kumar, G. V. P.; Bouhelier, A., Modal and wavelength conversions in plasmonic nanowires. *Opt. Exp.* 2021, *29* (10), 15366–15381.

88. Lu, G.; De Keersmaecker, H.; Su, L.; Kenens, B.; Rocha, S.; Fron, E.; Chen, C.; Van Dorpe, P.; Mizuno, H.; Hofkens, J.; Hutchison, J. A.; Uji-i, H., Live-cell SERS endoscopy using plasmonic nanowire waveguides. *Adv. Mater.* 2014, *26*, 5124.

89. Shanmukh, S.; Jones, L.; Driskell, J.; Zhao, Y.; Dluhy, R.; Tripp, R. A., Rapid and sensitive detection of respiratory virus molecular signatures using a silver nanorod array SERS-substrate. *Nano Lett.* 2006, *6*, 2630.

90. Jarvis, R. M.; Brooker, A.; Goodacre, R., Surface-enhanced Raman spectroscopy for bacterial discrimination utilizing a scanning electron microscope with a Raman spectroscopy interface. *Anal. Chem.* 2004, *76*, 5198.

91. Jarvis, R. M.; Goodacre, R., Rapid discrimination of bacteria using surface enhanced Raman spectroscopy. *Anal. Chem.* 2004, *76*, 40.

92. Dina, N. E.; Raluca Gherman, A. M.; Chiş, V.; Sârbu, C.; Wieser, A.; Bauer, D.; Haisch, C., Characterization of clinically relevant fungi via SERS fingerprinting assisted by novel chemometric models. *Anal. Chem.* 2018, *90*, 2484.

93. Alvarez-Puebla, R. A.; Agarwal, A.; Khanal, B. P.; Aldeanueva-Potel, P.; Carbó-Argibay, E.; Pazos-Pérez, N.; Zubarev, E. R.; Kotov, N. A.; Liz-Marzán, L. M., Real-time detection of scrambled prions on 3D supercrystals of gold nanorods. *Proc. Natl. Acad. Sci. USA* 2011, *108*, 8157.

94. Lomonaco, S., Identification methods | Application of single nucleotide polymorphisms–based typing for DNA fingerprinting of foodborne bacteria. In *Encyclopedia of Food Microbiology* (2nd ed.), Batt, C. A.; Tortorello, M. L., Eds. Oxford: Academic Press, 2014; pp. 289–294.

95. Kehn-Hall, K.; Bavari, S., Chapter 28: Detection of highly pathogenic viral agents: Implications for therapeutics, vaccines and biodefense. In *Molecular Diagnostics* (2nd ed.), Patrinos, G. P.; Ansorge, W. J., Eds. San Diego: Academic Press, 2010; pp. 417–429.

96. Kahraman, M.; Mullen, E. R.; Korkmaz, A.; Wachsmann-Hogiu, S., Fundamentals and applications of SERS-based bioanalytical sensing. *Nanophotonics* 2017, *6* (5), 831–852.

97. Ngo, H. T.; Wang, H. N.; Fales, A. M.; Vo-Dinh, T., Label-free DNA biosensor based on SERS molecular sentinel on nanowave chip. *Anal. Chem.* 2013, *85*, 6378.

98. Ngo, H. T.; Wang, H. N.; Burke, T.; Ginsburg, G. S.; Vo-Dinh, T., Multiplex detection of disease biomarkers using SERS molecular sentinel-on-chip. *Anal. Bioanal. Chem.* 2014, *406*, 3335.

99. Wang, H. N.; Crawford, B. M.; Fales, A. M.; Bowie, M. L.; Seewaldt, V. L.; Vo-Dinh, T., Multiplexed detection of microRNA biomarkers using SERS-based inverse molecular sentinel (iMS) nanoprobes. *J. Phys. Chem. C* 2016, *120*, 21047.

100. Zakel, S.; Rienitz, O.; Güttler, B.; Stosch, R., Double isotope dilution surface-enhanced Raman scattering as a reference procedure for the quantification of biomarkers in human serum. *Analyst* 2011, *136*, 3956.

101. Gong, T.; Hong, Z. Y.; Chen, C. H.; Tsai, C. Y.; Liao, L. D.; Kong, K. V., Optical Interference-free surface-enhanced Raman scattering CO-nanotags for logical multiplex detection of vascular disease-related biomarkers. *ACS Nano* 2017, *11*, 3365.

102. Bhamidipati, M.; Cho, H. Y.; Lee, K. B.; Fabris, L., SERS-based quantification of biomarker expression at the single cell level enabled by gold nanostars and truncated aptamers. *Bioconjugate Chem.* 2018, *29*, 2970.

103. Lauridsen, R. K.; Sommer, L. M.; Johansen, H. K.; Rindzevicius, T.; Molin, S.; Jelsbak, L.; Engelsen, S. B.; Boisen, A., SERS detection of the biomarker hydrogen cyanide from pseudomonas aeruginosa cultures isolated from cystic fibrosis patients. *Sci. Rep.* 2017, *7*, 45264.

104. Garcia-Algar, M.; Fernandez-Carrascal, A.; Olano-Daza, A.; Guerrini, L.; Feliu, N.; Parak, W. J.; Guimera, R.; Garcia-Rico, E.; Alvarez-Puebla, R. A., Adaptive metabolic pattern biomarker for disease monitoring and staging of lung cancer with liquid biopsy. *NPJ Precis. Oncol.* 2018, *2*, 16.

105. Qu, A.; Wu, X.; Xu, L.; Liu, L.; Ma, W.; Kuang, H.; Xu, C., SERS- and luminescence-active Au-Au-UCNP trimers for attomolar detection of two cancer biomarkers. *Nanoscale* 2017, *9*, 3865.

106. Xu, L.; Yan, W.; Ma, W.; Kuang, H.; Wu, X.; Liu, L.; Zhao, Y.; Wang, L.; Xu, C., SERS encoded silver pyramids for attomolar detection of multiplexed disease biomarkers. *Adv. Mater.* 2015, *27*, 1706.

107. Drescher, D.; Guttmann, P.; Buchner, T.; Werner, S.; Laube, G.; Hornemann, A.; Tarek, B.; Schneider, G.; Kneipp, J., Specific biomolecule corona is associated with ring-shaped organization of silver nanoparticles in cells. *Nanoscale* 2013, *5*, 9193.

108. Zhang, X.; Young, M. A.; Lyandres, O.; Van Duyne, R. P., Rapid detection of an anthrax biomarker by surface-enhanced Raman spectroscopy. *J. Am. Chem. Soc.* 2005, *127*, 4484.

109. Chou, I. H.; Benford, M.; Beier, H. T.; Coté, G. L.; Wang, M.; Jing, N.; Kameoka, J.; Good, T. A., Nanofluidic biosensing for β-amyloid detection using surface enhanced Raman spectroscopy. *Nano Lett.* 2008, *8* (6), 1729–1735.

110. Pavan Kumar, G. V.; Ashok Reddy, B. A.; Arif, M.; Kundu, T. K.; Narayana, C., Surface-enhanced Raman scattering studies of human transcriptional coactivator p300. *J. Phys. Chem. B* 2006, *110* (33), 16787–16792.

111. Choi, I.; Huh, Y. S.; Erickson, D., Ultra-sensitive, label-free probing of the conformational characteristics of amyloid beta aggregates with a SERS active nanofluidic device. *Microfluid. Nanofluid.* 2012, *12* (1), 663–669.

112. Siddhanta, S.; Narayana, C., Surface enhanced Raman spectroscopy of proteins: Implications for drug designing. *Nanomater. Nanotechnol.* 2012, *2*, 1.

113. Feliu, N.; Hassan, M.; Garcia Rico, E.; Cui, D.; Parak, W.; Alvarez-Puebla, R., SERS quantification and characterization of proteins and other biomolecules. *Langmuir* 2017, *33* (38), 9711–9730.

114. Bruzas, I.; Lum, W.; Gorunmez, Z.; Sagle, L., Advances in surface-enhanced Raman spectroscopy (SERS) substrates for lipid and protein characterization: Sensing and beyond. *Analyst* 2018, *143* (17), 3990–4008.

115. Karthigeyan, D.; Siddhanta, S.; Kishore, A. H.; Perumal, S. S. R. R.; Ågren, H.; Sudevan, S.; Bhat, A. V.; Balasubramanyam, K.; Subbegowda, R. K.; Kundu, T. K.; Narayana, C., SERS and MD simulation studies of a kinase inhibitor demonstrate the emergence of a potential drug discovery tool. *Proc. Natl. Acad. Sci. USA* 2014, *2014*, 02695.

116. Aggarwal, S.; Mondal, S.; Siddhanta, S.; Bharat, E.; Nagamalleswari, E.; Nagaraja, V.; Narayana, C., Divalent ion-induced switch in DNA cleavage of KpnI endonuclease probed through surface-enhanced Raman spectroscopy. *J. Phys. Chem. B* 2021, *125* (9), 2241–2250.

117. Ip, S.; MacLaughlin, C. M.; Gunari, N.; Walker, G. C., Phospholipid membrane encapsulation of nanoparticles for surface-enhanced Raman scattering. *Langmuir* 2011, *27*, 7024.

118. Alipour, E.; Halverson, D.; McWhirter, S.; Walker, G. C., Phospholipid bilayers: Stability and encapsulation of nanoparticles. *Annu. Rev. Phys. Chem.* 2017, *68*, 261.

119. Carroll, S.; Al-Rubeai, M., ACSD labelling and magnetic cell separation: A rapid method of separating antibody secreting cells from non-secreting cells. *J. Immunol. Methods* 2005, *296*, 171.

120. Pallaoro, A.; Hoonejani, M. R.; Braun, G. B.; Meinhart, C. D.; Moskovits, M., Rapid identification by surface-enhanced Raman spectroscopy of cancer cells at low concentrations flowing in a microfluidic channel. *ACS Nano* 2015, *9*, 4328.

121. Wu, X.; Luo, L.; Yang, S.; Ma, X.; Li, Y.; Dong, C.; Tian, Y.; Zhang, L.; Shen, Z.; Wu, A., Improved SERS nanoparticles for direct detection of circulating tumor cells in the blood. *ACS Appl. Mater. Interf.* 2015, *7*, 9965.

122. Théry, C., Exosomes: Secreted vesicles and intercellular communications. *F1000 Biol Rep* 2011, *3*, 15–15.

123. Kalra, H.; Drummen, G. P.; Mathivanan, S., Focus on extracellular vesicles: Introducing the next small big thing. *Int. J. Mol. Sci.* 2016, *17* (2), 170.

124. Jeppesen, D. K.; Fenix, A. M.; Franklin, J. L.; Higginbotham, J. N.; Zhang, Q.; Zimmerman, L. J.; Liebler, D. C.; Ping, J.; Liu, Q.; Evans, R.; Fissell, W. H.; Patton, J. G.; Rome, L. H.; Burnette, D. T.; Coffey, R. J., Reassessment of exosome composition. *Cell* 2019, *177* (2), 428–445.e18.

125. Merdalimova, A.; Chernyshev, V.; Nozdriukhin, D.; Rudakovskaya, P.; Gorin, D.; Yashchenok, A., Identification and analysis of exosomes by surface-enhanced Raman spectroscopy. *Appl. Sci.* 2019, *9* (6), 1135.

126. Stremersch, S.; Marro, M.; Pinchasik, B. E.; Baatsen, P.; Hendrix, A.; De Smedt, S. C.; Loza-Alvarez, P.; Skirtach, A. G.; Raemdonck, K.; Braeckmans, K., Identification of individual exosome-like vesicles by surface enhanced Raman spectroscopy. *Small* 2016, *12* (24), 3292–301.

127. Park, J.; Hwang, M.; Choi, B.; Jeong, H.; Jung, J.-H.; Kim, H. K.; Hong, S.; Park, J.-H.; Choi, Y., Exosome classification by pattern analysis of surface-enhanced Raman spectroscopy data for lung cancer diagnosis. *Anal. Chem.* 2017, *89* (12), 6695–6701.

128. Chalapathi, D.; Padmanabhan, S.; Manjithaya, R.; Narayana, C., Surface-enhanced Raman spectroscopy as a tool for distinguishing extracellular vesicles under autophagic conditions: A marker for disease diagnostics. *J. Phys. Chem. B* 2020, *124* (48), 10952–10960.

129. Zong, S.; Wang, L.; Chen, C.; Lu, J.; Zhu, D.; Zhang, Y.; Wang, Z.; Cui, Y., Facile detection of tumor-derived exosomes using magnetic nanobeads and SERS nanoprobes. *Anal. Meth.* 2016, *8* (25), 5001–5008.

130. Lim, C. Y.; Granger, J. H.; Porter, M. D., SERS detection of clostridium botulinum neurotoxin serotypes A and B in buffer and serum: Towards the development of a biodefense test platform. *Anal. Chim. Acta X* 2019, *1*, 100002.

131. Zengin, A.; Tamer, U.; Caykara, T., Fabrication of a SERS based aptasensor for detection of ricin B toxin. *J. Mater. Chem. B* 2015, *3*, 306.

132. Zhu, Y.; Kuang, H.; Xu, L.; Ma, W.; Peng, C.; Hua, Y.; Wang, L.; Xu, C., Gold nanorod assembly based approach to toxin detection by SERS. *J. Mater. Chem.* 2012, *22*, 2387.

133. Granger, J. H.; Porter, M. D., The case for human serum as a highly preferable sample matrix for detection of anthrax toxins. *ACS Sens.* 2018, *3*, 2303.

134. Gao, F.; Feng, S.; Chen, Z.; Li-Chan, E. C.; Grant, E.; Lu, X., Detection and quantification of chloramphenicol in milk and honey using molecularly imprinted polymers: Canadian Penny-based SERS nanobiosensor. *J. Food Sci.* 2014, *79*, N2542.

135. Gao, F.; Hu, Y.; Chen, D.; Li-Chan, E. C. Y.; Grant, E.; Lu, X., Determination of Sudan I in Paprika powder by molecularly imprinted polymers-thin layer chromatography-surface enhanced Raman spectroscopic biosensor. *Talanta* 2015, *143*, 344.

136. Radu, A. I.; Ryabchykov, O.; Bocklitz, T. W.; Huebner, U.; Weber, K.; Cialla-May, D.; Popp, J., Toward food analytics: Fast estimation of lycopene and B-carotene content in tomatoes based on surface enhanced Raman spectroscopy (SERS). *Analyst* 2016, *141*, 4447.

137. Radu, A. I.; Kuellmer, M.; Giese, B.; Huebner, U.; Weber, K.; Cialla-May, D.; Popp, J., Surface-enhanced Raman spectroscopy (SERS) in food analytics: Detection of vitamins B2 and B12 in cereals. *Talanta* 2016, *160*, 289–297.

138. Patze, S.; Huebner, U.; Liebold, F.; Weber, K.; Cialla-May, D.; Popp, J., SERS as an analytical tool in environmental science: The detection of sulfamethoxazole in water in the nanomolar range by applying a microfluidic cartridge setup. *Anal. Chim. Acta* 2017, *949*, 1.

139. Dugandžić, V.; Kupfer, S.; Jahn, M.; Henkel, T.; Weber, K.; Cialla-May, D.; Popp, J., A SERS-based molecular sensor for selective detection and quantification of copper (II) ions. *Sens. Actuators, B* 2019, *279*, 230.

140. Mueller, M.; Tebbe, M.; Andreeva, D. V.; Karg, M.; Alvarez-Puebla, R. A.; Pazos-Perez, N.; Fery, A., Large-area organization of pNIPAM-coated nanostars as SERS platforms for polycyclic aromatic hydrocarbons sensing in gas phase. *Langmuir* 2012, *28*, 9168.

141. Alvarez-Puebla, R. A.; Contreras-Caceres, R.; Pastoriza-Santos, I.; Perez-Juste, J.; Liz-Marzan, L. M., Au@pNIPAM colloids as molecular traps for surface-enhanced, spectroscopic, ultra-sensitive analysis. *Angew. Chem., Int. Ed.* 2009, *48*, 138.

6 Near-Field Nanospectroscopy and Tip-Enhanced Raman Spectroscopy (TERS)

Andrey Krayev, Jeremy F. Schultz, Nan Jiang,
Sreetosh Goswami, Agnès Tempez, Sharad Ambardar,
Dmitri V. Voronine, Naresh Kumar, Kaiyuan Yao,
Shuai Zhang, Emanuil Yanev, Kathleen McCreary,
Hsun-Jen Chuang, Matthew R. Rosenberger,
Thomas Darlington, Berend T. Jonker, James C. Hone,
D. N. Basov, P. James Schuck, and Avinash Patsha

CONTENTS

DOI: 10.1201/9781003248323-6

6.1   TERS: BASICS, HISTORY, APPLICATIONS, AND PROSPECTS

Tip-enhanced Raman scattering (TERS) is a hyphenated spectroscopic imaging method that combines the advantages of two techniques developed in the fourth quarter of the 20th century: scanning probe microscopy (SPM) and surface-enhanced Raman scattering (SERS). In order to understand the unique advantages and challenges of TERS, we'll briefly discuss the techniques on which it was founded.

Scanning probe microscopy allows characterization of various materials at the nanoscale by maintaining certain interactions between a sharp tip and the sample surface which is scanned in a controlled manner relative to the tip (Figure 6.1). It allows imaging of the topographical features down to atomic resolution in XY and Z. In addition to mere topography, SPM may provide information on many other properties of the samples like distribution of the density of states, surface potential, contact resistance, mechanical properties, local friction, etc.

Scanning probe microscopy started with the invention of scanning tunneling microscopy (STM) back in 1981 by Gerd Binnig and Heinrich Rohrer [1,2], for which they were awarded a Nobel prize in physics in 1986. Atomic force microscopy (AFM) invented later in 1986 by the same Gerd Binnig, Calvin Quate, and Christoph Gerber [3] complemented STM and became an extremely versatile and easy-to-use technique which popularity in various research communities by far exceeded the popularity of the STM. The capability to provide three-dimensional information on the sample's topography along with the flexibility in terms of the type of the samples that SPM could address and with its relative ease of use (compared to SEM or TEM) due to the "table top" design of many SPM instruments, made this technique a significant contributor to the broad field of nanotechnology.

Surface-enhanced Raman scattering was discovered by Martin Fleishmann, James McQuilian, and Patrick Hendra back in 1974 [4] in the course of measurements of Raman response from pyridine molecules absorbed on silver electrode subjected to cyclic potential sweeping. Surprisingly, the authors of this original work did not fully appreciate the scale and the nature of their findings. SERS was "rediscovered" in 1977 when two publications appeared that clearly showed that strong Raman signal coming from the molecules absorbed on roughened silver electrodes could by no means be attributed to the mere increase of the electrodes' surface area since the enhancement factors were estimated to be at least 10^5–10^6. Albrecht and Creighton [5] suggested a mechanism of enhancement related to the charge transfer between silver and the absorbed molecules and the interaction of this modified molecule with the surface plasmons as proposed earlier by Philpott [6]. Jeanmaire and Van Duyne [7] proposed the influence of local electric fields appearing at the roughened silver electrodes to be responsible for the dramatic enhancement of the Raman response from the absorbate. A bit later in 1978 Moskovits [8] proposed a mechanism of SERS enhancement based on the appearance of the optical resonances

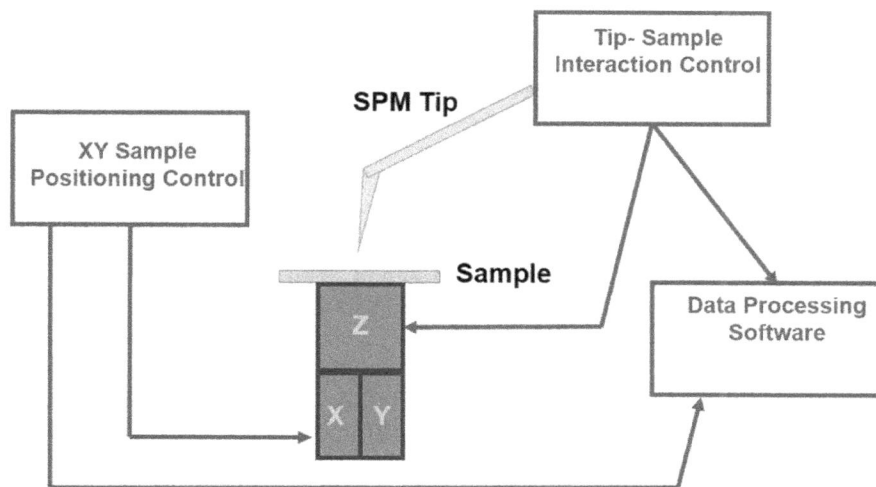

FIGURE 6.1 Principle scheme a scanning probe microscope. The tip-sample interaction control system signal is fed back to the Z scanner in order to maintain the interaction at desired preset level. XY positioning system allows controlled positioning of the sample relative to the tip. Data processing software allows plotting various SPM signals as the function of the sample position.

in two-dimensional array of colloidal metallic nanoparticles which de facto form as a result of electrochemical cycling of silver electrodes. Later the mechanism of SERS enhancement was confirmed to be due to two concurrently acting contributions: electromagnetic enhancement related to the extremely high local electric fields of the surface plasmons excited in colloidal metallic particles confined to the nanometer-scale volume (electromagnetic mechanism) and the molecular polarizability enhancement and appearance of resonant conditions as the result of the charge transfer between the absorbed molecule and supporting metallic particle (chemical mechanism). It has been generally accepted that the electromagnetic mechanism is responsible for the majority of the SERS effect, while chemical contribution is limited to the enhancement factors of <1,000 [9].)

In 1997, two independent groups demonstrated the single molecule sensitivity of SERS [10,11] which led to the explosion of interest toward this technique and development of methods of efficient synthesis of the noble metal nanoparticles with controlled shape and size as well as nano-fabricated structures which rapidly replaced roughened silver electrodes as SERS substrates. Such ultimate sensitivity suggested that the combined enhancement factor in SERS may be as high as ten orders of magnitude [12]. Dramatic variation of SERS signal in time when the intensity of some SERS spectra in the recorded time sequence exceeded the average value by many orders of magnitude (so-called blinking) led to the inception of the concept of so-called hot spots which were responsible for the majority of SERS effect. A junction between two plasmonic particles (a dimer) with an optical electric field strongly localized in the gap is one of the examples of a possible hot spot construct. Fractal nanoparticle aggregates can represent a somewhat alternative hot spot structure [13]. For more detailed description of the origins and developments of SERS, we can refer the reader to several recent comprehensive reviews [14,15].

Though demonstrated sensitivity of SERS was very impressive, this method, at least applied as is, could not improve the spatial resolution of Raman imaging beyond the diffraction limit of regular optical microscopy. To break the diffraction limit it was necessary to create a deterministic hot spot and be able to move it across the sample with high precision. Raman imaging beyond diffraction limit naturally led to the idea of combining SERS and SPM: if a plasmonic structure capable of SERS-like enhancement is formed at the apex of an SPM probe, thus serving as a well-defined hot spot, it would become possible to scan a sample under such tip illuminated with an external laser simultaneously collecting enhanced Raman signal, topography and other physical channels provided by SPM with very similar nanoscale resolution, thus breaking the spatial resolution limitations of conventional Raman microscopy (Figure 6.2). Such an apparatus was proposed by Wessel in 1985 [16].

It is interesting to note that the idea to use a single metallic nanoparticle or a metallic surface with a nanoscale aperture for sub-diffractional optical microscopy was proposed by Synge back in 1928 [17]. Like many other revolutionary ideas that appeared well before their time, it was not embraced by the research community, though in the retrospective, his design proved to be amazingly accurate. The first publication reporting experimental TERS measurements was published in 2000 by Stokle, Suh, Deckert, and Zenobi [18], and was quickly followed same year by the reports by Anderson [19], Hayazawa et al. [20], and Pettinger et al. [21]. Thus, the era of TERS began.

With demonstrated Angstrom-scale spatial resolution of the SPM and single-molecule sensitivity of SERS, the promises of analytical applications for TERS were enormous, but as great were the challenges associated with this technique, both the instrumentational and fundamental. Standard AFMs of late 1990s and up to 2010 were not designed to be coupled with Raman spectroscopy. Most of the AFM instruments had red feedback laser which excluded the possibility to use that spectral range for Raman spectroscopy. There was no standard solution for precise (better than 100 nm) and long-term (10 min) alignment of the apex of the TERS active tip and the focus of Raman laser. Combining high numerical aperture (NA) objectives with SPM instrumentation, except for the case of the bottom excitation/collection schemes was fairly challenging. Reliable and reproducible TERS

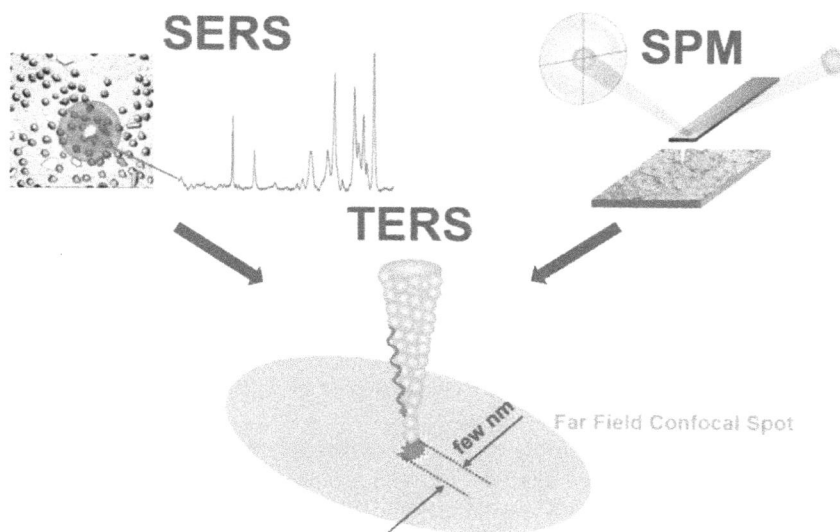

FIGURE 6.2 Combination of the ideas of SERS and SPM by creating a deterministic plasmonic structure at the apex of the SPM tip leads to the inception of TERS.

probes also remained a problem until 2015–2016 when several types of commercial TERS probes with guaranteed near 100% performance became available. The methodology of the TERS measurements also had to be developed and optimized for each specific feedback control mechanism of the SPM.

Currently, there are three major mechanisms of the control of the tip-sample interaction in SPM: in scanning tunneling microscopy the tip-sample interaction is control by maintaining the tunneling current between the tip and a conductive sample. The latter condition is absolutely mandatory – STM feedback can not be utilized on dielectric samples. AFM provides more freedom in this respect as in this technique the tip-sample interaction is either controlled by maintaining a certain deflection of a soft cantilever to which a sharp tip is attached (contact mode AFM), or by maintaining the amplitude or the oscillating frequency of either a similar cantilever or a quartz tuning fork to which a sharp tip is glued. AFM feedback control does not necessarily require the sample to be conductive.

Pettinger et al. [21] used an STM setup with an etched silver wire as a TERS probe and a monolayer of brilliant cresyl blue (BCB) molecules on 12 nm smooth gold film as the substrate. Silver tip and its electrostatic reflection in the gold substrate de facto formed a hot spot similar to the junction of two metallic nanoparticles. Such measurement condition is called a gap-mode TERS which highlights that the sample is positioned in the gap between the plasmonic tip and a plasmonic or at least highly reflective [22] metallic substrate. Hayazawa et al. [20] also used similar gap-mode TERS with a soft silicon cantilever coated with silver, since the Rhodamine 6G molecules they used as the Raman probe were absorbed on top of silver island film. Stokle et al. [18] in their pioneering work deposited the analytes (BCB and C_{60} fullerenes) directly onto the glass substrate, thus implementing non-gap-mode TERS conditions where the enhancement was enabled by the tip only.

The gap-mode TERS always provides significantly stronger enhancement as compared to the tip-only TERS measurements, and therefore the samples that can fit into the few-to subnanometer gap between the tip and the substrate benefit the most from strong enhancement: molecular monolayers or individual molecules, 0-(fullerenes), 1-[carbon nanotubes (CNTs)] or 2-dimensional [graphene, transition metal dichalcogenides (TMDs), etc.] materials deposited on metallic surfaces. This consideration explains the choice of the samples used in the early days of TERS.

The progress of the field during the first decade was very impressive – after the single point measurements, TERS imaging was demonstrated on CNTs, first – at fixed wavelength [23] and later – hyperspectral [24] when a full Raman spectrum within certain spectral range was collected in each pixel. This was a very important step as collection of hyperspectral TERS maps allowed characterization of nanoscale features in the samples such as defects [25] or local strain [26]. Sensitivity wise – a single molecule sensing level of TERS was demonstrated by the group of Richard Van Duyne [27], while the group led by Zhenchao Dong demonstrated submolecular resolution of TERS measurements [28] when they showed that different parts of the same porphyrin molecule absorbed on silver that was imaged using ultrahigh-vacuum (UHV) STM-based TERS probe produced noticeably different TERS spectra (Figure 6.3), thus fulfilling an ultimate dream of the analytical chemists of being able to see an individual molecule and identify its chemical nature by means of vibrational spectroscopy. TERS measurements of several biological molecules were demonstrated [29–31], expanding TERS applications to biology.

It's important to note that the progress of the field during the first decade since the discovery of TERS was enabled by a very limited number of groups around the world. Experimental difficulties of the technique and the lack of readily available highly and reproducibly performing TERS probes were prohibitive for broader applications.

The history and achievements of the first decade of TERS have been well covered in number of early and relatively recent reviews [32–37] to which we will refer the readers for more detailed information.

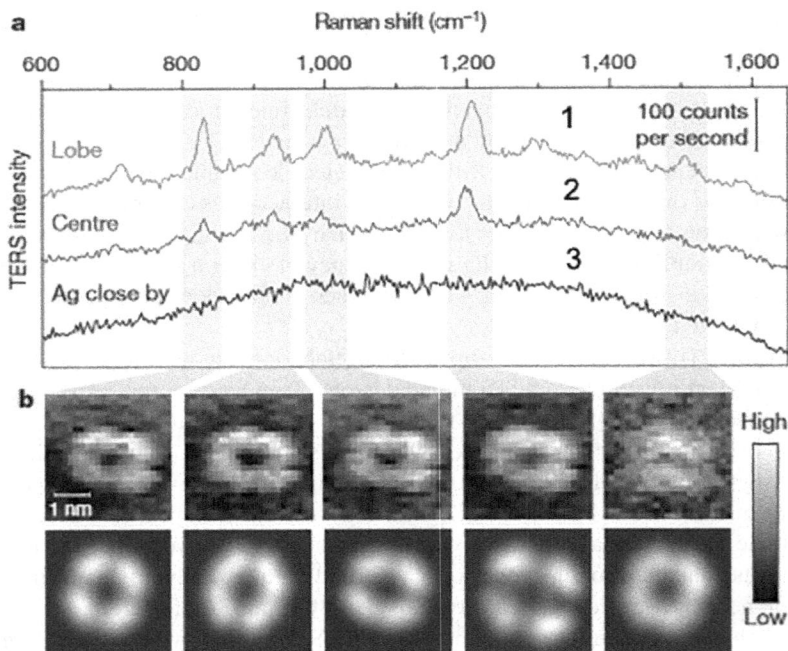

FIGURE 6.3 (a) Representative single-molecule TERS spectra on the lobe (red, 1) and center (blue, 2) of a flat-lying molecule on Ag(111). The TERS spectrum on the bare Ag about 1 nm away from the molecule is also shown in (black, 3) (120 mV, 1 nA, 3 s). (b) The top panels show experimental TERS mapping of a single molecule for different Raman peaks (23×23 pixels^2, 0.16 nm/pixel), processed from all individual TERS spectra acquired at each pixel (120 mV, 1 nA, 0.3 s; image size: 3.6×3.6 nm^2). The bottom panels show the theoretical simulation of the TERS mapping. Used with permission from Ref. [28].

The situation in the field started to change in 2010 when a startup company AIST-NT came up with an SPM instrument that was designed from the scratch to be coupled with Raman spectroscopy. This new instrumentation addressed many of the problems that existed in early, often – home-built TERS setups. For example, the AFM control was done using 1,300 nm feedback laser which left the whole visible range available for Raman/TERS. The AFM feedback laser's optical path was completely decoupled from the Raman laser optical path which dramatically improved the stability of both the SPM and the Raman side and the ease of use of the combined instrument. For precise, long-term alignment of the focus of the Raman laser and the apex of the TERS probe a piezo-controlled, sensor-equipped objective scanner was used for the first time [38]. Mechanically rigid flexure-based design of this objective scanner and the fact that it was positioning with the nanometer precision the very last optical element (the objective itself) in front of the TERS probe, combined with a fairly straightforward alignment procedure greatly improved the speed and stability of the tip-to-Raman laser alignment procedure. The use of multiple independently controlled objective scanners also allowed for performing split excitation/collection experiments [39,40]. In addition to much-improved hardware, an efficient methodology of TERS measurements was discovered [41]. The core idea of this methodology as far as the AFM-cantilever-based TERS probes were concerned, was that in order to achieve significant TERS enhancement, it was necessary to get in gentle direct contact with the sample's surface, a condition the importance of which was somehow underestimated in the community despite the fact that many of early TERS experiments were conducted with the tip in direct contact with the sample. The transition between the adjacent pixels was proposed to be done in semicontact mode thus minimizing the lateral drag on the sample which enabled TERS imaging of loosely attached nanoscale objects and improved the tip's lifetime. All the mentioned design and methodology improvements combined with more readily available well-performing TERS probes, and reproducible stable test samples of graphene oxide and CNTs co-deposited on gold substrate, resulted in bringing TERS measurements from the lab to the exhibition floor. Out-of-the-lab TERS measurement was demonstrated for the first time in 2013 at the TERS-III conference in Zurich on an AIST-NT OmegaScope-R system combined with HORIBA XploRA Raman microscope. In the following years, TERS imaging on combined AIST-NT/HORIBA systems was routinely demonstrated in the exhibition halls of multiple international conferences, even demonstrating near record-breaking spatial resolution, thus dramatically democratizing TERS field.

The pace of the progress was so high that within just 1 year in 2014 application scientists of AIST-NT and HORIBA generated comparable number of TERS images to what was published by the whole scientific community in prior decade. The fact that the TERS measurements could be routinely performed within a commercial company allowed fast developing of sophisticated TERS imaging protocols and, importantly, efficient operation training procedures which allowed even the newcomers in this field who never worked with an SPM or Raman before to get high-quality TERS data within 1–2 days of training. Now, that the hurdles of mastering TERS instrumentation were to a great degree overcome, it became possible to apply this technique for nanoscale spectroscopic characterization of multiple scientifically relevant samples *on demand*.

Quite often the names of engineers who enable just another technical revolution stay unknown to broad public, which is not really fare. The TERS community deserves to know the names of AIST-NT principle design engineers who stood behind the TERS breakthroughs of 2013–2014: Vasily Gavrilyuk, a true Renaissance man – an artist, an extremely knowledgeable optical physicist and a talented engineer; Alexey Belyaev and Vladimir Zhizhimontov – two perfectionists of mechanical engineering and Alexander Yagovkin who streamlined fast implementation of the breakthrough designs into actual parts and assemblies; Vladimir "Kirilych" Ivanov who developed a unique technology of assembly of ultra-stiff AFM and objective scanners; Dr. Alexey Temiryazev who created multiple new AFM imaging modes including the one that became the precursor of the patented TERS imaging mode; Sergey Kacur – a perfectionist of electronic engineering; Sergey

Bashkirov – a software and firmware engineer of encyclopedic knowledge of programming languages; Dmitry Evplov – another Renaissance man in AIST-NT team – a bright physicist, a perfectionist software engineer, whose efforts made it possible to fully utilize the advances of new electronics and mechanical designs, all the TERS imaging modes were programmed by him; Yury Turlapov, the developer of very powerful and user-friendly AFM and TERS image processing software. And of course all other members of the AIST-NT engineering and production team who stood behind the development and manufacturing of the new generation of TERS instrumentation.

Tip-enhanced Raman spectroscopy proved to be particularly beneficial for the characterization of nanoscale heterogeneities of various natures in 2D materials such as graphene and – to considerably greater extent – TMDs and other 2D semiconductors or even 2D metals such as MXenes. The first publication on the TERS imaging of MoS_2 [42] was followed by publications on this and other TMDs from multiple groups around the world which will be discussed in detail in corresponding chapters on TERS and tip-enhanced photoluminescence (TEPL) imaging of 2D materials. Though 2D materials and their heterostructures benefited the most from the progress in TERS and TEPL imaging so far, there were numbers of other exciting applied publications in recent years where TERS characterization of the real world, not the proof-of-concept samples provided a direct spectroscopic information on the structure of the nanoscale objects like, for example, RDX-TNT coreshell nanoparticles [43], CL20-HMX nano co-crystals [44] or chemical reorganization within a single molecular layer as a result of electrical switching [45].

To summarize this introductory chapter, we can safely state that thanks to the efforts of the early adopters of TERS and TEPL which were supported by the progress with corresponding instrumentation, methodology, and appearance of the commercial reproducibly performing probes, TERS is starting to fulfill its promises as an ultimate nanoscale spectroscopic imaging technique that can be applied to characterization of various scientifically or industrially relevant materials and structures. Further progress with the TERS probes design that should improve the current level of enhancement by at least 10–100 times, as well as the broader use of pulsed lasers and thus – nonlinear effects in TERS or performing the TERS imaging in an electrochemical environment open even wider horizons for the advanced research and industrial applications of this technique.

REFERENCES

1. Binnig, G.; Rohrer, H.; Gerber, Ch.; Weibel, E. Tunneling through a controllable vacuum gap. *Applied Physics Letters*, 1982, *40*(2). https://doi.org/10.1063/1.92999.
2. Binnig, G.; Rohrer, H.; Gerber, Ch.; Weibel, E. Surface studies by scanning tunneling microscopy. *Physical Review Letters*, 1982, *49*(1). https://doi.org/10.1103/PhysRevLett.49.57.
3. Binnig, G.; Quate, C. F.; Gerber, Ch. Atomic force microscope. *Physical Review Letters*, 1986, *56*(9). https://doi.org/10.1103/PhysRevLett.56.930.
4. Fleischmann, M.; Hendra, P. J.; McQuillan, A. J. Raman spectra of pyridine adsorbed at a silver electrode. *Chemical Physics Letters*, 1974, *26*(2). https://doi.org/10.1016/0009-2614(74)85388-1.
5. Albrecht, M. G.; Creighton, J. A. Anomalously intense Raman spectra of pyridine at a silver electrode. *Journal of the American Chemical Society*, 1977, *99*(15). https://doi.org/10.1021/ja00457a071.
6. Philpott, M. R. Effect of surface plasmons on transitions in molecules. *The Journal of Chemical Physics*, 1975, *62*(5). https://doi.org/10.1063/1.430708.
7. Jeanmaire, D. L.; van Duyne, R. P. Surface Raman spectroelectrochemistry. *Journal of Electroanalytical Chemistry and Interfacial Electrochemistry*, 1977, *84*(1). https://doi.org/10.1016/S0022-0728(77)80224-6.
8. Moskovits, M. Surface roughness and the enhanced intensity of Raman scattering by molecules adsorbed on metals. *The Journal of Chemical Physics*, 1978, *69*(9). https://doi.org/10.1063/1.437095.
9. Moore, J. E.; Morton, S. M.; Jensen, L. Importance of correctly describing charge-transfer excitations for understanding the chemical effect in SERS. *The Journal of Physical Chemistry Letters*, 2012, *3*(17). https://doi.org/10.1021/jz300492p.
10. Kneipp, K.; Wang, Y.; Kneipp, H.; Perelman, L. T.; Itzkan, I.; Dasari, R. R.; Feld, M. S. Single molecule detection using surface-enhanced Raman scattering (SERS). *Physical Review Letters*, 1997, *78*(9). https://doi.org/10.1103/PhysRevLett.78.1667.

11. Nie, S. Probing single molecules and single nanoparticles by surface-enhanced Raman scattering. *Science*, 1997, *275*(5303). https://doi.org/10.1126/science.275.5303.1102.

12. le Ru, E. C.; Etchegoin, P. G. Quantifying SERS enhancements. *MRS Bulletin*, 2013, *38*(8). https://doi.org/10.1557/mrs.2013.158.

13. Tsai, D. P.; Kovacs, J.; Wang, Z.; Moskovits, M.; Shalaev, V. M.; Suh, J. S.; Botet, R. Photon scanning tunneling microscopy images of optical excitations of fractal metal colloid clusters. *Physical Review Letters*, 1994, *72*(26). https://doi.org/10.1103/PhysRevLett.72.4149.

14. Pilot, R.; Signorini, R.; Durante, C.; Orian, L.; Bhamidipati, M.; Fabris, L. A review on surface-enhanced Raman scattering. *Biosensors*, 2019, *9*(2). https://doi.org/10.3390/bios9020057.

15. Langer, J.; Jimenez de Aberasturi, D.; Aizpurua, J.; Alvarez-Puebla, R. A.; Auguié, B.; Baumberg, J. J.; Bazan, G. C.; Bell, S. E. J.; Boisen, A.; Brolo, A. G.; Choo, J.; Cialla-May, D.; Deckert, V.; Fabris, L.; Faulds, K.; García de Abajo, F. J.; Goodacre, R.; Graham, D.; Haes, A. J.; Haynes, C. L.; Huck, C.; Itoh, T.; Käll, M.; Kneipp, J.; Kotov, N. A.; Kuang, H.; le Ru, E. C.; Lee, H. K.; Li, J.-F.; Ling, X. Y.; Maier, S. A.; Mayerhöfer, T.; Moskovits, M.; Murakoshi, K.; Nam, J.-M.; Nie, S.; Ozaki, Y.; Pastoriza-Santos, I.; Perez-Juste, J.; Popp, J.; Pucci, A.; Reich, S.; Ren, B.; Schatz, G. C.; Shegai, T.; Schlücker, S.; Tay, L.-L.; Thomas, K. G.; Tian, Z.-Q.; van Duyne, R. P.; Vo-Dinh, T.; Wang, Y.; Willets, K. A.; Xu, C.; Xu, H.; Xu, Y.; Yamamoto, Y. S.; Zhao, B.; Liz-Marzán, L. M. Present and future of surface-enhanced Raman scattering. *ACS Nano*, 2020, *14*(1). https://doi.org/10.1021/acsnano.9b04224.

16. Wessel, J. Surface-enhanced optical microscopy. *Journal of the Optical Society of America B*, 1985, *2*(9). https://doi.org/10.1364/JOSAB.2.001538.

17. Synge, E. H. XXXVIII. A suggested method for extending microscopic resolution into the ultra-microscopic region. *The London, Edinburgh, and Dublin Philosophical Magazine and Journal of Science*, 1928, *6*(35). https://doi.org/10.1080/14786440808564615.

18. Stöckle, R. M.; Suh, Y. D.; Deckert, V.; Zenobi, R. Nanoscale chemical analysis by tip-enhanced Raman spectroscopy. *Chemical Physics Letters*, 2000, *318*(1–3). https://doi.org/10.1016/S0009-2614(99)01451-7.

19. Anderson, M. S. Locally enhanced Raman spectroscopy with an atomic force microscope. *Applied Physics Letters*, 2000, *76*(21). https://doi.org/10.1063/1.126546.

20. Hayazawa, N.; Inouye, Y.; Sekkat, Z.; Kawata, S. Metallized tip amplification of near-field Raman scattering. *Optics Communications*, 2000, *183*(1–4). https://doi.org/10.1016/S0030-4018(00)00894-4.

21. Pettinger, B.; Picardi, G.; Schuster, R.; Ertl, G. Surface enhanced Raman spectroscopy: Towards single molecule spectroscopy. *Electrochemistry*, 2000, *68*(12). https://doi.org/10.5796/electrochemistry.68.942.

22. Krayev, A.; Krylyuk, S.; Ilic, R.; Hight Walker, A. R.; Bhattarai, A.; Joly, A. G.; Velický, M.; Davydov, A. V.; El-Khoury, P. Z. Comparable enhancement of TERS signals from WSe$_2$ on chromium and gold. *The Journal of Physical Chemistry C*, 2020, *124*(16). https://doi.org/10.1021/acs.jpcc.0c01298.

23. Hartschuh, A.; Sánchez, E. J.; Xie, X. S.; Novotny, L. High-resolution near-field Raman microscopy of single-walled carbon nanotubes. *Physical Review Letters*, 2003, *90*(9). https://doi.org/10.1103/PhysRevLett.90.095503.

24. Anderson, N.; Hartschuh, A.; Cronin, S.; Novotny, L. Nanoscale vibrational analysis of single-walled carbon nanotubes. *Journal of the American Chemical Society*, 2005, *127*(8). https://doi.org/10.1021/ja045190i.

25. Stadler, J.; Schmid, T.; Zenobi, R. Nanoscale chemical imaging of single-layer graphene. *ACS Nano*, 2011, *5*(10). https://doi.org/10.1021/nn2035523.

26. Yano, T.; Ichimura, T.; Kuwahara, S.; H'Dhili, F.; Uetsuki, K.; Okuno, Y.; Verma, P.; Kawata, S. Tip-enhanced nano-Raman analytical imaging of locally induced strain distribution in carbon nanotubes. *Nature Communications*, 2013, *4*(1). https://doi.org/10.1038/ncomms3592.

27. Sonntag, M. D.; Klingsporn, J. M.; Garibay, L. K.; Roberts, J. M.; Dieringer, J. A.; Seideman, T.; Scheidt, K. A.; Jensen, L.; Schatz, G. C.; van Duyne, R. P. Single-molecule tip-enhanced Raman spectroscopy. *The Journal of Physical Chemistry C*, 2012, *116*(1). https://doi.org/10.1021/jp209982h.

28. Zhang, R.; Zhang, Y.; Dong, Z. C.; Jiang, S.; Zhang, C.; Chen, L. G.; Zhang, L.; Liao, Y.; Aizpurua, J.; Luo, Y.; Yang, J. L.; Hou, J. G. Chemical mapping of a single molecule by plasmon-enhanced Raman scattering. *Nature*, 2013, *498*(7452). https://doi.org/10.1038/nature12151.

29. Ichimura, T.; Hayazawa, N.; Hashimoto, M.; Inouye, Y.; Kawata, S. Tip-enhanced coherent anti-stokes Raman scattering for vibrational nanoimaging. *Physical Review Letters*, 2004, *92*(22). https://doi.org/10.1103/PhysRevLett.92.220801.

30. Bailo, E.; Deckert, V. Tip-enhanced Raman spectroscopy of single RNA strands: Towards a novel direct-sequencing method. *Angewandte Chemie International Edition*, 2008, *47*(9). https://doi.org/10.1002/anie.200704054.

31. Alexander, K. D.; Schultz, Z. D. Tip-enhanced Raman detection of antibody conjugated nanoparticles on cellular membranes. *Analytical Chemistry*, 2012, *84*(17). https://doi.org/10.1021/ac301739k.

32. Deckert-Gaudig, T.; Taguchi, A.; Kawata, S.; Deckert, V. Tip-enhanced Raman spectroscopy – From early developments to recent advances. *Chemical Society Reviews*, 2017, *46*(13). https://doi.org/10.1039/C7CS00209B.

33. Shao, F.; Zenobi, R. Tip-enhanced Raman spectroscopy: Principles, practice, and applications to nano-spectroscopic imaging of 2D materials. *Analytical and Bioanalytical Chemistry*, 2019, *411*(1). https://doi.org/10.1007/s00216-018-1392-0.

34. Verma, P. Tip-enhanced Raman spectroscopy: Technique and recent advances. *Chemical Reviews*, 2017, *117*(9). https://doi.org/10.1021/acs.chemrev.6b00821.

35. Pozzi, E. A.; Goubert, G.; Chiang, N.; Jiang, N.; Chapman, C. T.; McAnally, M. O.; Henry, A.-I.; Seideman, T.; Schatz, G. C.; Hersam, M. C.; Van Duyne, R. P. Ultrahigh-vacuum tip-enhanced Raman spectroscopy. *Chemical Reviews*, 2017, *117*(7). https://doi.org/10.1021/acs.chemrev.6b00343.

36. Novotny, L. *Chapter 5: The History of Near-Field Optics, Progress in Optics*, 2007, *50* https://doi.org/10.1016/S0079-6638(07)50005-3.

37. Kumar, N.; Mignuzzi, S.; Su, W.; Roy, D. Tip-enhanced Raman spectroscopy: Principles and applications. *EPJ Techniques and Instrumentation*, 2015, *2*(1). https://doi.org/10.1140/epjti/s40485-015-0019-5.

38. Nicklaus, M.; Nauenheim, C.; Krayev, A.; Gavrilyuk, V.; Belyaev, A.; Ruediger, A. Note: Tip enhanced Raman spectroscopy with objective scanner on opaque samples. *Review of Scientific Instruments*, 2012, *83*(6). https://doi.org/10.1063/1.4725528.

39. Bhattarai, A.; Crampton, K. T.; Joly, A. G.; Wang, C.-F.; Schultz, Z. D.; El-Khoury, P. Z. A closer look at corrugated Au tips. *The Journal of Physical Chemistry Letters*, 2020, *11*(5). https://doi.org/10.1021/acs.jpclett.0c00305.

40. Wang, C.-F.; El-Khoury, P. Z. Imaging plasmons with sub-2 nm spatial resolution via tip-enhanced four-wave mixing. *The Journal of Physical Chemistry Letters*, 2021, *12*(14). https://doi.org/10.1021/acs.jpclett.1c00763.

41. Saunin S. A., Krayev A. V., Zhishimontov V. V., Gavrilyuk V. V., Grigorov L. N., Belyaev A. V., & Evplov D. A. *Systems and Methods for Non-Destructive Surface Chemical Analysis of Samples*, Patent Number US20150338439A1, USA, 2018.

42. Zhang, Y.; Voronine, D. V.; Qiu, S.; Sinyukov, A. M.; Hamilton, M.; Liege, Z.; Sokolov, A. V.; Zhang, Z.; Scully, M. O. Improving resolution in quantum subnanometre-gap tip-enhanced Raman nanoimaging. *Scientific Reports*, 2016, *6*(1). https://doi.org/10.1038/srep25788.

43. Deckert-Gaudig, T.; Pichot, V.; Spitzer, D.; Deckert, V. High-resolution Raman spectroscopy for the nanostructural characterization of explosive nanodiamond precursors. *ChemPhysChem*, 2017, *18*(2). https://doi.org/10.1002/cphc.201601276.

44. Hübner, J.; Deckert-Gaudig, T.; Glorian, J.; Deckert, V.; Spitzer, D. Surface characterization of nanoscale co-crystals enabled through tip enhanced Raman spectroscopy. *Nanoscale*, 2020, *12*(18). https://doi.org/10.1039/D0NR00397B.

45. Goswami, S.; Deb, D.; Tempez, A.; Chaigneau, M.; Rath, S. P.; Lal, M.; Ariando; Williams, R. S.; Goswami, S.; Venkatesan, T. Nanometer-scale uniform conductance switching in molecular memristors. *Advanced Materials*, 2020, *32*(42). https://doi.org/10.1002/adma.202004370.

6.2 DEVELOPMENT OF TERS INSTRUMENTATION AND METHODOLOGY

Before getting into the details of the more common and some exotic configurations of the TERS instrumentation, let's discuss several general principles which are common for all these configurations.

It's a rather obvious statement that in order to improve the signal-to-noise ratio in TERS spectra, it's important to collect as many scattered photons as possible. This simple statement leads to the conclusion that increasing the NA of the collecting optics improves the performance of a TERS setup. Polarization control is another important factor in TERS experiments. In order to have the "hot spot" located at the apex of the TERS probe facing the sample, electric field of the excitation laser should be directed along the TERS probe which is usually almost perpendicular to the sample plane. The technical means of implementation of this requirement is another important factor in assessment of different optical schemes of TERS instruments. Finally, optical elements that guide and collect the light to/from the apex of the TERS probe should minimally affect the performance of the SPM side of the experimental setup. The latter is an important requirement since in order to keep the laser focus on the

SPM probe alignment stable, in most of the TERS setups the scanning in X, Y, and Z directions is usually performed on the sample side. As we will see further in this section, some illumination/collection schemes may affect the SPM performance stronger compared to the alternatives.

Tip-enhanced Raman spectroscopy instruments based on standard, preferably – infinity-corrected objectives that focus the excitation laser at the apex of a TERS probe and collect the scattered Raman photons, comprise the overwhelming majority of existing TERS setups. There are three possible configurations within this family.

6.2.1 Transmission (Bottom) Illumination/Collection

In this optical scheme, laser beam is focused on the SPM tip through the objective located below the transparent sample, Raman signal is collected with the same objective. In most cases, the sample is prepared on the surface of a thin (ca 170 μm) cover glass slip (Figure 6.4). This configuration was very popular in the early days of TERS as it provides the highest NA possible, up to 1.4 when oil or water immersion objectives are used.

Unfortunately, high NA is realistically the only advantage of this approach, while it creates number of limitations. First – in this scheme, it's impossible to work with non-transparent samples. Organization of the Z-polarized light, in the laser focus (a mandatory requirement for efficient TERS) requires using radial polarizers [1] or careful manipulation of the optical beam when the central part is blocked [2]. The arrangement leads to the condition when only the outer part of the aperture that comes to the focal point almost along the sample surface and which intrinsically has a higher Z component in the focus, is being used. The presence of immersion oil between the face of the objective and a thin cover glass slip leads to certain drag on the sample which significantly decreases the stability of the feedback and decreases the quality of the SPM images. Also, it should

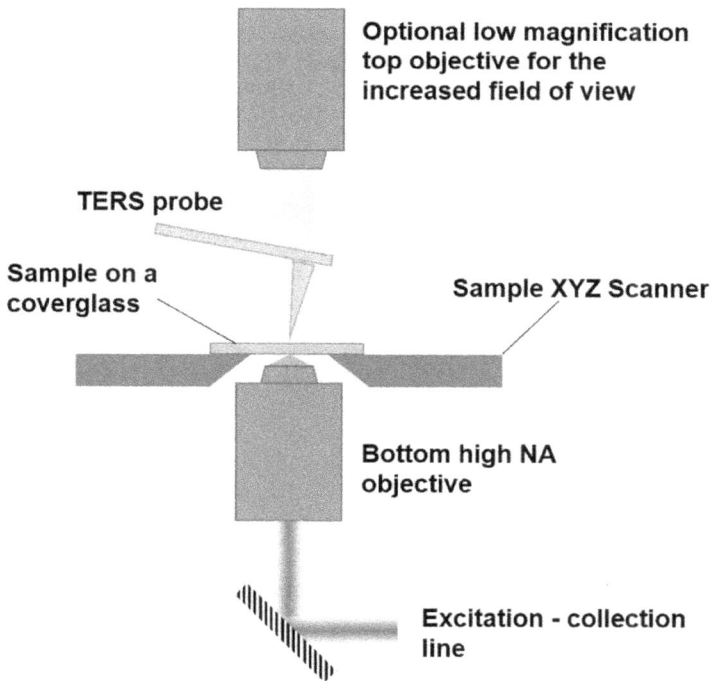

FIGURE 6.4 Scheme of the bottom illumination/collection TERS setup.

be mentioned that in order to accommodate the bottom objective, it's necessary to use fairly bulky sample scanners with large aperture in the middle. Such scanners have naturally low resonant frequencies due to their large mass, which in its turn affects the stability and the noise floor of the SPM measurements.

6.2.2 Top Illumination/Collection

In this optical scheme (Figure 6.5) the laser is focused on the tip with a long working distance (6–20 mm) objective located above and as in case of the bottom illumination/collection scheme, the Raman signal is collected with the same objective.

In order to be able to see the apex of the probe in this scheme, it's mandatory to use protruding, top-visible probes like, for example, an Access-type (Applied Nanostructures Inc.) probe presented in Figure 6.6a.

The apex of such probe can be clearly seen with the top objective despite inevitable shading of the portion of the aperture by the front portion of the cantilever and the base of the needle. An example of the top view of an access-type TERS probe is shown in Figure 6.6b.

Obviously, top configuration allows working with non-transparent samples, though at the expense of the NA of the objective which is in most instruments is 0.7 or less. In principle, top illumination also requires careful preparation of the laser beam for increased Z polarization in the laser focus, though in practice it's less picky compared to the bottom illumination and can produce decent TERS enhancement with linear polarization oriented along the cantilever Y axis [3,4]. Since no immersion liquid is involved, in top illumination/collection scheme the objective does not affect the SPM performance, especially if the AFM feedback laser is *not* coming through this objective. The most important drawbacks of the top illumination scheme apart from the decreased NA compared to the bottom objectives are the following:

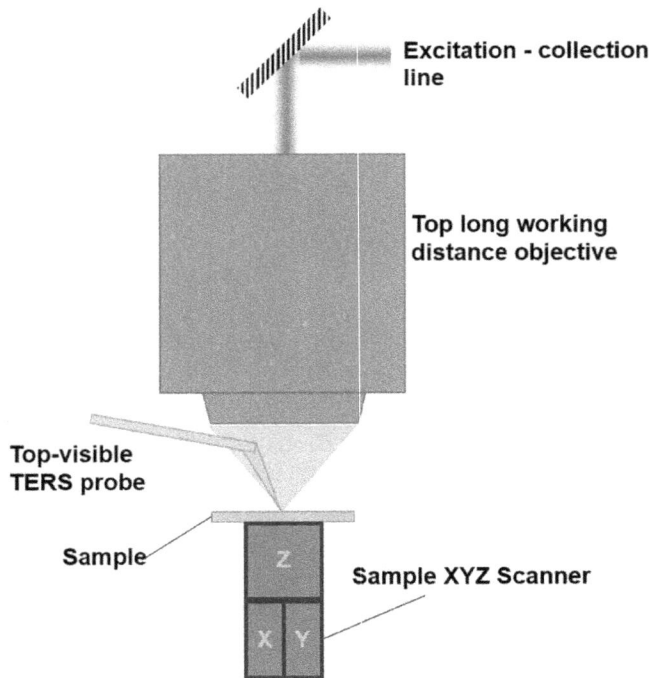

FIGURE 6.5 Scheme of the top illumination/collection TERS setup.

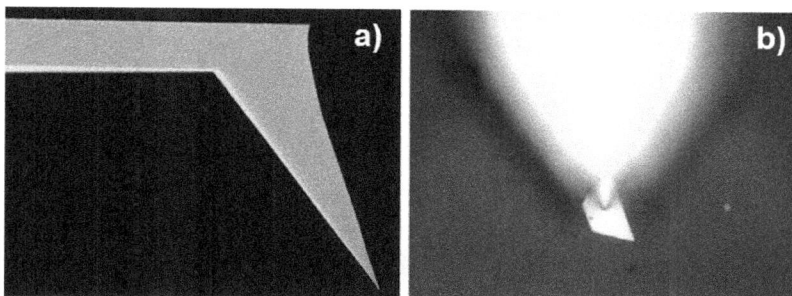

FIGURE 6.6 (a) SEM image of an access-type AFM probe (Applied Nanostructures Inc.). (b) Optical view of the access-type TERS probe as seen through 100×, 0.7 NA top objective. Images used with permission from Applied Nanostructures Inc. and HORIBA Scientific correspondingly.

- Limited field of view of the high magnification objectives, and therefore, somewhat complicated sample navigation.
- Alignment of the laser focus and the apex of the TERS probe can be realistically done only after the tip is engaged with the sample when perfect focusing of the laser on the sample surface guarantees that the apex of the tip and the laser focus is in the same Z plane.
- In instruments where the AFM feedback laser is directed through the same top objective as the Raman laser, the noise level of the AFM registration system increases noticeably because of relatively poor mechanical stiffness of the assembly that holds a heavy top objective. In addition, inevitable use of the beam splitters in case when the top objective is shared between the AFM registration system and the Raman excitation/collection line leads to limitations of the spectral range of Raman/TERS response and overall decrease of the intensity of collected TERS signal.

6.2.3 SIDE ILLUMINATION/COLLECTION

In this optical scheme that was first implemented back in 2005–2007 [5–7] (Figure 6.7) the excitation and collection of scattered light are executed through the side objective whose optical axis is inclined 15°–30° to the sample plane.

Long working distance (over 6 mm) with NA up to 0.75 can be used in this illumination/collection scheme, same as the ones used for the top excitation/collection. Sample observation is done through a different, smaller magnification top objective which greatly improves the ease of sample navigation, a very important advantage when a specific feature of only a few microns in size has to be found on the sample, which is usually in the order of few tens to few hundreds of square millimeters.

Side illumination/collection provides the simplest means of arranging significant Z component of the electric field in the focal point as vertically (P) polarized light has a very significant projection on the axis of the TERS probe. Another extremely useful advantage of the side excitation/collection is the possibility to perform near-perfect alignment of the TERS probe apex and the Raman laser focus away from the sample due to very good visibility of the probe's needle in the side objective (Figure 6.8a).

Though the top-visible AFM probes are the most convenient for the side-illumination/collection operation, other types of the probes can be used as well, like the ones showed in Figure 6.9.

The limiting factor of a specific AFM cantilever type is the ratio of the depth of the recess of the needle from the end of the cantilever and the height of the needle itself, which determines the degree to which the apex of the probe is shaded by the overhanging part of the cantilever beam.

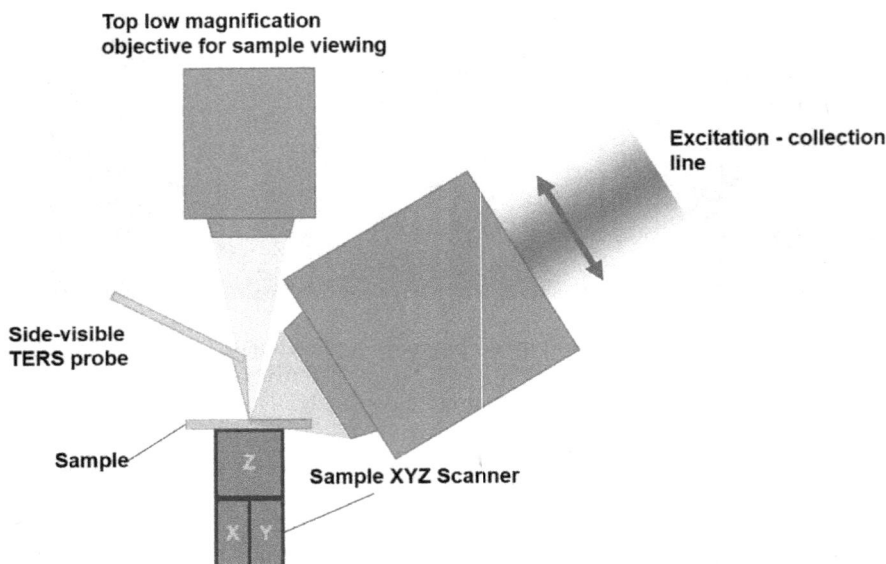

FIGURE 6.7 Side-illumination-collection optical scheme. *P* polarization of the excitation laser results in significant projection of the optical electric field to the *Z* axis.

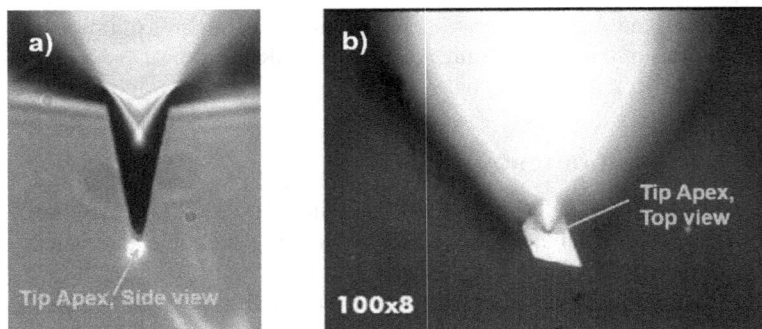

FIGURE 6.8 (a) Optical image of an access-type TERS probe through the side 100×, 0.7 NA objective when the apex of the probe is aligned with the focus of the Raman laser. (b) Optical image of a similar probe with the same objective in the top position. Obviously, the side view is way more intuitive and easy to interpret.

Significant progress with the on-demand TERS and TEPL imaging of the nanoscale heterogeneities in 2D semiconductors seen in recent years [8–20] is to great extent due to the considerably broader use of the side-illumination/collection optical scheme in TERS setups in combination with advanced TERS/TEPL imaging methodology and highly reproducible strongly enhancing commercial TERS AFM probes.

Of course, the side-illumination collection scheme is not drawback-free. In addition to smaller NA objectives used in this scheme as compared to the bottom illumination, we can mention the limitation on the sample size along the direction parallel to the objective's optical axis due to finite (and not very large) working distance of the objectives.

As was already mentioned in the Section 6.1, using several independently controlled objective scanners allows co-localization of the focal spots of two different objectives with the precision

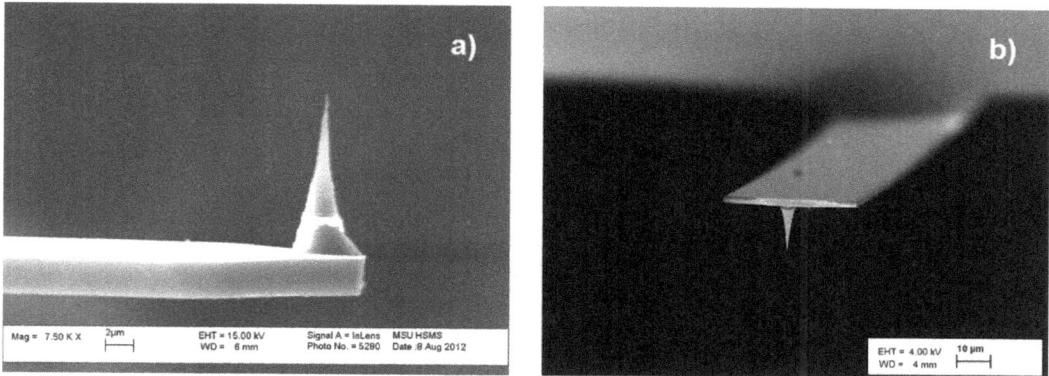

FIGURE 6.9   (a) Side view SEM image of the single crystal diamond-based AFM probe that can be used for the side-access TERS experiments, though it is not top-visible. (b) Front view SEM image of a similar probe. SEM images courtesy of Artech Carbon OU.

down to few tens of nanometers. This in its turn enables implementation of the split excitation/ collection TERS experiments when for example the excitation is performed through the side objective and the collection – through the bottom one [21] (Figure 6.10), thus combining the advantages of the ease of arrangement of Z polarization in the side channel and high NA in the bottom collection channel.

6.2.4   MORE EXOTIC ILLUMINATION/COLLECTION SCHEMES

In addition to the three common schemes based on the standard microscope objectives, there are several alternatives.

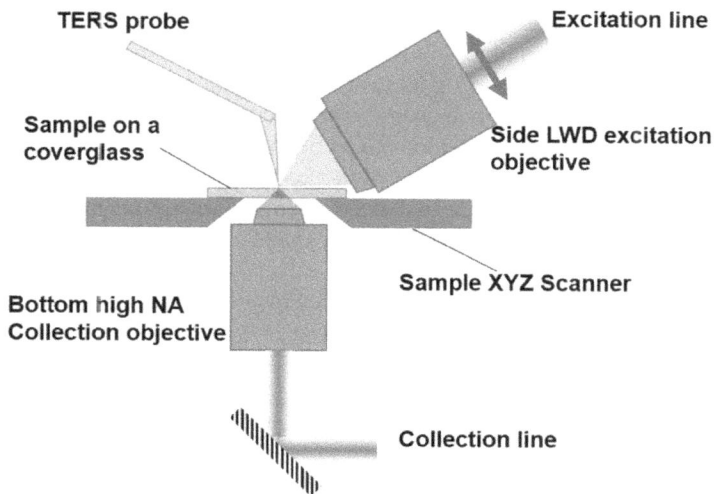

FIGURE 6.10   Split excitation–collection setup scheme when excitation is done with P-polarized beam focused on the TERS probe's apex through the side objective, while the collection of the signal is done through the high NA bottom objective.

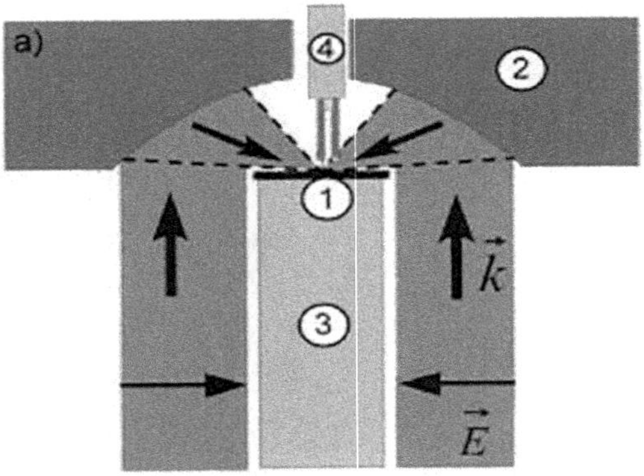

FIGURE 6.11 TERS setup based on parabolic mirror with NA of 1. Adopted with permission from Ref. [22]. Such an optical scheme combines the advantage of the de facto top illumination, and therefore – the possibility to work with non-transparent samples and the high numerical aperture of the collection optics.

Using an on-axis parabolic mirror instead of the standard, lens or even mirror-based objective was one of the earliest advances in TERS instrumentation [22,23] (Figure 6.11).

The use of the parabolic mirror with NA of 1.0 enables efficient collection of Raman signal while still allowing working with non-transparent samples. Despite its advantages, this illumination scheme did not gain significant popularity, mainly due to two reasons – high-quality parabolic mirrors with high NA are not broadly available and considerably more expensive compared to the lens-based long working distance objectives. In addition, the on-axis parabolic mirror creates significant complications for the use of standard laser feedback AFM detection and limits such TERS setups to STM or tuning-fork-based SPM feedback which does not improve the ease of use of such instrumentation.

An alternative approach to the TERS schemes described above was used by several groups [24–29] who combined the near-field excitation with the efficient far-field collection. In this family of TERS experimental setups excitation is achieved by the near-field emission from either an apex of an optical fiber with a plasmonic structure attached to it, for example, a bow-type antenna [28,29] or a single crystal silver wire that serves as the plasmonic waveguide. Collection of the signal is usually performed with a standard objective with as high NA as the specific implementation of the TERS setup can allow, though efficient signal collection efficiency has been demonstrated with the same fiber-based probe [26]. In addition to improved spatial resolution, the near-field excitation eliminates the parasitic far-field signal which has to be measured and subtracted in case when the TERS/TEPL signal is comparable or a fraction of the far-field signal. An alternative way of exciting near-field emission from a TERS probe is to convert a far field into propagating plasmons in a solid or metallized TERS probes using a grating [27] or a single slit [25,30] cut in the upper part of the tip needle (Figure 6.12).

Complicated manufacturing of such probes that involves the focused ion beam milling (FIB) (in case of the single slit or the grating-based probes) or tedious and not perfectly reproducible attachment of the single crystal gold or silver nanowires to the apex of the probe are hampering their broader use at the moment, though breakthroughs in the mass production technology of such sophisticated near-field probes are not impossible.

6.2.5 TERS PROBES

As important as the efficiency and convenience of the optical excitation–collection part of a TERS setup is, efficient and stable TERS probes will always remain to be the heart of any

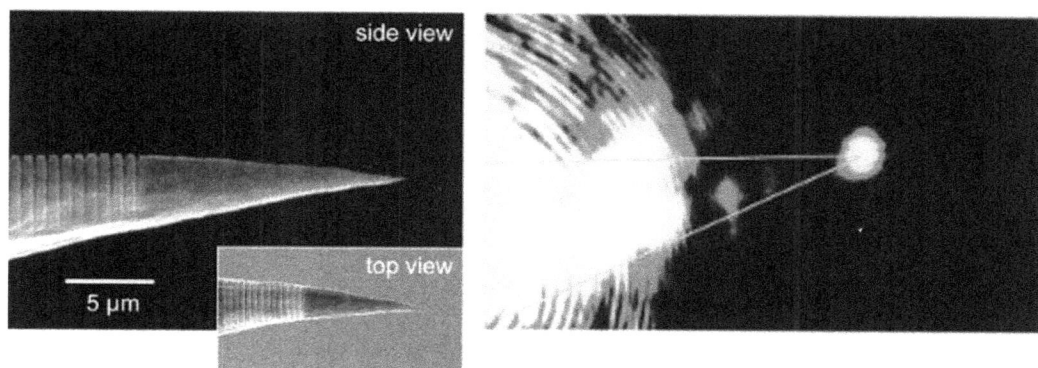

FIGURE 6.12 Near-field excitation etched gold TERS probe with a coupling grating cut few microns away from the tip apex. Image of the light emitted from the apex when excitation laser is properly aligned with the grating. Adopted with permission from Ref. [27].

TERS instrument. Again, before we discuss different types of the TERS probes, let's review the requirements that an efficient TERS probe should meet:

- It should enable efficient coupling of the far field into localized plasmons at the apex of the tip and vice versa.
- An efficient TERS probe should feature a reasonably broad spectral range of the enhancement – ideally – covering the whole visible – NIR range of 450–900 nm.
- The tip should normally operate under focused laser power of up to a few hundreds of μW and remain active for at least 10^4–10^5 measurement cycles for efficient TERS/TEPL imaging.
- TERS probes should be reasonably sharp to enable high spatial resolution of TERS/TEPL imaging.
- Tip exchange and re-alignment with Raman laser have to be simple and fast.
- TERS probe should be compatible with as many SPM techniques as possible – topography, scanning Kelvin microscopy (surface potential imaging), conductivity, etc. This automatically implies that the probe should be able to operate both in the intermittent contact (oscillating mode) and direct contact.
- Probes should feature consistent enhancement and be readily available at a reasonable cost so that researchers can truly treat them as consumables not that much different from conventional SPM probes.

Taking into account the above criteria, let's review the TERS probes that have been developed over the last two decades.

6.2.6 Etched Gold/Silver Wire Probes

Electrochemically etched gold wire that tapers down to the radius of about 10–20 nm was proposed as the TERS probe in the original TERS publication [31]. This type of TERS probes (Figure 6.13) became the most popular during the first 10–15 years thanks to the relative ease of fabrication at low cost and rapid progress in the development of sophisticated etching methods leading to

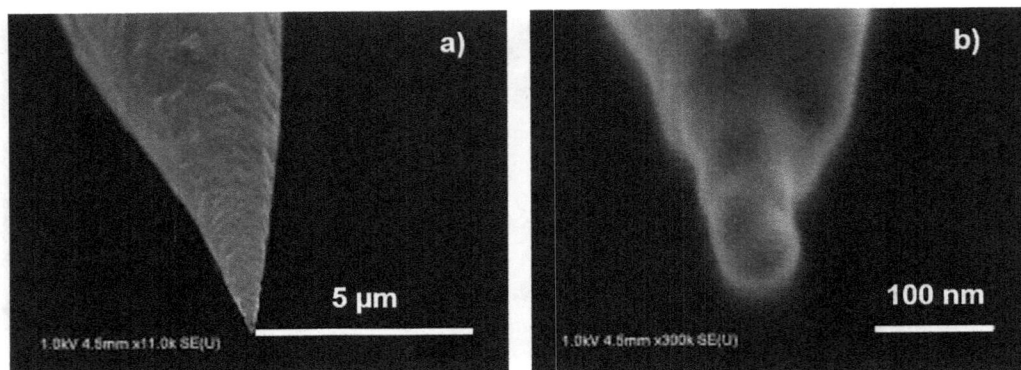

FIGURE 6.13 SEM images of an etched gold probe. Certain roughness of the tip cone is clearly seen in the low magnification image in panel (a). Zoom-in image in panel (b) shows the tip radius of about 20 nm. Images courtesy of HORIBA Scientific.

reasonably good reproducibility of the tip radius and TERS enhancement. Both silver and gold, the two metals that feature the best combination of the real and imaginary parts of dielectric functions in visible-to NIR range for efficient excitation of the surface plasmons, were used for manufacturing of such tips. Due to significant chemical activity of silver which in ambient results in its quick oxidation or sulfidation, etched TERS probes based on this metal are mostly used in UHV cryogenic TERS systems, and these probes demonstrated the best spatial resolution achieved in TERS so far [32,33]. Gold, thanks to its chemical inertness has been broadly used in TERS experiments conducted in air at room temperature (RT), and early progress in TERS was made mostly thanks to such etched gold wire probes that could be used in setups with various tip-sample distance control mechanism – STM [34,35], tuning fork, where a small piece of etched gold wire is glued to one of the prongs of a quartz tuning fork [36,37] or even AFM for which etched gold or silver wires were flattened and bent de facto forming a solid metallic AFM cantilever [38,39]. The latter etched gold AFM probes played an important role in the development of advanced TERS imaging methodology back in 2013–2014 that enabled on-demand high pixel density TERS imaging [40] (Figure 6.14).

Etched TERS probes demonstrate strong enhancement in broad range from ~470 to 900 nm for silver and 590–900 nm for gold. They can tolerate fairly high power in the Raman laser focus up to several mW without losing the performance. In case of picking up organic contamination, gold wires can be cleaned in concentrated sulfuric acid without the loss of TERS enhancement [41]. However, despite all this and the fact that the etched gold and silver wire TERS probes have been the most popular type of TERS probes during first 10–15 years, and still remain the champions in terms of the spatial resolution of TERS imaging, they do not meet several important criteria of a perfect TERS probe. Limited commercial availability of such probes leads to their preferential in-lab manufacturing which naturally limits their broad use across the TERS community. Etched gold/silver wires used with the STM – or tuning fork-based tip-sample distance control are not as flexible as conventional AFM probes in terms of the number of different SPM channels they can probe. The necessity to manually glue the etched wires to the tuning forks makes the probe manufacturing slow, tedious, and extremely dependent on the skill level of an experimentalist. Due to rather significant variation of the geometry of such probes, their exchange and re-alignment are considerably more complicated and slow compared to the TERS probes based on standard silicon or silicon nitride AFM cantilevers which we'll discuss next.

Coating of conventional AFM probes with a thin layer of silver or gold was also proposed in the original TERS publication, though for a long time performance of coated probes remained

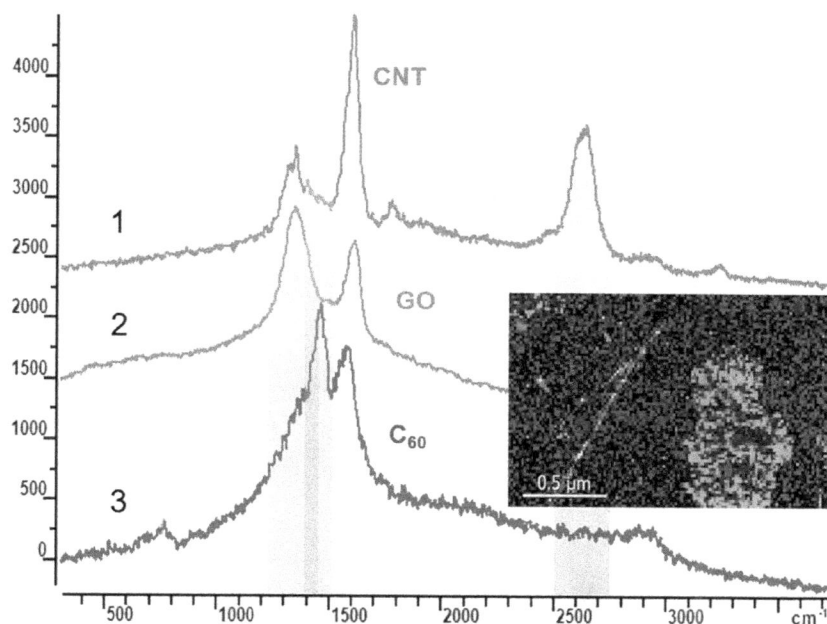

FIGURE 6.14 Averaged TERS spectra of CNTs, graphene oxide and C_{60} fullerene co-deposited on gold substrate. In the insert – 128 pixels/line TERS map showing the distribution of the intensity of D band (green, 2), 2D band (red, 1) and 1,450 cm^{-1}, a characteristic peak of C_{60} (blue, 3). Map was recorded with a solid gold etched AFM probe using the SpecTop™ TERS imaging mode (HORIBA Scientific). Adopted with permission from Ref. [40].

extremely poorly reproducible and during the first decade of TERS development, the only group that consistently used silver-coated AFM probes was the group of Volker Deckert in Germany.

Only in 2015, a comprehensive work appeared that thoroughly investigated the physics of coated TERS probes [42] both theoretically and experimentally. It was shown that a group of closely located, preferably non-touching plasmonic nanoparticles at the apex of the AFM tip couple the far field into the localized plasmons and vice versa significantly more efficiently compared to the smooth coating with the same plasmonic material, thus highlighting the importance of creating a certain degree of nanoscale roughness in the coating of AFM-based TERS probes for efficient performance. Though separated silver nanoparticles showed the strongest plasmonic response in the 450–500 nm range, an identical number of connected nanoparticles was calculated to outperform the disconnected ones in 750–800 nm range. Superior performance of the tips with particle-like coating over the tips with smooth coating was demonstrated experimentally on a graphene sample when the tip with disconnected particles was shown to outperform the smoothly coated probe by almost an order of magnitude (Figure 6.15).

Though separated nanoparticles showed the best TERS enhancement, probes with such coating may not necessarily be the most desirable since the lack of electrical connection between adjacent particles at the tip apex makes conductivity or Kelvin probe microscopy (SKM) measurements with such probes impossible.

Modern commercially available TERS probes, mostly based on protruding type silicon AFM probes like the one presented in Figure 6.6 feature continuous gold or protected silver coating with certain nanoscale roughness, though not as high as in work of Taguchi, Verma, and Kawata [42] (Figure 6.16).

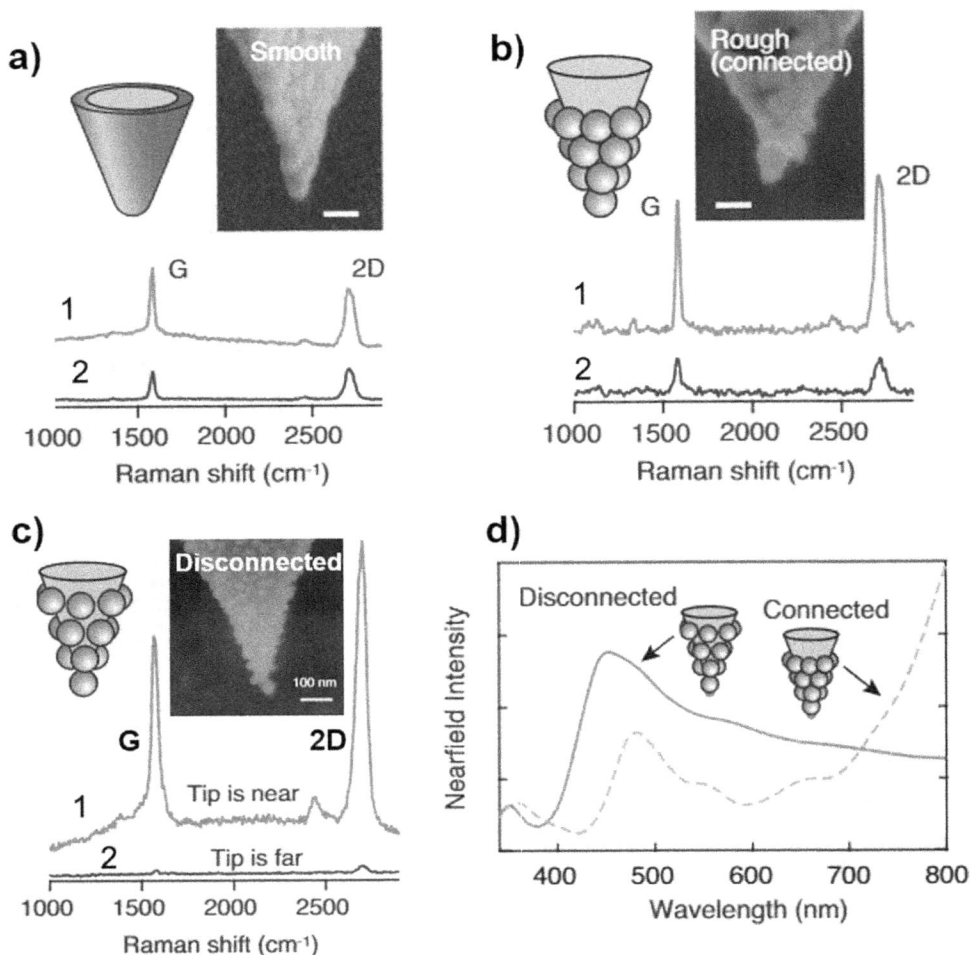

FIGURE 6.15 TERS enhancement of multiple grain probe. (a) Enhancement of Raman signal from graphene with smooth TERS probe. (b) Enhancement of the Raman signal from graphene with the TERS probe coated with merged silver nanoparticles. (c) Enhancement of the Raman signal from graphene with the TERS probe coated with disconnected silver nanoparticles. In (a)–(c) the dark blue (2) spectrum corresponds to the retracted tip, red spectrum (1) – the tip is near the sample. (d) Spectral dependence of the near-field intensity for disconnected (red line) and connected (dotted cyan curve) silver particles. Connected particles perform better in NIR. Adopted with permission from Ref. [42].

Practically 100% reliability and consistency of modern commercial TERS probes played a significant role in democratization of TERS measurements and the use of TERS for on-demand measurements of scientifically relevant samples, even in out-of-lab conditions. As an example in Figure 6.17, three TERS maps collected with three different silver AFM probes OMNI-TERS-FM-Ag (Applied Nanostructures Inc.) of the same graphene oxide monolayer on gold are presented. Despite slight variations in the spatial resolution caused by the variation of the tip radius, the reproducibility of TERS maps is fairly impressive.

An alternative approach to engineering a plasmonic nanostructure at the apex of an SPM probe for efficient far- to near-field conversion was employed by the groups of Ado Jorio and Luiz Gustavo Concado [43,44]. They proposed to use the "monopole" nano-pyramid gold probes which can be fine-tuned for specific excitation laser. This design is to some extent reminiscent of the earlier

FIGURE 6.16 Helium ion microscopy image of gold-coated access-NC-Au TERS probe from Applied Nanostructures Inc. Image courtesy of Patrick El-Khoury and Bruce Arey (PNNL).

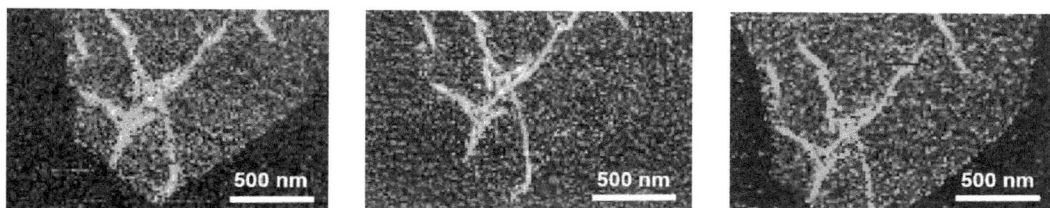

FIGURE 6.17 TERS maps of the same graphene oxide monolayer crystal on gold collected with three different OMNBI-TERS-FM-Ag AFM probes (Applied Nanostructures Inc).

published idea of the "stepped pyramid" design [45], but it seems to be simpler in manufacturing as in this case only one nano-pyramid needs to be etched into the silicon mold. These "monopole" nano-pyramid gold probes demonstrated amazing performance producing not only strong enhancement on graphene and carbon nanotubes even without the gap mode [43,44], but proved to be able to resolve the reconstructed Moire patterns in twisted bilayer graphene [46], which certainly became a milestone achievement for the TERS community (Figure 6.18).

Summarizing this part devoted to the TERS probes, we should say that despite obvious progress with the TERS probes performance, reproducibility, and commercial availability, there is quite some room for further improvement in TERS tips performance. Future progress in TERS/TEPL field will greatly depend on the pace at which the performance of the TERS probes will be improving.

6.2.7 TERS Imaging Methodology

We'll start this part with two rather self-obvious statements: TERS imaging, when the sample is scanned under illuminated TERS probe and in every pixel of the map a full TERS spectrum is recorded, is always preferred over the single point TERS measurement as the TERS map plotted as the distribution of the intensity/spectral position/linewidth of one or several Raman bands across the scanned area allows assessment of the structural or compositional heterogeneities of the sample. TERS enhancement strongly depends on the tip-sample distance, and the strongest enhancement occurs when this distance is below 1 nm or even when the tip is in direct contact with the sample (Figure 6.19).

Maintaining this distance optimized for maximal TERS enhancement, without inflicting damage to the tip or the sample in the process of scanning is the most essential part of TERS imaging. Let's consider three major feedback control mechanisms in SPM in this respect.

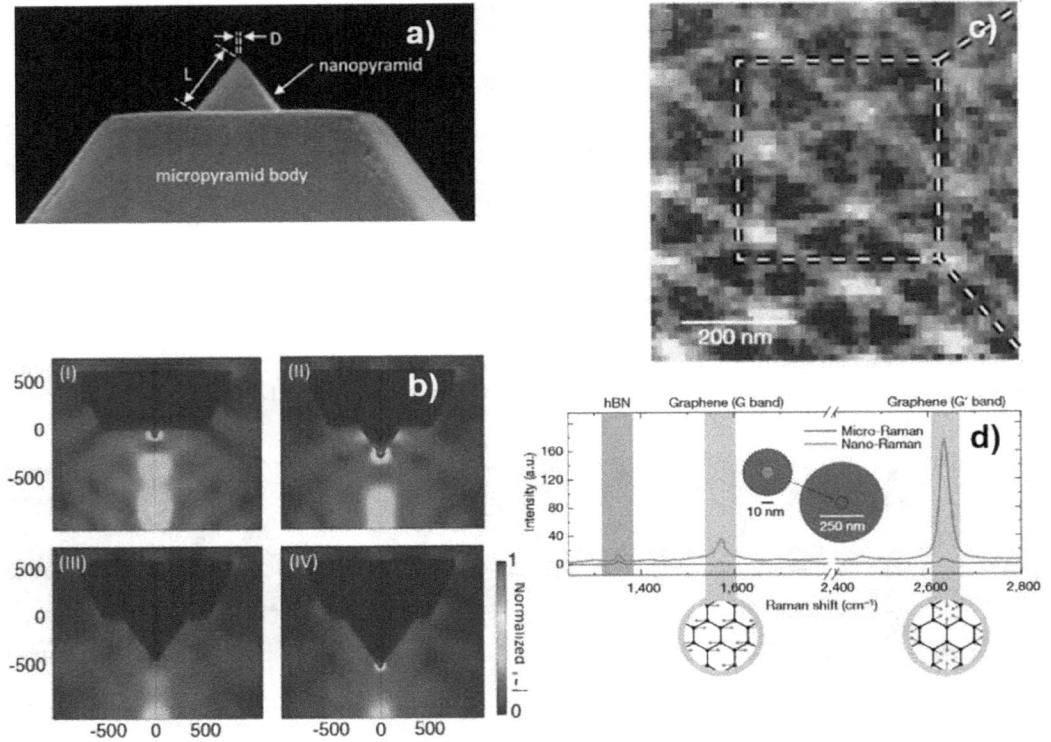

FIGURE 6.18 (a) SEM image of the monopole nano-pyramid TERS probe. (b) Electric field dependence on the nano-pyramid size L: (i) $L=125$ nm; (ii) $L=425$ nm; (iii) $L=650$ nm; and (iv) $L=750$ nm. The color code on panels numbered (i–iv) displays the electric field module $|E|$, all normalized to the highest value on tip with $L=425$ nm. Adopted with permission from Ref. [44]. (c) TERS image of the reconstructed Moire pattern in twisted bilayer graphene based on the intensity of G′(2D) band. (d) Comparative far-field (green) and TERS (red) spectra from the same twisted bilayer graphene showing strong enhancement of the Raman signal by the nano-pyramid TERS probe. Adopted with permission from Ref. [46].

In STM and Tuning fork feedback the tip-sample distance is automatically maintained at the subnanometer level which is dictated by the condition of maintaining a reasonably strong tunneling current in case of STM or reasonably large frequency or amplitude shift in case of the tuning fork control, so optimization for successful TERS imaging is automatically built-in these modes. At the same time, it's important to remember that both for STM and tuning fork, the mechanical stiffness of the probe is substantially higher compared to the AFM cantilevers, which puts fairly strict requirements on the maintenance of the tip-sample distance as even short-term imperfection in the tip-sample distance control may result in severe damage both to the tip and the sample.

With AFM feedback control things are more complicated. Intermittent contact more commonly known as the tapping mode is the standard AFM operation these days when the tip-sample distance is maintained based on maintaining certain amplitude or (considerably less frequently) frequency shift of the AFM probe which oscillates near its resonant frequency. In this case, an average distance between the tip and the sample is roughly equal to the value of the oscillation amplitude, which is usually in the range of 20–40 nm. Such distance is way too big for efficient TERS enhancement of most of the TERS probes currently available, though some exceptionally well-performing probes may demonstrate noticeable enhancement even in tapping, when the majority of TERS signal is generated during the short portion of the oscillation cycle when the tip-sample distance is at its minimum. In order to make TERS measurements more consistent with the commercially available

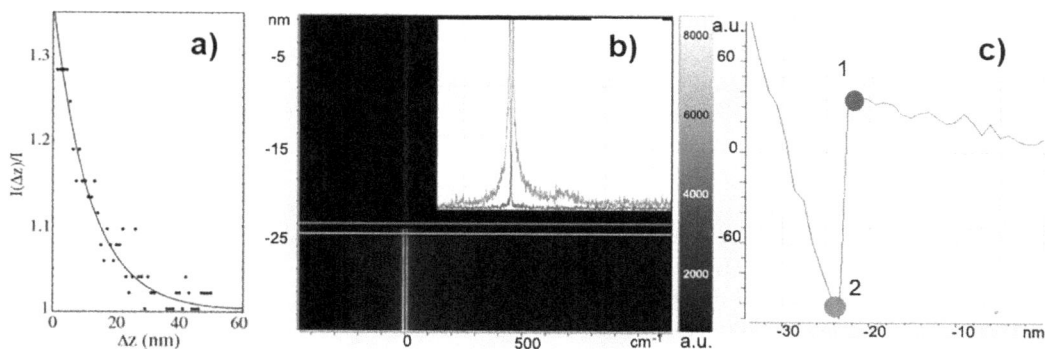

FIGURE 6.19 (a) Dependence of the TERS enhancement on the tip-sample distance for an etched gold probe operating in share force (tuning fork) feedback. Adopted with permission from Ref. [47]. (b) CCD camera snapshot sequence as the function of the tip-sample distance demonstrating sharp rise of the TERS signal at the moment when soft AFM-TERS probe snaps into contact with the sample due to capillary forces. In the insert – two Raman spectra corresponding to similarly colored lines in the sequence. (c) Force-distance curve recorded concurrently with the spectra in (b). Blue (1) and red (2) marks correspond to the blue and red spectra in the insert in (b).

AFM-TERS probes, they need to be conducted in contact mode. In this case, the tip-sample distance is optimized for TERS automatically, like in case of STM/tuning fork control, but such scanning or TERS imaging may not be compatible with loosely attached [CNTs or quantum dots (QD)] or fragile samples. Of course, one can argue that decreasing the spring constant of the cantilevers used for gold or silver deposition down to 0.05 N/m or less may be an answer to this problem, but in reality even very soft cantilevers can exert significant force on the sample in ambient conditions which is caused by the surface water capillary attraction which can easily be in the order of 5–10 nN. In addition to this, extremely soft cantilevers can only be operated in contact mode which excludes the use of such important SPM imaging modes as the SKM or scanning capacitance (dC/dZ or d^2C/dZ^2) microscopy (SCM).

An elegant way of addressing this problem was proposed back in 2014–2015 [48]. In this approach, the TERS probes may be based on AFM cantilevers with reasonably large spring constants in the range of 1–100 N/m, which are perfectly suitable for standard intermittent contact operations including SKM and SCM.

TERS imaging is performed by bringing the AFM probe in direct contact with the sample surface in each pixel of TERS maps for preset time, while the transition between the adjacent pixels is performed in intermittent contact which drastically decreases lateral forces exerted on the sample. Raman spectra can be collected either only during the contact mode portion of the imaging cycle, which is usually done when the TERS signal is overwhelmingly stronger than the far-field signal (SpecTop™ mode) or twice – during the direct and the intermittent contact portions of the cycle (DualSpec™ mode), when the Raman signal collected in intermittent contact de facto corresponds to the far-field signal. In this latter mode, two maps are collected concurrently and the far-field map can be subtracted later in order to extract the pure near-field component when the TERS/TEPL enhancement is not very strong.

Broad use of these two TERS imaging modes which allowed 10^4–10^5 TERS measurements cycles during the useful life of a TERS probe, enabled fast and efficient on-demand TERS and/or TEPL imaging of many scientifically relevant samples, especially 2D semiconductors [8,9,11–20].

It should be noted that even in case when an AFM-TERS probe demonstrates significant enhancement in intermittent contact, its enhancement in direct contact will always be stronger, while the TERS spectra obtained under contact and non-contact conditions may actually differ due to various effects related to the charge transfer from the tip to the sample [49].

It's practically inevitable that in the course of collecting TERS data the probe may pick up some contamination that can reduce or completely destroy TERS enhancement, this is particularly true for the AFM probes that generate TERS data in direct contact with the sample. Luckily, there is a simple method that allows to clean the probe and restore its performance – usually, it's sufficient to bring the tip in contact with a clean area on the sample with the typical force of 15–30 nN and rub it against the surface by scanning small, few hundreds of nanometers across, area in contact mode while monitoring the TERS response. When the contamination particle is removed, the background Raman signal will dramatically increase which serves as an indication that the tip is restored. Such cleaning can be repeated multiple times during the life cycle of the TERS probe, though at the expense of a slight increase of the tip radius.

To conclude this section of the chapter we can state that despite a very significant progress in the development of TERS instrumentation, probes, and methodology, there remains plenty of room for improvement. Increasing the enhancement of the TERS probes by another 1–3 orders of magnitude over the current state of the art as well as advancing the TERS imaging methodology to minimize the tip contamination and the tip-sample interaction force would move the TERS field to the next level, further expanding its general applicability.

REFERENCES

1. Yano, T.; Ichimura, T.; Kuwahara, S.; H'Dhili, F.; Uetsuki, K.; Okuno, Y.; Verma, P.; Kawata, S. Tip-enhanced nano-Raman analytical imaging of locally induced strain distribution in carbon nanotubes. *Nature Communications*, 2013, *4*(1). https://doi.org/10.1038/ncomms3592.

2. Hayazawa, N.; Inouye, Y.; Sekkat, Z.; Kawata, S. Metallized tip amplification of near-field Raman scattering. *Optics Communications*, 2000, *183*(1–4). https://doi.org/10.1016/S0030-4018(00)00894-4.

3. Huang, T.-X.; Cong, X.; Wu, S.-S.; Lin, K.-Q.; Yao, X.; He, Y.-H.; Wu, J.-B.; Bao, Y.-F.; Huang, S.-C.; Wang, X.; Tan, P.-H.; Ren, B. Probing the edge-related properties of atomically thin MoS_2 at nanoscale. *Nature Communications*, 2019, *10*(1). https://doi.org/10.1038/s41467-019-13486-7.

4. Chan, K. L. A.; Kazarian, S. G. Tip-enhanced Raman mapping with top-illumination AFM. *Nanotechnology*, 2011, *22*(17). https://doi.org/10.1088/0957-4484/22/17/175701.

5. Mehtani, D.; Lee, N.; Hartschuh, R. D.; Kisliuk, A.; Foster, M. D.; Sokolov, A. P.; Maguire, J. F. Nano-Raman spectroscopy with side-illumination optics. *Journal of Raman Spectroscopy*, 2005, *36*(11). https://doi.org/10.1002/jrs.1409.

6. Saito, Y.; Motohashi, M.; Hayazawa, N.; Iyoki, M.; Kawata, S. Nanoscale characterization of strained silicon by tip-enhanced Raman spectroscope in reflection mode. *Applied Physics Letters*, 2006, *88*(14). https://doi.org/10.1063/1.2191949.

7. Lee, N.; Hartschuh, R. D.; Mehtani, D.; Kisliuk, A.; Maguire, J. F.; Green, M.; Foster, M. D.; Sokolov, A. P. High contrast scanning nano-Raman spectroscopy of silicon. *Journal of Raman Spectroscopy*, 2007, *38*(6). https://doi.org/10.1002/jrs.1698.

8. Velický, M.; Rodriguez, A.; Bouša, M.; Krayev, A. V.; Vondráček, M.; Honolka, J.; Ahmadi, M.; Donnelly, G. E.; Huang, F.; Abruña, H. D.; Novoselov, K. S.; Frank, O. Strain and charge doping fingerprints of the strong interaction between monolayer MoS_2 and gold. *The Journal of Physical Chemistry Letters*, 2020, *11*(15). https://doi.org/10.1021/acs.jpclett.0c01287.

9. Rahaman, M.; Rodriguez, R. D.; Plechinger, G.; Moras, S.; Schüller, C.; Korn, T.; Zahn, D. R. T. Highly localized strain in a MoS_2/Au heterostructure revealed by tip-enhanced Raman spectroscopy. *Nano Letters*, 2017, *17*(10). https://doi.org/10.1021/acs.nanolett.7b02322.

10. Sheremet, E.; Rodriguez, R. D.; Agapov, A. L.; Sokolov, A. P.; Hietschold, M.; Zahn, D. R. T. Nanoscale imaging and identification of a four-component carbon sample. *Carbon*, 2016, *96*. https://doi.org/10.1016/j.carbon.2015.09.104.

11. Su, W.; Kumar, N.; Krayev, A.; Chaigneau, M. In situ topographical chemical and electrical imaging of carboxyl graphene oxide at the nanoscale. *Nature Communications*, 2018, *9*(1). https://doi.org/10.1038/s41467-018-05307-0.

12. Tang, C.; He, Z.; Chen, W.; Jia, S.; Lou, J.; Voronine, D. V. Quantum plasmonic hot-electron injection in lateral WSe_2-$MoSe_2$ heterostructures. *Physical Review B*, 2018, *98*(4). https://doi.org/10.1103/PhysRevB.98.041402.

13. Krayev, A.; Bailey, C. S.; Jo, K.; Wang, S.; Singh, A.; Darlington, T.; Liu, G.-Y.; Gradecak, S.; Schuck, P. J.; Pop, E.; Jariwala, D. Dry transfer of van der Waals crystals to noble metal surfaces to enable characterization of buried interfaces. *ACS Applied Materials & Interfaces*, 2019, *11*(41). https://doi.org/10.1021/acsami.9b09798.

14. Smithe, K. K. H.; Krayev, A. V.; Bailey, C. S.; Lee, H. R.; Yalon, E.; Aslan, Ö. B.; Muñoz Rojo, M.; Krylyuk, S.; Taheri, P.; Davydov, A. V.; Heinz, T. F.; Pop, E. Nanoscale heterogeneities in monolayer MoSe$_2$ revealed by correlated scanning probe microscopy and tip-enhanced Raman spectroscopy. *ACS Applied Nano Materials*, 2018, *1*(2). https://doi.org/10.1021/acsanm.7b00083.

15. Darlington, T. P.; Carmesin, C.; Florian, M.; Yanev, E.; Ajayi, O.; Ardelean, J.; Rhodes, D. A.; Ghiotto, A.; Krayev, A.; Watanabe, K.; Taniguchi, T.; Kysar, J. W.; Pasupathy, A. N.; Hone, J. C.; Jahnke, F.; Borys, N. J.; Schuck, P. J. Imaging strain-localized excitons in nanoscale bubbles of monolayer WSe$_2$ at room temperature. *Nature Nanotechnology*, 2020, *15*(10). https://doi.org/10.1038/s41565-020-0730-5.

16. Jariwala, D.; Krayev. A.; Wong, J.; Robinson, A. E.; Sherrott, M. C.; Wang, S.; Liu, G.-Y.; Terrones, M.; Atwater, H. A. Nanoscale doping heterogeneity in few-layer WSe$_2$ exfoliated onto noble metals revealed by correlated SPM and TERS imaging. *2D Materials*, 2018, *5*(3). https://doi.org/10.1088/2053-1583/aab7bc.

17. Darlington, T. P.; Krayev, A.; Venkatesh, V.; Saxena, R.; Kysar, J. W.; Borys, N. J.; Jariwala, D.; Schuck, P. J. Facile and quantitative estimation of strain in nanobubbles with arbitrary symmetry in 2D semiconductors verified using hyperspectral nano-optical imaging. *The Journal of Chemical Physics*, 2020, *153*(2). https://doi.org/10.1063/5.0012817.

18. Bhattarai, A.; Krayev, A.; Temiryazev, A.; Evplov, D.; Crampton, K. T.; Hess, W. P.; El-Khoury, P. Z. Tip-enhanced Raman scattering from nanopatterned graphene and graphene oxide. *Nano Letters*, 2018, *18*(6). https://doi.org/10.1021/acs.nanolett.8b01690.

19. Su, W.; Esfandiar, A.; Lancry, O.; Shao, J.; Kumar, N.; Chaigneau, M. Visualising structural modification of patterned graphene nanoribbons using tip-enhanced Raman spectroscopy. *Chemical Communications*, 2021, *57*(56). https://doi.org/10.1039/D1CC01769A.

20. Zhang, Y.; Voronine, D. V.; Qiu, S.; Sinyukov, A. M.; Hamilton, M.; Liege, Z.; Sokolov, A. V.; Zhang, Z.; Scully, M. O. Improving resolution in quantum subnanometre-gap tip-enhanced Raman nanoimaging. *Scientific Reports*, 2016, *6*(1). https://doi.org/10.1038/srep25788.

21. Wang, C.-F.; El-Khoury, P. Z. Imaging plasmons with sub-2nm spatial resolution via tip-enhanced four-wave mixing. *The Journal of Physical Chemistry Letters*, 2021, *12*(14). https://doi.org/10.1021/acs.jpclett.1c00763.

22. Anger, P.; Feltz, A.; Berghaus, T.; Meixner, A. J. Near-field and confocal surface-enhanced resonance Raman spectroscopy at cryogenic temperatures. *Journal of Microscopy*, 2003, *209*(3). https://doi.org/10.1046/j.1365-2818.2003.01089.x.

23. Sackrow, M.; Stanciu, C.; Lieb, M. A.; Meixner, A. J. Imaging nanometre-sized hot spots on smooth au films with high-resolution tip-enhanced luminescence and Raman near-field optical microscopy. *ChemPhysChem*, 2008, *9*(2). https://doi.org/10.1002/cphc.200700723.

24. Walke, P.; Fujita, Y.; Peeters, W.; Toyouchi, S.; Frederickx, W.; de Feyter, S.; Uji-i, H. Silver nanowires for highly reproducible cantilever based AFM-TERS microscopy: Towards a universal TERS probe. *Nanoscale*, 2018, *10*(16). https://doi.org/10.1039/C8NR02225A.

25. Taguchi, K.; Umakoshi, T.; Inoue, S.; Verma, P. Broadband plasmon nanofocusing: Comprehensive study of broadband nanoscale light source. *The Journal of Physical Chemistry C*, 2021, *125*(11). https://doi.org/10.1021/acs.jpcc.0c11541.

26. Kim, S.; Yu, N.; Ma, X.; Zhu, Y.; Liu, Q.; Liu, M.; Yan, R. High external-efficiency nanofocusing for lens-free near-field optical nanoscopy. *Nature Photonics*, 2019, *13*(9). https://doi.org/10.1038/s41566-019-0456-9.

27. Ropers, C.; Neacsu, C. C.; Elsaesser, T.; Albrecht, M.; Raschke, M. B.; Lienau, C. Grating-coupling of surface plasmons onto metallic tips: A nanoconfined light source. *Nano Letters*, 2007, *7*(9). https://doi.org/10.1021/nl071340m.

28. Bao, W.; Borys, N. J.; Ko, C.; Suh, J.; Fan, W.; Thron, A.; Zhang, Y.; Buyanin, A.; Zhang, J.; Cabrini, S.; Ashby, P. D.; Weber-Bargioni, A.; Tongay, S.; Aloni, S.; Ogletree, D. F.; Wu, J.; Salmeron, M. B.; Schuck, P. J. Visualizing nanoscale excitonic relaxation properties of disordered edges and grain boundaries in monolayer molybdenum disulfide. *Nature Communications*, 2015, *6*(1). https://doi.org/10.1038/ncomms8993.

29. Calafiore, G.; Koshelev, A.; Darlington, T. P.; Borys, N. J.; Melli, M.; Polyakov, A.; Cantarella, G.; Allen, F. I.; Lum, P.; Wong, E.; Sassolini, S.; Weber-Bargioni, A.; Schuck, P. J.; Cabrini, S.; Munechika, K. Campanile near-field probes fabricated by nanoimprint lithography on the facet of an optical fiber. *Scientific Reports*, 2017, *7*(1). https://doi.org/10.1038/s41598-017-01871-5.

30. Umakoshi, T.; Tanaka, M.; Saito, Y.; Verma, P. White nanolight source for optical nanoimaging. *Science Advances*, 2020, *6*(23). https://doi.org/10.1126/sciadv.aba4179.

31. Stöckle, R. M.; Suh, Y. D.; Deckert, V.; Zenobi, R. Nanoscale chemical analysis by tip-enhanced Raman spectroscopy. *Chemical Physics Letters*, 2000, *318*(1–3). https://doi.org/10.1016/S0009-2614(99)01451-7.

32. Tallarida, N.; Lee, J.; Apkarian, V. A. Tip-enhanced Raman spectromicroscopy on the Angstrom scale: Bare and CO-terminated Ag tips. *ACS Nano*, 2017, *11*(11). https://doi.org/10.1021/acsnano.7b06022.

33. Zhang, R.; Zhang, Y.; Dong, Z. C.; Jiang, S.; Zhang, C.; Chen, L. G.; Zhang, L.; Liao, Y.; Aizpurua, J.; Luo, Y.; Yang, J. L.; Hou, J. G. Chemical mapping of a single molecule by plasmon-enhanced Raman scattering. *Nature*, 2013, *498*(7452). https://doi.org/10.1038/nature12151.

34. Chen, C.; Hayazawa, N.; Kawata, S. A 1.7 nm resolution chemical analysis of carbon nanotubes by tip-enhanced Raman imaging in the ambient. *Nature Communications*, 2014, *5*(1). https://doi.org/10.1038/ncomms4312.

35. Pettinger, B.; Picardi, G.; Schuster, R.; Ertl, G. Surface enhanced Raman spectroscopy: Towards single molecule spectroscopy. *Electrochemistry*, 2000, *68*(12). https://doi.org/10.5796/electrochemistry.68.942.

36. Anderson, N.; Hartschuh, A.; Cronin, S.; Novotny, L. Nanoscale vibrational analysis of single-walled carbon nanotubes. *Journal of the American Chemical Society*, 2005, *127*(8). https://doi.org/10.1021/ja045190i.

37. Park, K.-D.; Khatib, O.; Kravtsov, V.; Clark, G.; Xu, X.; Raschke, M. B. Hybrid tip-enhanced nano-spectroscopy and nanoimaging of monolayer WSe_2 with local strain control. *Nano Letters*, 2016, *16*(4). https://doi.org/10.1021/acs.nanolett.6b00238.

38. Snitka, V.; Rodriguez, R. D.; Lendraitis, V. Novel gold cantilever for nano-Raman spectroscopy of graphene. *Microelectronic Engineering*, 2011, *88*(8). https://doi.org/10.1016/j.mee.2011.02.046.

39. Rodriguez, R. D.; Sheremet, E.; Müller, S.; Gordan, O. D.; Villabona, A.; Schulze, S.; Hietschold, M.; Zahn, D. R. T. Compact metal probes: A solution for atomic force microscopy based tip-enhanced Raman spectroscopy. *Review of Scientific Instruments*, 2012, *83*(12). https://doi.org/10.1063/1.4770140.

40. Sheremet, E.; Rodriguez, R. D.; Agapov, A. L.; Sokolov, A. P.; Hietschold, M.; Zahn, D. R. T. Nanoscale imaging and identification of a four-component carbon sample. *Carbon*, 2016, *96*, 588–593. https://doi.org/10.1016/j.carbon.2015.09.104.

41. Liu, Z.; Wang, X.; Dai, K.; Jin, S.; Zeng, Z.-C.; Zhuang, M.-D.; Yang, Z.-L.; Wu, D.-Y.; Ren, B.; Tian, Z.-Q. Tip-enhanced Raman spectroscopy for investigating adsorbed nonresonant molecules on single-crystal surfaces: Tip regeneration, probe molecule, and enhancement effect. *Journal of Raman Spectroscopy*, 2009, *40*(10). https://doi.org/10.1002/jrs.2431.

42. Taguchi, A.; Yu, J.; Verma, P.; Kawata, S. Optical antennas with multiple plasmonic nanoparticles for tip-enhanced Raman microscopy. *Nanoscale*, 2015, *7*(41). https://doi.org/10.1039/C5NR05022G.

43. Vasconcelos, T. L.; Archanjo, B. S.; Oliveira, B. S.; Valaski, R.; Cordeiro, R. C.; Medeiros, H. G.; Rabelo, C.; Ribeiro, A.; Ercius, P.; Achete, C. A.; Jorio, A.; Cançado, L. G. Plasmon-tunable tip pyramids: Monopole nanoantennas for near-field scanning optical microscopy. *Advanced Optical Materials*, 2018, *6*(20). https://doi.org/10.1002/adom.201800528.

44. Miranda, H.; Rabelo, C.; Vasconcelos, T. L.; Cançado, L. G.; Jorio, A. Optical properties of plasmon-tunable tip pyramids for tip-enhanced Raman spectroscopy. *Physica Status Solidi (RRL) – Rapid Research Letters*, 2020, *14*(9). https://doi.org/10.1002/pssr.202000212.

45. Lindquist, N. C.; Nagpal, P.; Lesuffleur, A.; Norris, D. J.; Oh, S.-H. Three-dimensional plasmonic nano-focusing. *Nano Letters*, 2010, *10*(4). https://doi.org/10.1021/nl904294u.

46. Gadelha, A. C.; Ohlberg, D. A. A.; Rabelo, C.; Neto, E. G. S.; Vasconcelos, T. L.; Campos, J. L.; Lemos, J. S.; Ornelas, V.; Miranda, D.; Nadas, R.; Santana, F. C.; Watanabe, K.; Taniguchi, T.; van Troeye, B.; Lamparski, M.; Meunier, V.; Nguyen, V.-H.; Paszko, D.; Charlier, J.-C.; Campos, L. C.; Cançado, L. G.; Medeiros-Ribeiro, G.; Jorio, A. Localization of lattice dynamics in low-angle twisted bilayer graphene. *Nature*, 2021, *590*(7846). https://doi.org/10.1038/s41586-021-03252-5.

47. Hartschuh, A.; Sánchez, E. J.; Xie, X. S.; Novotny, L. High-resolution near-field Raman microscopy of single-walled carbon nanotubes. *Physical Review Letters*, 2003, *90*(9). https://doi.org/10.1103/PhysRevLett.90.095503.

48. Saunin, S. A.; Krayev, A. V.; Zhishimontov, V. V.; Gavrilyuk, V. V.; Grigorov, L. N.; Belyaev, A. V.; Evplov, D. A. Systems and devices for non-destructive surface chemical analysis of samples. US 9,989,556 B2, 2018.

49. Wang, C.-F.; O'Callahan, B. T.; Kurouski, D.; Krayev, A.; Schultz, Z. D.; El-Khoury, P. Z. Suppressing molecular charging, nanochemistry, and optical rectification in the tip-enhanced Raman geometry. *The Journal of Physical Chemistry Letters*, 2020, *11*(15). https://doi.org/10.1021/acs.jpclett.0c01413.

6.3 DEVELOPMENT OF TERS INSTRUMENTATION: PUSHING THE LIMITS AND EXPANDING THE APPLICATIONS

6.3.1 Atomic Force Microscopy (AFM) Based TERS Study

In this chapter, we'll review design peculiarities of the TERS systems based on standard AFM cantilevers and the results obtained with their help.

Of three major SPM feedback control methodologies, STM, tuning forks, and the AFM cantilever deflection, the latter proved to be the most universal and easy to use. Also, it should be noted that the tip-sample interaction with AFM cantilevers used as the probe is the most gentle. Resolution-wise, a decent AFM operating in ambient condition can *easily* detect monoatomic steps in graphite (about 0.3 nm) in Z direction with molecular resolution in XY when features as small as 1 nm or even less can be resolved (Figure 6.20).

Even though the ambient AFM resolution can not be as good as the true atomic resolution achieved in UHV with STM or tuning-fork-based AFM imaging [1], it's good enough for the majority of practical applications. Let's start with the discussion of a TERS setup based on the AFM cantilevers and peculiarities of its operation.

In practically all the modern AFMs the system that detects the deflection of the cantilever is based on the optical lever when a feedback laser operating at 5–10 mW is focused on the backside of the cantilever. The beam reflected from the cantilever is directed onto a four-section photodiode. Vertical or torsional bending of the cantilever results in a misbalance between the top and bottom or left and right sections correspondingly (Figure 6.21), thus providing the signal based on which a certain tip-sample interaction can be maintained.

For efficient TERS operation, it's important that the feedback laser does not interfere with Raman measurements and does not alter the sample through photoexcitation of charge carriers. This requirement inevitably leads to the necessity to move the wavelength of the feedback laser out of visible range to NIR. Which in its turn results in the necessity to have fully automated feedback laser-to-cantilever alignment since the light with a wavelength over 1 micron is invisible to the naked eye or even silicon-based video CCDs. Moreover, for the sake of convenient re-alignment of the, TERS probes with the Raman laser it's important that the relative position of the focus of Raman and the feedback lasers stays the same, meaning that it's the AFM cantilever that has to be

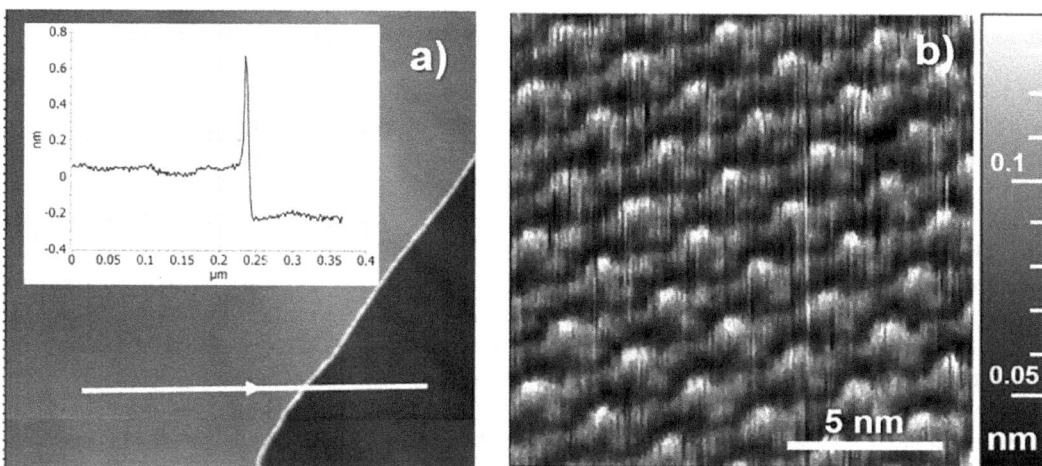

FIGURE 6.20 (a) Topography image of HOPG showing a monoatomic step of about 0.3 nm as can be seen from the section curve in the insert. (b) Topography image showing individual molecules of cholesteryl stearate self-assembled on HOPG. Image was collected in intermittent contact in ambient conditions with conventional HQ NSC-14-AL/BS probe (Mikromasch). Both images courtesy of HORIBA Scientific.

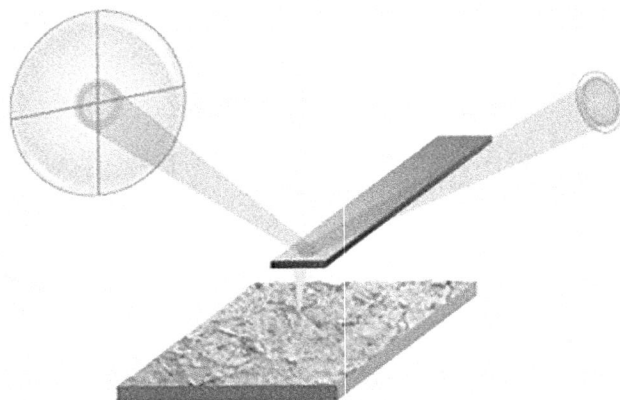

FIGURE 6.21 Illustration of the optical lever detection system in AFMs. Feedback laser is being focused on the back side of the AFM cantilever. Reflected beam is directed to the four-section photodiode. Misbalance between the top/bottom or left/right sections represents vertical or torsional bending of the cantilever.

moved relative to the fixed feedback laser, rather than the other way around, as is still done in practically all the stand-alone AFMs not designed specifically for TERS operation.

It's important to keep the feedback and the Raman optical channels completely decoupled, meaning that no objectives are supposed to be shared by these two systems. Such decoupling improves the noise floor of the AFM feedback system since the feedback laser does not have to pass through heavy objective which is susceptible to low-frequency vibrations. For the Raman/TERS side, complete decoupling with the feedback line means that there is no need to use beamsplitters in the registration channels or sacrifice a portion of the Raman range to allow the feedback laser, which improves the collection efficiency and spectral versatility of TERS experiments.

Precise long-term alignment of the focus of the Raman laser and the apex of the TERS probe is another crucial condition for efficient TERS measurements. In order to maintain such alignment, the tip has to be stationary, which automatically means that the scanning in XY and Z should be performed on the sample side, though a shallow, 10–20 nm, Z displacement of the tip within the focus of Raman laser is acceptable. There are several approaches to the nanometer-precise alignment of the focus of the Raman laser and the apex of a TERS probe:

6.3.1.1 Dual Scan AFMs

In AFM setups where there exists the possibility to scan by both the tip and the sample, precise positioning of the apex of the TERS probe within the focus of Raman laser can be achieved by moving the tip scanner to the desired position [2,3]. Such an approach has several limitations: the tip XYZ scanner must be a closed-loop scanner, meaning that inevitable creeps of the piezo actuators, which occur even when the voltage applied to them remains constant, are being constantly measured with some internal sensors and properly compensated in real time. Open loop scanners by design can't maintain the desired position of the tip for prolonged time. Operation of a closed-loop tip scanner inevitably decreases the XY noise characteristic of the overall AFM instrument. In addition to these physical factors, adding a fully capable additional scanner in the AFM system is fairly expensive. Such tip-sample alignment mechanism has not become very popular in the TERS community and is not currently being used in the overwhelming majority of the commercial TERS instruments.

6.3.1.2 Steering Mirror

In alternative approaches, it's the Raman laser focus which is being precisely positioned relative to the tip. Introduction of a steering mirror attached to a closed-loop scanner can address precise

XY positioning of the laser focus within the field of view of corresponding objective, though the displacement of the focus along the optical axis of the objective has to be done separately, which complicates the alignment procedure (Figure 6.22).

Also, it's important to note that significant lengths of the optical path between the steering mirror and the TERS probe's apex limit the stability of the alignment, being susceptible to thermal drifts in the optical setup.

6.3.1.3 Objective Scanner

The most appropriate currently available mechanism of precise and long-term alignment is based on a mechanically stiff closed-loop objective scanner which moves the objective itself, the very last optical element in front of the TERS probe, with nanometer-scale precision in X, Y, and Z [4]. This approach towards the TERS tip-to-Raman laser alignment is gaining the popularity with multiple TERS and TEPL papers being published each year [5–17]. The compact design of such an objective scanner allows concurrent use of multiple units in the same TERS setup which makes possible sophisticated experiments with split excitation and collection optical channels [18].

6.3.1.3.1 TERS Imaging in AFM-Based Setups

As was noted already in the previous chapter, intermittent contact also known as tapping mode has become the standard operating technique for cantilever-based AFMs. It allows imaging of fairly rough samples with topography of up to few hundreds of nanometers or even a few microns without damaging the tip or the sample, which is almost impossible to achieve with STM or tuning fork feedback. Additional information provided by AFM like the surface potential, capacitance, friction, etc can come extremely helpful for characterization of the real-world samples when certain degree of inhomogeneity is almost inevitably present either by design or through the defects of various natures. Therefore, when a TERS probe is based on conventional AFM cantilever, it can provide comprehensive cross-correlated information about the sample.

TERS imaging methodology with AFM cantilever-based TERS probes has been described in detail in the previous chapter, so here will address in more detail the experimental results paying particular attention to the spatial resolution of AFM-based TERS measurements, and cross-correlation of TERS data with other SPM channels provided with the same cantilever.

Unfortunately, currently (late 2021) still there are no commercially available AFM cantilever – based TERS probes that would demonstrate consistent batch-to-batch enhancement in intermittent contact, though such performance may happen in particularly successful batches (Figure 6.23).

In most cases in order to get reproducible TERS imaging results, it's necessary to collect TERS data either in direct contact or in one of specialized TERS imaging modes where "TERS" cycles with direct contact between the tip and the sample are alternating with cycles when the tip is in intermittent contact, like for example SpecTop mode (HORIBA Scientific). Spatial resolution that can be achieved with AFM cantilever-based TERSprobes depends on both the tip and the sample.

FIGURE 6.22 Optical scheme of possible implementation of the XY positioning of the Raman laser focus using a steering mirror. In such an optical scheme a separate control of the focus of Raman laser (Z axis control) is required. Image courtesy of Vassili Gavrilyuk, HORIBA Scientific.

Asperities on the TERS probes, even at the near-atomic level, can enhance the spatial resolution of TERS images [19,20], which according to calculations should enable sub-nm XY spatial resolution in TERS imaging in gap mode under other favorable conditions.

Since TERS effect can be considered as a manifestation of local optical electric fields [21–23], favorable geometry of the sample, like for example ribs of monocrystalline silver nanocubes functionalized with Raman active molecules, may additionally enhance the spatial resolution in TERS images, even collected with non-ideal TERS probes, to the level of 2–3 nm provided the pixel size of the map is small enough [24] (panel e) in Figure 6.24.

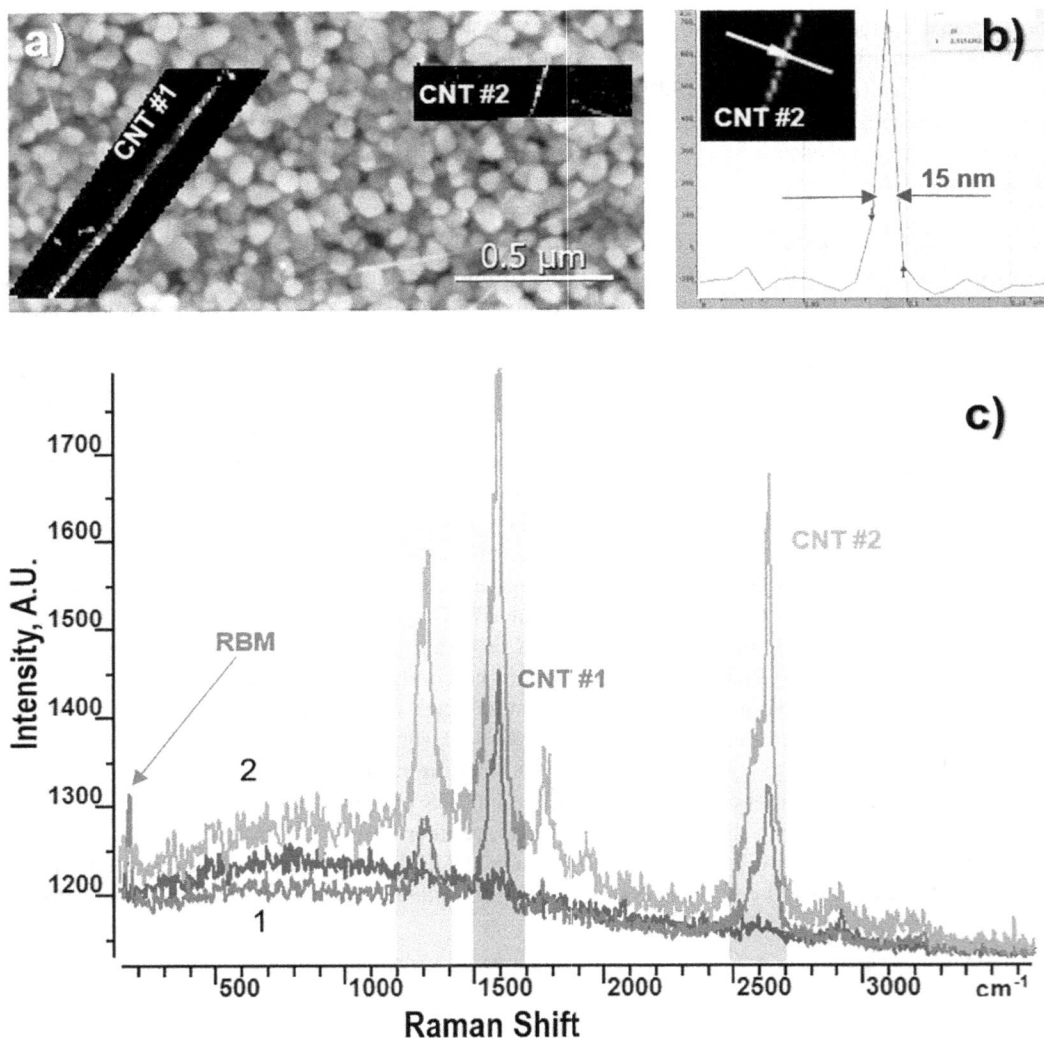

FIGURE 6.23 (a) Combined TERS map (intensity of correspondingly colored bands) of CNTs on gold overlaid on top of the topography image. TERS map was collected in intermittent contact with the amplitude of the TERS cantilever (OMNI-TERS-NC-Au, Applied Nanostructures Inc.) oscillation of about 10 nm. 100 ms/pixel integration time. (b) Section curve of TERS image of CNT#2 showing the apparent width of about 15 nm, which was to certain extent limited by the pixel size of 7 nm/pixel. (c) TERS spectra averaged over the CNTs#1 and #2, red (1) and cyan (2) spectra correspondingly, and the background signal averaged over the gold area immediately adjacent to the CNT#1. Please note a prominent RBM band in CNT#1 which is absent in TERS spectra of CNT#2 even though the overall signal from this CNT was stronger. Images courtesy of HORIBA Scientific.

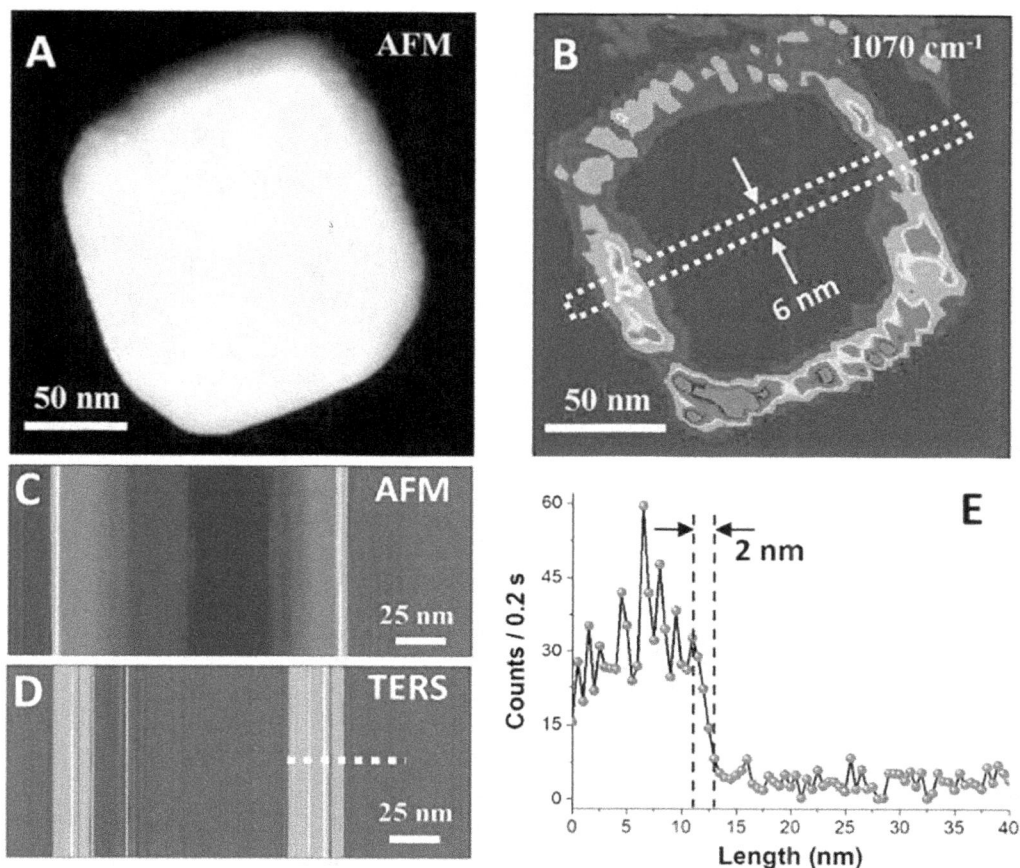

FIGURE 6.24 (a) AFM image of a silver nanocube functionalized with 4-mercaptobenzonitrile. (b) TERS image (intensity of the 1,070 cm^{-1} band) of the same nanocube showing strongly increased response from the ribs where the optical electric field is the highest. (c) High-resolution AFM topography. (d) TERS images of the area within white dotted rectangle in (b). (e) TERS profile along the white dotted line shown in panel (d). Adopted with permission from Ref. [24].

Similar spatial resolution has been achieved in the AFM cantilever-based TERS imaging of domains in CNTs, even though apparent width of the CNT was about 8 nm, which by itself was still a fairly decent result, the borderline between adjacent domains along the tube with drastically different ratios between the D and G bands was pixel-limited at about 1.4 nm. This spatial resolution was even more impressive taking into account that it was obtained in out-of-lab conditions in the exhibition hall of SPIE conference in San Diego (Figure 6.25).

One of the most important features of the AFM cantilever TERS probes-based experiments is the capability to collect multiple AFM channels like topography, phase shift, surface potential, photocurrent, etc. with the same TERS probe and then cross-correlate these data with TERS imaging results. This capability proved to be particularly useful for characterization of the nanoscale heterogeneities in 2D semiconductors.

In the course of investigation of a mono-to a few-layer-thick crystals of WSe$_2$ exfoliated to gold or silver, SKM characterization of these crystals showed noticeable variations of the surface potential, which was not totally surprising [8]. What was more surprising and rather unexpected, TERS imaging of these crystals showed that there were two types of Raman response, resonant, as expected with 638 nm excitation used in the experiments, and non-resonant, while the TERS

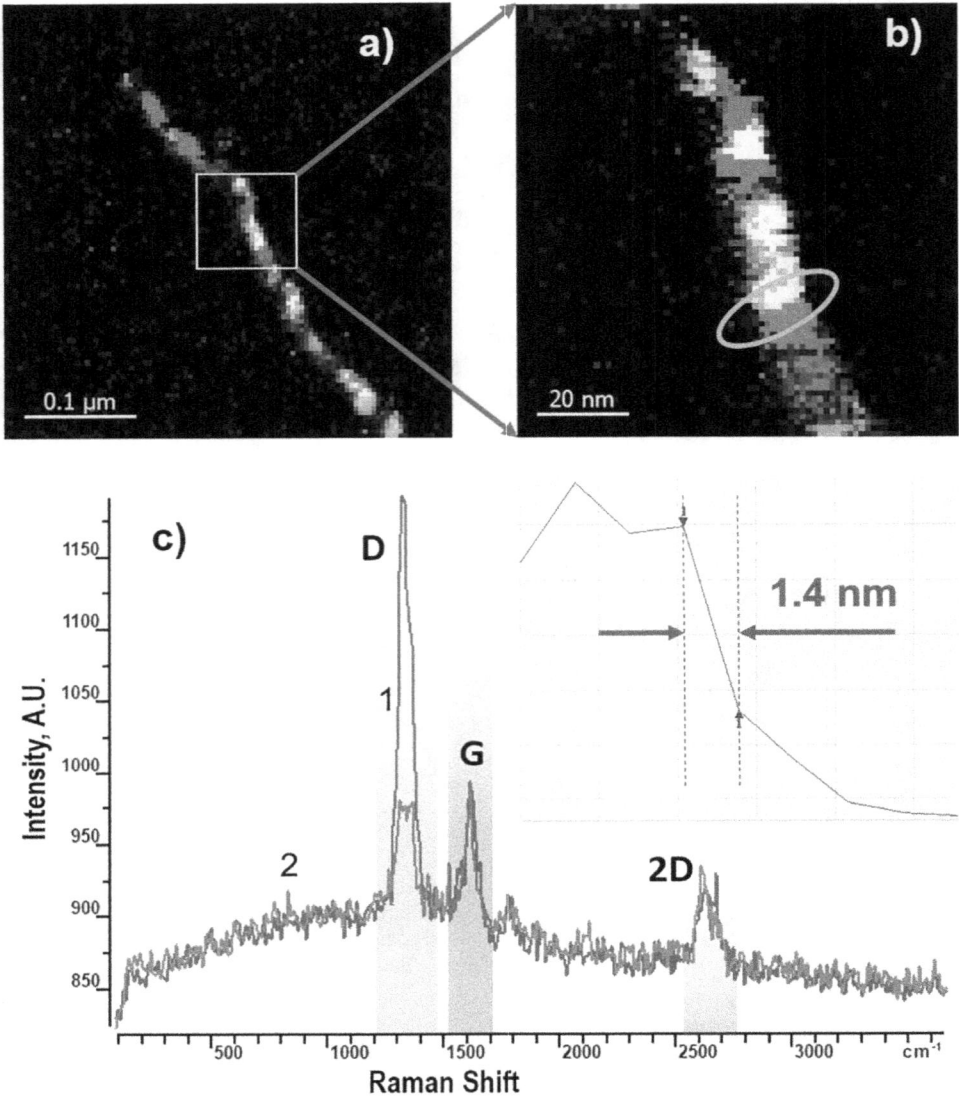

FIGURE 6.25 (a) and (b) Combined TERS maps (intensities of the correspondingly colored bands) of the same CNT on gold collected in SpecTop™ TERS imaging mode with 4 nm/pixel (a) and 1.3 nm/pixel (b) resolution. (c) TERS spectra averaged over the single pixel wide line inside (blue spectrum, 1) and immediately outside (red spectrum, 2) the high *D* band intensity domain borderline highlighted with the cyan circle in (b). In the insert – a section of the *D* band intensity map across the domain borderline showing strong drop at the distance of 1.4 nm. Images courtesy of HORIBA Scientific.

map based on the intensity of the peaks at around 250 cm^{-1} correlated surprisingly well with the distribution of the surface potential, but it was not directly related to the thickness of the crystals (Figure 6.26).

Across-the-crystal (along *Z* axis) photocurrent maps recorded in the course of the TERS imaging of these crystals showed that areas with higher surface potential and resonant TERS response generated photocurrent that was the opposite polarity of the photocurrent generated in areas with lower surface potential and non-resonant TERS response. This triple cross-correlation of the SKM,

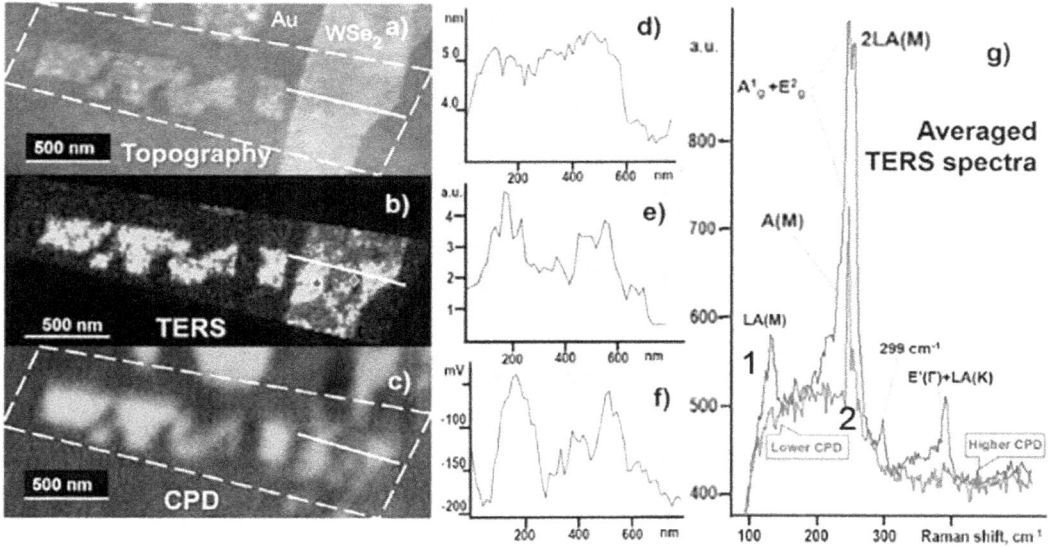

FIGURE 6.26 WSe$_2$ exfoliated to template-stripped gold. Topography (a), TERS map (intensity of complex TERS peak at around 250 cm^{-1}) (b), corresponding surface potential map (c). Dashed line in (a) and (c) marks the area where TERS mapping was performed. Section analysis of the maps in (a)–(c), correspondingly (d)–(f), shows that TERS map correlates well with the distribution of the surface potential, while the thickness of the flakes remains almost constant. Averaged spectra from location with high (blue spectrum, 1) and low (red spectrum, 2) surface potential (g). Used with permission from Ref. [8].

TERS, and photocurrent maps allowed the authors to come to the conclusion that in exfoliated crystals of WSe$_2$ there were nanoscale (few tens to few hundreds of nanometers) domains with different doping (Figure 6.27).

Another example of the power of cross-correlation of the SKM and TERS imaging is a study of WS$_2$–WSe$_2$ vertical/lateral heterostructures. In this set of experiments WSe$_2$ layer was CVD grown on top of the mono- to a few-layer CVD-grown WS$_2$ crystals on Si/SiO$_2$. In order to improve the quality of the TERS and SKM imaging, crystals were transferred from the growth substrate with the help of thin (<100 nm) gold layer evaporated on top, following the dry transfer procedure published earlier [11]. Ideally, such transfer procedure leads to extremely smooth sample surface which is de facto a replica of the Si/SiO$_2$ wafer surface used for the crystal growth, and therefore a mere topography image of the transferred sample is of little help for identification of the components. But, cross-correlating the results of the SKM and TERS imaging allowed to conclude that the outer edges of the WS$_2$ monolayer crystals or the bilayer WS$_2$ islands that occasionally grew in the middle of the WS$_2$ monolayers, served as the seeding location for the growth of the WSe$_2$ layer, which in the latter case formed peculiar lateral-vertical heterostructure (Figure 6.28).

The presence of such a heterostructure was evident in the SKM (contact potential difference, CPD) image, but its chemical nature was clearly confirmed by TERS imaging with 785 nm excitation. The monolayer WS$_2$ showed a prominent 410 cm^{-1} A′ peak which was about 6 cm^{-1} red-shifted relative to its normal position, most probably due to the strong interaction with gold. The bilayer WS$_2$ showed the presence of both the red-shifted (410 cm^{-1}) and more regular (416 cm^{-1}) A$_{1g}$ peaks, as should be expected as only one layer was experiencing intimate connection with gold substrate. TERS spectra of the WS$_2$–WSe$_2$ vertical/lateral heterobilayer showed the features of both components – strong peak at around 250 cm^{-1} characteristic for WSe$_2$ and A$_{1g}$ peak at 416 cm^{-1} typical for WS$_2$.

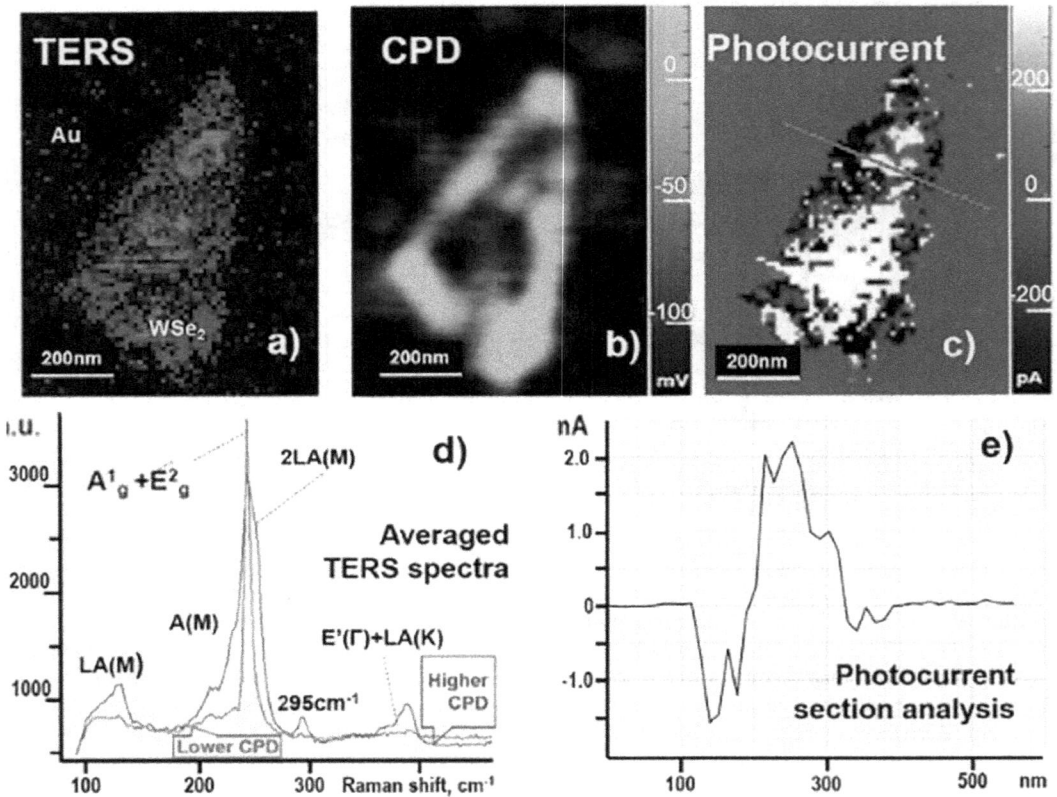

FIGURE 6.27 WSe$_2$ exfoliated to template-stripped gold. Combined TERS map (intensity of complex resonant $E'(\Gamma)+L'K$) peak at around 390 cm^{-1}, (blue) and non-resonant peak at around 250 cm^{-1} A$_{1g}$+E$_{2g}$, red) (a); corresponding surface potential map (b), photocurrent map measured at zero bias voltage simultaneously with TERS measurements (c). Averaged TERS spectra taken in areas with resonant (blue) and non-resonant response (red) (d); section analysis of the photocurrent map showing that the photocurrent generated by the domains with lower and higher surface potential was of similar value, but of opposite direction, which implies the semiconducting properties of both types of these domains, but with the charge carriers of opposite sign. Used with permission from Ref. [8].

Since AFM cantilever-based TERS probes are considerably softer compared to stiff STM or tuning fork-based TERS probes, they are particularly well suited for characterization of soft and biological materials.

As an illustration of this aspect, we'll discuss the results of the TERS study of TNT-RDX nanoparticles [25] already mentioned in Chapter 6.1. Such particles are used as precursors for manufacturing of nanodiamonds with the diameter smaller than 5 nm through controlled detonation [26,27]. TNT-RDX particles in this study were obtained by flash spray evaporation when the mixture of these chemicals in a common solvent was sprayed on the substrate resulting in the formation of the particles in approximately 40–115 nm diameter range. Such sphere-like particles can be readily imaged using AFM cantilever-based TERS probes, but are impossible to image using STM feedback due to the lack of electric conductivity and are extremely challenging for the tuning fork-based feedback SPM imaging because of their significant height and relatively weak attachment to the substrate. TERS imaging of these particles which were naturally sensitive to the chemical compound located at the surface of the particles clearly showed the signature of TNT in overwhelming majority of analyzed particles, which suggested their RDX core-TNT shell structure. In control sample which was obtained by physical mixture of pre-fabricated TNT – or RDX-only

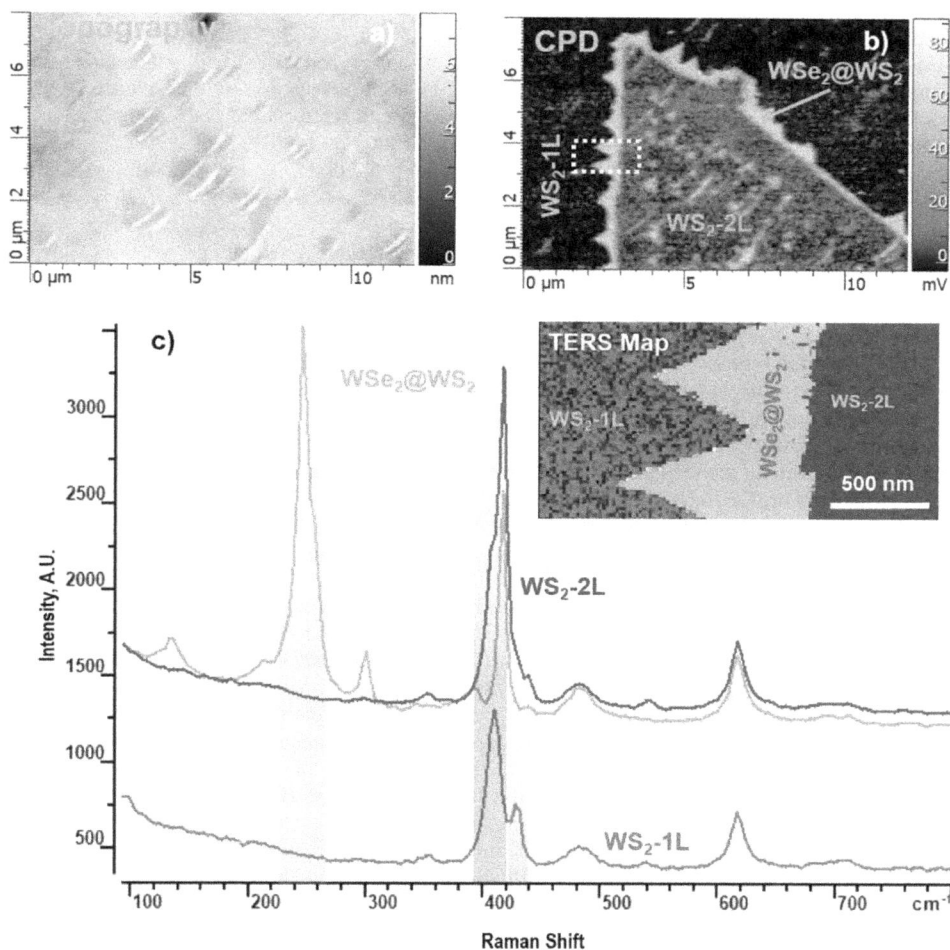

FIGURE 6.28 (a) AFM topography image of the vertical–lateral WS_2–WSe_2 heterostructure after gold-assisted transfer. The surface was very smooth with the exception of several defects probably induced in the course of stripping the crystals from the growth substrate. (b) Corresponding CPD image clearly showing three different materials within the scan area. (c) Averaged TERS spectra collected with 785 nm excitation from the monolayer WS_2 (red spectrum), bilayer WS_2 (blue spectrum), and WS_2–WSe_2 vertical heterobilayer (cyan spectrum). In the insert – combined TERS map of the area within white dotted rectangle in panel (b), showing the distribution of the intensity of the ca 430 cm^{-1} band (red color), peak position within 390–420 cm^{-1} range (blue color) and the intensity of the ca 250 cm^{-1} peak characteristic for WSe_2 (green color). Sample courtesy of Prof. Peng Chen (Southern University of Science and Technology, China) and Prof. Xiangfeng Duan (UCLA), images courtesy of HORIBA Scientific.

nanoparticles, TERS imaging results was evenly split between the Raman signatures of the TNT and RDX.

More examples of the AFM-cantilever-based TERS probe characterization of inorganic and organic or biological materials will be discussed in corresponding dedicated chapters later in the book. It will be safe to state, summarizing this section, that the AFM cantilever-based TERS characterization is currently the most user-friendly and versatile in terms of the types of the samples it can address. Despite the fact that the best spatial resolution of TERS imaging demonstrated with this version of TERS instrumentation is not as high as the one obtained with STM-TERS probes under UHV conditions, it's still sufficient for many practical applications.

REFERENCES

1. Gross, L.; Mohn, F.; Moll, N.; Liljeroth, P.; Meyer, G. The chemical structure of a molecule resolved by atomic force microscopy. *Science*, (1979) 2009, *325*(5944). https://doi.org/10.1126/science.1176210.
2. Rasmussen, A.; Deckert, V. Surface- and tip-enhanced raman scattering of DNA components. *Journal of Raman Spectroscopy*, 2006, *37*(1–3). https://doi.org/10.1002/jrs.1480.
3. Owen, R. J.; Heyes, C. D.; Knebel, D.; Röcker, C.; Nienhaus, G. U. An integrated instrumental setup for the combination of atomic force microscopy with optical spectroscopy. *Biopolymers*, 2006, *82*(4). https://doi.org/10.1002/bip.20414.
4. Nicklaus, M.; Nauenheim, C.; Krayev, A.; Gavrilyuk, V.; Belyaev, A.; Ruediger, A. Note: Tip enhanced Raman spectroscopy with objective scanner on opaque samples. *Review of Scientific Instruments*, 2012, *83*(6). https://doi.org/10.1063/1.4725528.
5. Darlington, T. P.; Krayev, A.; Venkatesh, V.; Saxena, R.; Kysar, J. W.; Borys, N. J.; Jariwala, D.; Schuck, P. J. Facile and quantitative estimation of strain in nanobubbles with arbitrary symmetry in 2D semi-conductors verified using hyperspectral nano-optical imaging. *The Journal of Chemical Physics*, 2020, *153*(2). https://doi.org/10.1063/5.0012817.
6. Bhattarai, A.; Krayev, A.; Temiryazev, A.; Evplov, D.; Crampton, K. T.; Hess, W. P.; El-Khoury, P. Z. Tip-enhanced Raman scattering from nanopatterned graphene and graphene oxide. *Nano Letters*, 2018, *18*(6). https://doi.org/10.1021/acs.nanolett.8b01690.
7. Zhang, Y.; Voronine, D. V.; Qiu, S.; Sinyukov, A. M.; Hamilton, M.; Liege, Z.; Sokolov, A. V.; Zhang, Z.; Scully, M. O. Improving resolution in quantum subnanometre-gap tip-enhanced Raman nanoimaging. *Scientific Reports*, 2016, *6*(1). https://doi.org/10.1038/srep25788.
8. Jariwala, D.; Krayev, A.; Wong, J.; Robinson, A. E.; Sherrott, M. C.; Wang, S.; Liu, G.-Y.; Terrones, M.; Atwater, H. A. Nanoscale doping heterogeneity in few-layer WSe$_2$ exfoliated onto noble metals revealed by correlated SPM and TERS imaging. *2D Materials*, 2018, *5*(3). https://doi.org/10.1088/2053-1583/aab7bc.
9. Darlington, T. P.; Carmesin, C.; Florian, M.; Yanev, E.; Ajayi, O.; Ardelean, J.; Rhodes, D. A.; Ghiotto, A.; Krayev, A.; Watanabe, K.; Taniguchi, T.; Kysar, J. W.; Pasupathy, A. N.; Hone, J. C.; Jahnke, F.; Borys, N. J.; Schuck, P. J. Imaging strain-localized excitons in nanoscale bubbles of monolayer WSe$_2$ at room temperature. *Nature Nanotechnology*, 2020, *15*(10). https://doi.org/10.1038/s41565-020-0730-5.
10. Smithe, K. K. H.; Krayev, A. V.; Bailey, C. S.; Lee, H. R.; Yalon, E.; Aslan, Ö. B.; Muñoz Rojo, M.; Krylyuk, S.; Taheri, P.; Davydov, A. V.; Heinz, T. F.; Pop, E. Nanoscale heterogeneities in monolayer MoSe$_2$ revealed by correlated scanning probe microscopy and tip-enhanced raman spectroscopy. *ACS Applied Nano Materials*, 2018, *1*(2). https://doi.org/10.1021/acsanm.7b00083.
11. Krayev, A.; Bailey, C. S.; Jo, K.; Wang, S.; Singh, A.; Darlington, T.; Liu, G.-Y.; Gradecak, S.; Schuck, P. J.; Pop, E.; Jariwala, D. Dry transfer of van der Waals crystals to noble metal surfaces to enable characterization of buried interfaces. *ACS Applied Materials & Interfaces*, 2019, *11*(41). https://doi.org/10.1021/acsami.9b09798.
12. Tang, C.; He, Z.; Chen, W.; Jia, S.; Lou, J.; Voronine, D. V. Quantum plasmonic hot-electron injection in lateral WSe$_2$–MoSe$_2$ heterostructures. *Physical Review B*, 2018, *98*(4). https://doi.org/10.1103/PhysRevB.98.041402.
13. Su, W.; Kumar, N.; Krayev, A.; Chaigneau, M. In situ topographical chemical and electrical imaging of carboxyl graphene oxide at the nanoscale. *Nature Communications*, 2018, *9*(1). https://doi.org/10.1038/s41467-018-05307-0.
14. Rahaman, M.; Rodriguez, R. D.; Plechinger, G.; Moras, S.; Schüller, C.; Korn, T.; Zahn, D. R. T. Highly localized strain in a MoS$_2$/Au heterostructure revealed by tip-enhanced Raman spectroscopy. *Nano Letters*, 2017, *17*(10). https://doi.org/10.1021/acs.nanolett.7b02322.
15. Sheremet, E.; Rodriguez, R. D.; Agapov, A. L.; Sokolov, A. P.; Hietschold, M.; Zahn, D. R. T. Nanoscale imaging and identification of a four-component carbon sample. *Carbon N Y*, 2016, *96*, 588–593. https://doi.org/10.1016/j.carbon.2015.09.104.
16. Velický, M.; Rodriguez, A.; Bouša, M.; Krayev, A. V.; Vondráček, M.; Honolka, J.; Ahmadi, M.; Donnelly, G. E.; Huang, F.; Abruña, H. D.; Novoselov, K. S.; Frank, O. Strain and charge doping finger-prints of the strong interaction between monolayer MoS$_2$ and gold. *The Journal of Physical Chemistry Letters*, 2020, *11*(15). https://doi.org/10.1021/acs.jpclett.0c01287.
17. Su, W.; Esfandiar, A.; Lancry, O.; Shao, J.; Kumar, N.; Chaigneau, M. Visualising structural modification of patterned graphene nanoribbons using tip-enhanced Raman spectroscopy. *Chemical Communications*, 2021, *57*(56). https://doi.org/10.1039/D1CC01769A.
18. Wang, C.-F.; El-Khoury, P. Z. Imaging plasmons with sub-2 nm spatial resolution via tip-enhanced four-wave mixing. *The Journal of Physical Chemistry Letters*, 2021, *12*(14). https://doi.org/10.1021/acs.jpclett.1c00763.

19. Trautmann, S.; Aizpurua, J.; Götz, I.; Undisz, A.; Dellith, J.; Schneidewind, H.; Rettenmayr, M.; Deckert, V. A Classical description of subnanometer resolution by atomic features in metallic structures. *Nanoscale*, 2017, *9*(1). https://doi.org/10.1039/C6NR07560F.

20. Bhattarai, A.; Crampton, K. T.; Joly, A. G.; Wang, C.-F.; Schultz, Z. D.; El-Khoury, P. Z. A closer look at corrugated Au tips. *The Journal of Physical Chemistry Letters*, 2020, *11*(5). https://doi.org/10.1021/acs.jpclett.0c00305.

21. Bhattarai, A.; Crampton, K. T.; Joly, A. G.; Kovarik, L.; Hess, W. P.; El-Khoury, P. Z. Imaging the optical fields of functionalized silver nanowires through molecular TERS. *The Journal of Physical Chemistry Letters*, 2018, *9*(24). https://doi.org/10.1021/acs.jpclett.8b03324.

22. El-Khoury, P. Z.; Aprà, E. Spatially resolved mapping of three-dimensional molecular orientations with ~2 nm spatial resolution through tip-enhanced Raman scattering. *The Journal of Physical Chemistry C*, 2020, *124*(31). https://doi.org/10.1021/acs.jpcc.0c04263.

23. El-Khoury, P. Z.; Schultz, Z. D. From SERS to TERS and beyond: Molecules as probes of nanoscopic optical fields. *The Journal of Physical Chemistry C*, 2020, *124*(50). https://doi.org/10.1021/acs.jpcc.0c08337.

24. Bhattarai, A.; Novikova, I. V.; El-Khoury, P. Z. Tip-enhanced Raman nanographs of plasmonic silver nanoparticles. *The Journal of Physical Chemistry C*, 2019, *123*(45). https://doi.org/10.1021/acs.jpcc.9b07811.

25. Deckert-Gaudig, T.; Pichot, V.; Spitzer, D.; Deckert, V. High-resolution Raman spectroscopy for the nanostructural characterization of explosive nanodiamond precursors. *ChemPhysChem*, 2017, *18*(2). https://doi.org/10.1002/cphc.201601276.

26. Pichot, V.; Risse, B.; Schnell, F.; Mory, J.; Spitzer, D. Understanding ultrafine nanodiamond formation using nanostructured explosives. *Scientific Reports*, 2013, *3*(1). https://doi.org/10.1038/srep02159.

27. Pichot, V.; Comet, M.; Risse, B.; Spitzer, D. Detonation of nanosized explosive: New mechanistic model for nanodiamond formation. *Diamond and Related Materials*, 2015, *54*. https://doi.org/10.1016/j.diamond.2014.09.013.

6.3.2 Scanning Tunneling Microscopy (STM)-based TERS Measurements

6.3.2.1 Introduction to the Scanning Tunneling Microscope

Most microscopes are optical in nature, relying upon a series of lenses to image objects and structures too small to be seen by the naked eye. As a result, their spatial resolution is diffraction-limited according to the Abbe diffraction limit, where the resolving distance is roughly proportional to $\dfrac{\lambda}{2}$ [1] where λ is the wavelength of light used to image a sample. Efforts to circumvent this limit have successfully used a particle beam of electrons to image samples or surfaces. This is the method used in the scanning electron microscope and transmission electron microscope. Since the wavelength of an electron is inversely related to its momentum via the de Broglie equation, it becomes necessary to use a high-energy beam to achieve the highest degree of spatial resolution. This high-energy electron beam can result in significant sample damage that prevents the characterization of delicate samples, such as organic thin films or certain 2D materials [2]. On the other hand, super-resolution fluorescence microscopy provides the ability to image 3D structures at the nanometer scale with light, but critically requires the use of fluorescent tags [3]. Alternatively, another branch of microscopy, SPM, relies on a physical probe to scan a surface, providing a method to nondestructively image individual molecules and even atoms without requiring light, therefore circumventing the diffraction limit.

The inherent nature of the scanning probe microscope (SPM) implies that the resolution of these instruments is limited only by the size of the probe-sample interaction volume. In the case of an atomically sharp tip, it becomes possible to resolve individual atoms. Although there is a wide range of the SPM techniques, two primary methods are AFM and STM. As mentioned in an earlier chapter, the AFM measures the interaction force between a tip and a surface to image topography [4], providing true atomic resolution [5]. This results in a general SPM method capable of imaging a wide range of materials, including metals and insulators as well as softer materials such as biological macromolecules, tissue samples, and self-assembled monolayers (SAMs). On the other hand, the STM relies on the quantum tunneling of electrons to image surfaces [6]. The resulting tunneling current is highly sensitive and can be used to yield atomic-resolution images of metal surfaces and 2D materials.

In STM, an atomically sharp metallic probe or tip is brought extremely close to conducting or semiconducting samples. A low bias (typically $|U| < 3\,\text{V}$) is applied between the sample and the tip resulting in the flow of electrons (tunneling current). This tunneling current (I_t) is exceedingly sensitive to the distance between the sample and tip according to a relationship that can be approximated by a rectangular potential barrier, which reduces to the following proportion:

$$I_t \propto e^{-2d} \frac{\sqrt{2m(V-E)}}{\hbar}$$

where V is the height of the potential barrier, d is the tip-sample distance, and E is the energy of the state into which tunneling occurs. This relationship implies that a change in the tip-sample gap distance of ~1 changes the tunneling current by an order of magnitude [7]. As a result, STM is exceedingly sensitive to the topography of a surface. As shown in Figure 6.29a, by raster scanning the STM tip across the surface and monitoring the tunneling current at each point, a surface can be imaged with atomic resolution. In the typical mode of operation, constant current mode, the tunneling current is maintained at a setpoint by a feedback loop that relies on a proportional-integral controller to adjust the gap distance dynamically. A piezoelectric scanner is used to scan the tip across the surface and simultaneously adjust the tip-sample gap distance, both with fine spatial resolution. An STM image, as shown in Figure 6.29b, is therefore a delineation of the tip height across the scanned area, where the tip height is represented in pseudo color.

Significantly, the previously mentioned approach, although a useful means to understand the fine spatial resolution of the STM, represents an oversimplification, and does not accurately describe the physical measurement that it captures. In 1985, Tersoff and Hamann presented a theory that provides a more general treatment due to the critical inclusion of the electronic structures of the tip and sample [8]. In brief, this theory found that the tunneling current is proportional to the local density of states (LDOS) of the surface at the position of the tip, which could be considered to be a convolution of the density of states of the STM tip and sample. Therefore, rather than simple topographic maps, STM images capture a complicated mixture of topological and electronic information. The LDOS can result in surface adsorbates appearing as either protrusions or depressions, obfuscating clear chemical identification.

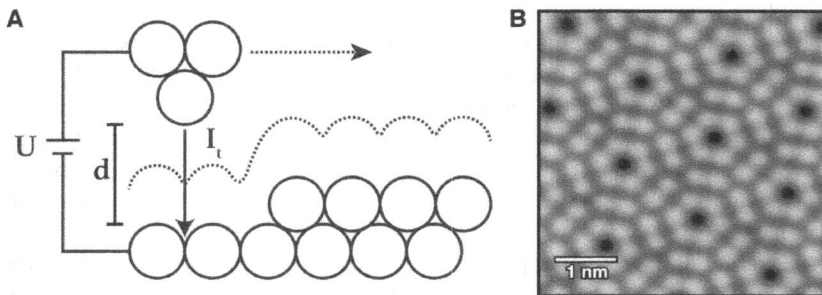

FIGURE 6.29 Schematic of the STM and a sample STM image. (a) Diagram of the principle underlying STM imaging. An applied voltage bias between the tip and sample (U) induces a tunneling current (I_t). In constant current mode, a feedback loop is used to maintain a constant tunneling current by adjusting the tip-sample gap distance (d) as the tip scans across the surface. (b) Sample STM image of the Si(111) 7×7 surface ($U = +2.0\,\text{V}$, $I_t = 250\,\text{pA}$).

Despite this potential complication, unlike any other instrument, the STM enables the visualization and manipulation of matter at the atomic scale [9]. And so, it has become the tool of choice to develop understandings of surfaces and on-surface chemistry, even at the single-molecule level [10]. The STM can even be used to induce individual steps of a chemical reaction and observe intermediates [11,12]. It is an extremely powerful tool within its own right, but it can be further developed through coupling optical spectroscopy into the tip-sample junction.

6.3.2.2   Early Efforts to Combine TERS with the STM

Edward Hutchinson Synge is commonly credited as first proposing the idea of scanning near-field optical microscopy (SNOM) circa 1928. He proposed a method to circumvent the diffraction limit of light and extend microscopic resolution into the ultra-microscopic resolution. However, he only presented the idea of the method, identified a few obvious technical problems, and then hoped that a clever experimentalist would solve the various associated difficulties. He identified two of the most obvious technical difficulties as the need for a very intense source of illumination and the ability to make very small adjustments in the position of a sample, on the order of 10^{-7} cm [13]. But it would be a number of years before new developments in equipment and instrumentation made near-field optical microscopy a viable method to perform Raman spectroscopy.

In 1985, shortly after the discovery of the STM, Wessel proposed surface-enhanced optical microscopy (SEOM), which builds upon the earlier discovery of surface-enhanced Raman spectroscopy in 1977 [14,15], to rely on a single well-defined hot spot as shown in Figure 6.30 [16]. This proposed method relied on the now widespread use of lasers for Raman spectroscopy and piezo-electric translators used in the STM to address the two most glaring technical difficulties identified by Synge significantly earlier. In SEOM, a laser beam is an incident on a submicrometer-sized plasmonic particle that is kept close to the sample surface, yielding a confined and enhanced electric field. This strong field arises from a plasmon resonance at the curved particle surface. By scanning the optical probe particle across the surface, Wessel expected the ability to obtain an excellent signal for enhanced Raman spectroscopic measurements with 5 nm spatial resolution. Significantly, Wessel recognized SEOM as closely related to STM, resembling the STM mechanically, and so it is no surprise that one of the first instances of tip-enhanced Raman spectroscopy (TERS) involved optical coupling into the STM.

Four separate groups realized what many consider the first experimental demonstrations of TERS in 2000. Three of the four first demonstrations combined an AFM with an inverted optical microscope where the probe tip was coated with a plasmonic metal [17–19]. However, one group coupled TERS with STM [20]. In this original demonstration of STM-TERS Pettinger, Picardi,

FIGURE 6.30   Proposed concept for SEOM. A laser beam is incident on an optical probe particle, which concentrates the field in a region adjacent to the sample surface. The Raman signal from the sample surface is then scattered. The surface can be scanned by moving the optically transparent sample holder with piezo-electric translators.

Schuster, and Ertl sought to extend earlier AFM-TERS studies of dye molecules, C_{60} molecules, and sulfur deposited onto glass substrates to the study of molecules at metal surfaces, modeling catalytic surfaces. Due to the necessity for a conductive substrate, they evaporated a smooth gold film of 12 nm thickness onto a 1 mm thick glass support. Critically, the substrate had to be as optically transparent as possible due to their use of an inverted microscope to introduce and collect light. Both the excitation and collected enhanced Raman signal had to pass through the substrate in this transmission mode setup. Importantly, despite the losses accrued due to this optical setup, Pettinger et al. found that the laser power had to be kept sufficiently low to avoid photo-bleaching of the dye molecules they investigated. The laser power was reduced from 5 to 0.05 mW at the sample. Ultimately, they were able to observe a strongly enhanced (500–1,200×) Raman signal of BCB molecules adsorbed on a smooth gold film.

Since these early experiments relied on inverted optical microscopes, both the excitation and Raman scattered photons had to transmit through the substrate. This required the use of transparent samples. The development of new optical setups would enable the use of opaque samples, improving the flexibility of STM-TERS to consider new types of systems. Specifically, the development of side illumination, as well as methods to use a parabolic mirror, was essential to the realization of coupling TERS with the STM, preserving the benefits of both methods to result in an exceptionally powerful tandem technique to investigate systems with chemical imaging at the nanoscale and below. Crucially, the combination with the STM includes the possibility to use TERS in nonambient conditions, such as in liquid, such as for the study of electrochemistry, or in vacuum where pristine and cryogenic conditions have resulted in new levels of spatial resolution. The development and applications of UHV STM-TERS and cryogenic UHV-STM-TERS will be the subject of the next sections. A focus will be dedicated to the improvements to the spatial resolution that arise from the pristine and ultrastable conditions.

6.3.2.3   Optimizing the STM Tip for TERS

The tip is perhaps the most important element in STM-TERS experiments and as a result, extensive work has been done to develop fabrication methods with a number of reviews available that focus on these efforts [21–27]. As a result, here we will provide a broad survey of the current methods with mentions of new or unusual advances. STM-TERS probes are typically Au [28–33] or Ag [34–39] with a sharp tip that results from an electrochemical etching process, as shown in Figure 6.31. Alternatively, plasmonic metals can be deposited onto a tip that would otherwise be plasmonically inactive [40], as can be found in AFM-TERS experiments that are performed under ambient conditions [41]. Two of the most common methods used to fabricate TERS tips are electrochemical etching and plasmonic metal deposition [22].

Electrochemically etched tips can be post-processed with Ar^+ ion sputtering or FIB in vacuum. Ar^+ ion sputtering can be used to clean a contaminated STM-TERS tip, therefore enabling the reuse and long-term preservation of the plasmonic properties of the probe [44]. It can also be used to smooth the tip, Tallarida et al. used field-directed sputter sharpening (FDSS) to fabricate silver tips that were found to exhibit nanoscale smoothness, as shown in Figures 6.31c and d [42]. A positive bias was applied to the tip during Ar^+ ion sputtering The resulting repulsive potential isolated to the localized electric field enhancement at the tip apex directs the sputtering process along the tip. This preserves and sharpens the apex, therefore, reducing the radius of curvature of the tip [45]. In contrast, Yang et al. relied on an ion beam sputtering technique to decorate a tip apex with a nanoneedle array, therefore, forming a "Tip-On-Tip" structure that significantly enhanced TERS signal of rhodamine 6G (R6G) [46].

Focused ion beam milling can be used to fabricate tips with nanoscale features providing unique methods to couple light into the tip-sample junction. Grooves and gratings in the body of the tip can manipulate the plasmonics of the tip. Experimental studies have shown that a single groove milled into the shaft of a tip can tune the STM-induced luminescence spectra

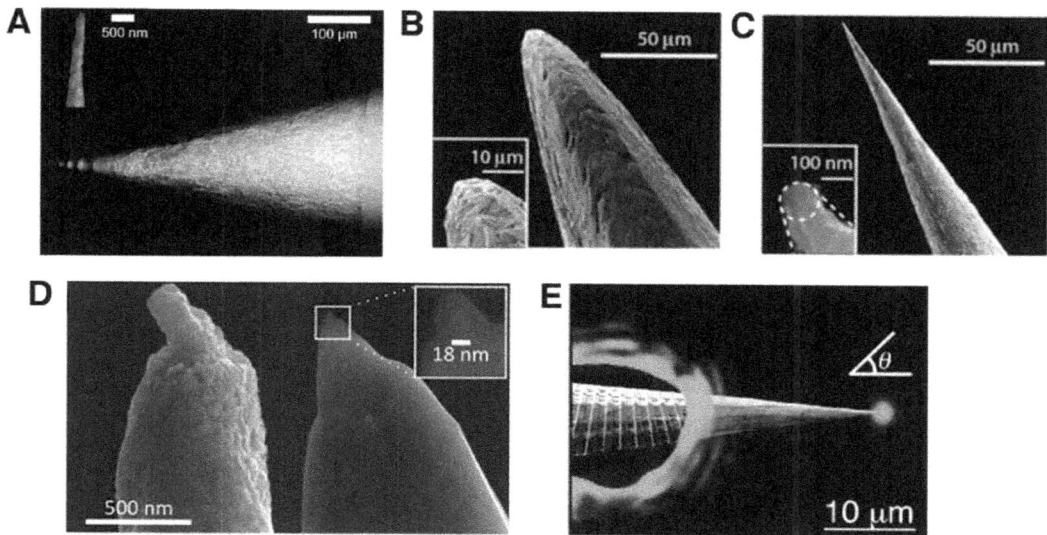

FIGURE 6.31 STM-TERS tips following fabrication and post-processing. (a) Scanning electron microscope (SEM) image of a Ag tip etched in a perchloric acid/methanol solution, with an enlarged inset SEM image. (Figure and caption reprinted from Ref. [34] with permission of AIP Publishing.) (b) and (c) Evolution of tip profiles as a result of a two-stage electrochemical process, involving an etching and subsequent polishing step. (Figure and caption reprinted from Ref. [39] with permission of AIP Publishing.) (d) SEM images of a two-stage electrochemically etched tip before and after batch processing via FDSS. (Figure and caption reprinted (adapted) under the terms of the ACS AuthorChoice License from Ref. [42], Copyright © 2017 American Chemical Society.) (e) SEM image of an electrochemically etched Au tip with a plasmonic grating fabricated using focused ion beam milling superimposed with an optical image of grating coupling and subsequent reradiation at the tip apex. (Figure and caption reprinted (adapted) with permission from Ref. [43], Copyright © 2012 American Chemical Society.)

due to interactions between localized and propagating surface plasmon modes [47]. Meanwhile, more intricate designs, such as linear nanoslit gratings, can result in a type of optical antenna. When the grating is illuminated a surface plasmon polariton results that propagate towards the tip apex, making it possible to use nonlocal excitation of the STM-TERS probe, as seen in Figure 6.31e [48,49]. Since this method avoids far-field Raman of the sample, such nonlocal excitation results in a decrease in background Raman signal during TERS measurements [43,50]. In a similar manner, a silver nanoparticle or nanostructure can be attached to the body of the tip to enable remote excitation [25,51]. Beyond enabling unique experimental setups, FIB has also been used to achieve spectral modulation through a Fabry-Pérot-type interference of surface plasmons. As shown in Figure 6.32, Böckmann et al. were able to use FIB to fabricate Au tips [47]. Figure 6.32b shows the Au tip following FIB, while the scanning tunneling microscope-induced luminescence (STML) spectra appear adjacent in Figure 6.32c. The single broad peak resulting from localized surface plasmon resonance (LSPR) excitation in the junction is in good agreement with electrodynamic simulations, with a strong dependence on the applied bias voltage as shown in Figure 6.32c [49]. Furthermore, the authors found that the STML spectra can be modulated by plasmonic Fabry-Pérot interference. The LSPR excitation in the STM junction also launches propagating surface plasmon polaritons (SPPs) on the tip shaft [52]. As a result, a single groove fabricated on the smooth tip shaft with FIB at a given distance from the tip apex, as shown in Figure 6.32d and e, was used to reflect the SPPs back to the apex. This results in coupling to the localized mode in the junction and causes spectral modulation of light emission

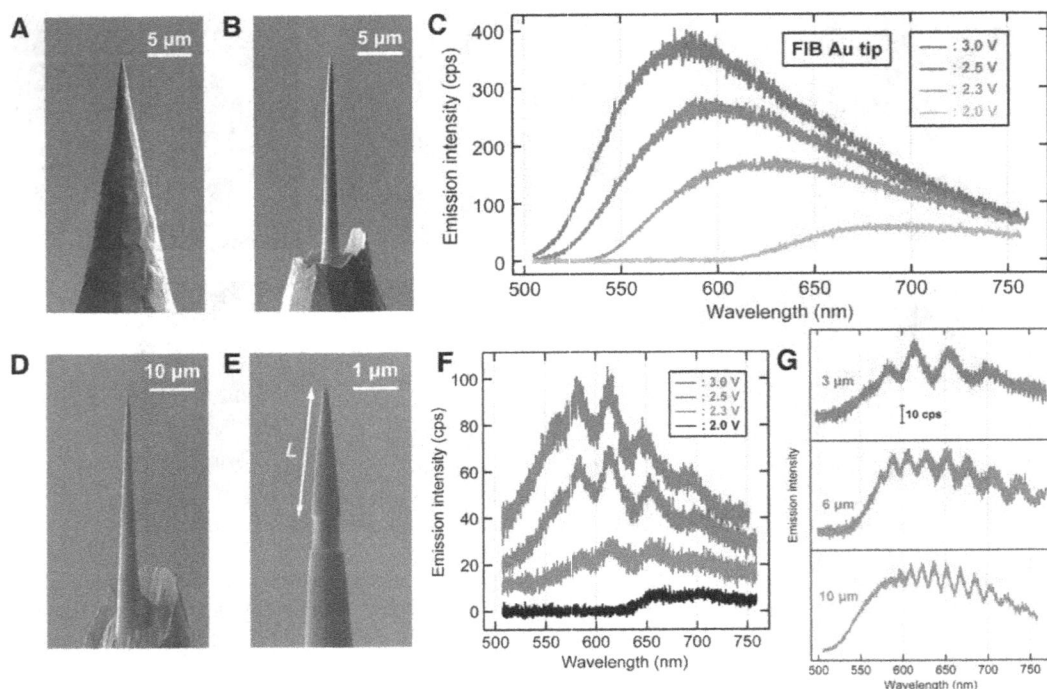

FIGURE 6.32 FIB fabrication of Au tips and STML spectra without and with a groove. (a) SEM image of an electrochemically etched Au tip. (b) SEM image of the same tip following FIB milling. (c) Bias voltage dependence of STML spectra measured over the Ag(111) surface with the FIB-tip ($I_t=9\,nA$, V_{bias} is indicated in the figure). (d) and (e) SEM images of the FIB processed Au tip with a groove located 3 µm away from the tip apex. (f) Bias voltage dependence of STML spectra measured over the Ag(111) surface with the grooved tip ($I_t=1\,nA$, V_{bias} is indicated in the figure). (g) STML spectra measured over the Ag(111) surface using Au tips with grooves located at 3, 6, and 10 µm from the tip apex ($V_{bias}=2.5\,V$, $I_t=9\,nA$). (Figure and caption reprinted (adapted) under a Creative Commons Attribution (CC-BY) License from Ref. [47], Copyright © 2019 American Chemical Society.)

from the junction. As shown in Figure 6.32f, STML spectra obtained with a grooved tip again exhibit strong voltage dependence, but also periodic oscillations are evident in the spectra due to the Fabry-Pérot-type interference on the tip shaft. By changing the distance between the groove and the tip apex, it was possible to control this spectral modulation. Such control over the plasmonic properties of the STM-TERS tip is highly desirable to the realization of TERS studies with the strongest signal and highest degree of spatial resolution. Although the current conventional methods rely on electrochemical etching to fabricate STM-TERS tips, the ongoing research that uses post-processing has demonstrated that there are potential avenues to improve the enhancement and reliability of STM-TERS probes. However, in the case of nanostructures or atomic apexes, the stability of the tip becomes increasingly important. In this manner, many of these highly sensitive fabrication methods have thus far only been successfully implemented in cryogenic UHV-STM systems, where the pristine conditions and supremely stable conditions can result in stable atomic-scale features.

Although it was first realized over 20 years ago, the experimental techniques and instrumentation necessary for STM-TERS experiments continue to develop. The development of new optical setups to couple light into the STM tip-sample junction has enabled the study of new systems, surmounting previous obstacles. Simultaneously, these ongoing improvements have pushed the spatial resolution of STM-TERS to the subnanomter regime. However, tip and sample stability become paramount for the realization of studies of single molecules with simultaneous high-resolution STM imaging and TERS. In this regard, the transition to incorporate STM-TERS into low temperatures and UHV has resulted in previously unprecedented levels of spatial resolution, including submolecular TERS images of single molecules that will be discussed later.

6.3.2.4   Single-molecule STM-TERS

Single-molecule detection and identification lie as an end goal for many analytical techniques, permitting the study of chemistry at the spatial limit [53]. This sensitivity results in the ability to interrogate highly localized chemical properties. However, single-molecule resolution with the STM is exceedingly difficult in ambient conditions, requiring alternative methods to validate single-molecule resolution. Zenobi and co-workers reported the first TERS measurements of a single molecule adsorbed on a flat gold substrate in 2007 [35]. The authors had to rely upon a number of observations to prove the single-molecule sensitivity of their measurements. They used extremely dilute samples. They reported spectroscopic fluctuations expected for the behavior of a single molecule. And they also observed discrete signal losses that suggested the decomposition of a single molecule. Contemporaneously, Raschke and co-workers reported TERS spectra of malachite green (MG) dye molecules that they described as corresponding to single molecules. Their assertion of single molecule sensitivity faced criticism that required response [54–56]. The Van Duyne group would later investigate rhodamine 6G (R6G) molecules, relying on isotopic substitution to confirm the single molecule nature of their TERS spectra. R6G–d_0 and a deuterated form, R6G–d_4, were both deposited onto a silver surface. The acquired TERS spectra showed vibrational fingerprints that correspond to only one isotopologue, with the ability to use TERS to distinguish between isotopologues based on the shift of a vibrational peak as shown in Figure 6.33. This observation was used to confirm that TERS was able to identify individual molecules [57].

Alternatively, another group attached a single molecule to the apex of the tip and monitored the line shape and peak position of different vibrational modes in TERS spectra, finding that changes in these spectral features correspond to conductance at the tip-sample junction. They referred to this method as fishing mode TERS (FM-TERS) [58]. Single-molecule sensitivity was verified based on the measured conductance of the tip-sample junction, observing attached (ON) states and detached (OFF) states. Ultimately, without the ability to observe individual molecules with STM topographic imaging, the assignment of single-molecule sensitivity becomes complicated. In contrast, when performed at low temperatures (LT) in ultrahigh vacuum, STM can routinely achieve studies of individual molecular adsorbates on a surface. In this manner, despite the technical difficulty of coupling optical spectroscopy into the LT-UHV-STM tip-sample junction, the increased stability becomes an increasingly attractive prospect. And indeed, early efforts to improve the spatial resolution of TERS relied on improving the experimental setup, specifically moving STM-TERS experiments into UHV conditions.

REFERENCES

1. Abbe, E., Beiträge zur Theorie des Mikroskops und der mikroskopischen Wahrnehmung. *Archiv für Mikroskopische Anatomie* 1873, *9*(1), 413–468.
2. Szczerbiński, J.; Gyr, L.; Kaeslin, J.; Zenobi, R., Plasmon-driven photocatalysis leads to products known from E-beam and X-ray-induced surface chemistry. *Nano Lett.* 2018, *18*(11), 6740–6749.
3. Huang, B.; Bates, M.; Zhuang, X., Super-resolution fluorescence microscopy. *Annu. Rev. Biochem* 2009, *78*(1), 993–1016.
4. Binnig, G.; Quate, C. F.; Gerber, C., Atomic force microscope. *Phys. Rev. Lett.* 1986, *56*(9), 930–933.

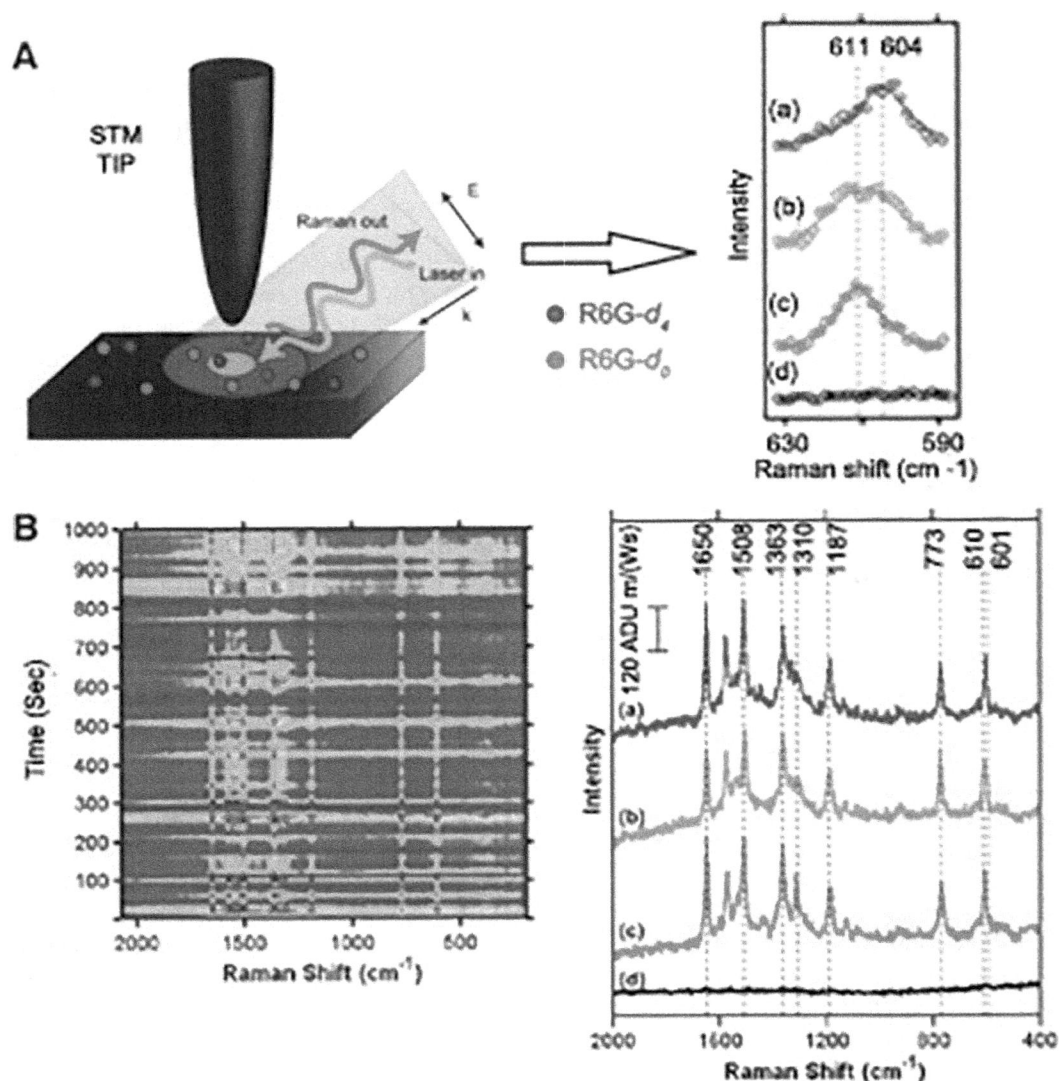

FIGURE 6.33 Single-molecule resolution TERS confirmed with isotopic substitution. (a) Diagram of the TERS isotope-based experiment used to validate the single-molecule nature of a TERS experiment (λ_{ex} = 532 nm). Two isotopologues of Rhodamine 6G were deposited onto the silver film, R6G–d_0 and R6G–d_4. The vibrational mode circa 600 cm^{-1}, which occurs at 611 cm^{-1} for R6G–d_0 and shifts to 611 cm^{-1} for R6G–d_4 allows for unambiguous isotopic identification. (b) Time series waterfall plot of spectra taken continuously. (c) Three-time slices extracted from the waterfall plot shown in panel (b), where (a) is R6G–d_4, (b) is both isotopologues, (c) is R6G–d_0, and (d) corresponds to retracted tip. (Figure and caption reprinted (adapted) with permission from Ref. [57], Copyright © 2007 American Chemical Society.)

5. Giessibl, F. J., Atomic resolution of the silicon (111)–(7×7) surface by atomic force microscopy. *Science* 1995, *267*(5194), 68–71.

6. Binnig, G.; Rohrer, H.; Gerber, C.; Weibel, E., Surface studies by scanning tunneling microscopy. *Phys. Rev. Lett.* 1982, *49*(1), 57–61.

7. Binnig, G.; Rohrer, H.; Gerber, C.; Weibel, E., Tunneling through a controllable vacuum gap. *Appl. Phys. Lett.* 1982, *40*(2), 178–180.

8. Tersoff, J.; Hamann, D. R., Theory of the scanning tunneling microscope. *Phys. Rev. B* 1985, *31*(2), 805–813.

9. Eigler, D. M.; Schweizer, E. K., Positioning single atoms with a scanning tunnelling microscope. *Nature* 1990, *344*(6266), 524–526.

10. Palma, C.-A.; Samorì, P., Blueprinting macromolecular electronics. *Nat. Chem.* 2011, *3*(6), 431–436.

11. Lee, H. J.; Ho, W., Single-bond formation and characterization with a scanning tunneling microscope. *Science* 1999, *286*(5445), 1719–1722.

12. Hla, S.-W.; Bartels, L.; Meyer, G.; Rieder, K.-H., Inducing all steps of a chemical reaction with the scanning tunneling microscope tip: Towards single molecule engineering. *Phys. Rev. Lett.* 2000, *85*(13), 2777–2780.

13. Synge, E. H., XXXVIII. A suggested method for extending microscopic resolution into the ultra-microscopic region. *The London, Edinburgh, and Dublin Philosophical Magazine and Journal of Science* 1928, *6*(35), 356–362.

14. Jeanmaire, D. L.; Vanduyne, R. P., Surface Raman spectroelectrochemistry. 1. Heterocyclic, aromatic, and aliphatic-amines adsorbed on anodized silver electrode. *J. Electroanal. Chem.* 1977, *84*(1), 1–20.

15. Albrecht, M. G.; Creighton, J. A., Anomalously intense Raman spectra of pyridine at a silver electrode. *J. Am. Chem. Soc.* 1977, *99*(15), 5215–5217.

16. Wessel, J., Surface-enhanced optical microscopy. *Journal of the Optical Society of America B-Optical Physics* 1985, *2*(9), 1538–1541.

17. Stöckle, R. M.; Suh, Y. D.; Deckert, V.; Zenobi, R., Nanoscale chemical analysis by tip-enhanced Raman spectroscopy. *Chem. Phys. Lett.* 2000, *318*(1–3), 131–136.

18. Anderson, M. S., Locally enhanced Raman spectroscopy with an atomic force microscope. *Appl. Phys. Lett.* 2000, *76*(21), 3130–3132.

19. Hayazawa, N.; Inouye, Y.; Sekkat, Z.; Kawata, S., Metallized tip amplification of near-field Raman scattering. *Opt. Commun.* 2000, *183*(1), 333–336.

20. Pettinger, B.; Picardi, G.; Schuster, R.; Ertl, G., Surface enhanced Raman spectroscopy: towards single molecule spectroscopy. *Electrochemistry* 2000, *68*, 942.

21. Ramanauskaite, L.; Xu, H.; Griskonis, E.; Batiuskaite, D.; Snitka, V., Comparison and evaluation of silver probe preparation techniques for tip-enhanced Raman spectroscopy. *Plasmonics* 2018, *13*(6), 1907–1919.

22. Huang, T.-X.; Huang, S.-C.; Li, M.-H.; Zeng, Z.-C.; Wang, X.; Ren, B., Tip-enhanced Raman spectroscopy: Tip-related issues. *Analytical and Bioanalytical Chemistry* 2015, *407*(27), 8177–8195.

23. Verma, P., Tip-enhanced Raman spectroscopy: Technique and recent advances. *Chem. Rev.* 2017, *117*(9), 6447–6466.

24. Deckert-Gaudig, T.; Taguchi, A.; Kawata, S.; Deckert, V., Tip-enhanced Raman spectroscopy – From early developments to recent advances. *Chem. Soc. Rev.* 2017, *46*(13), 4077–4110.

25. Yasuhiko, F.; Peter, W.; Steven De, F.; Hiroshi, U.-I., Tip-enhanced Raman scattering microscopy: Recent advance in tip production. *Jpn. J. Appl. Phys.* 2016, *55*(8S1), 08NA02.

26. Blum, C.; Opilik, L.; Atkin, J. M.; Braun, K.; Kämmer, S. B.; Kravtsov, V.; Kumar, N.; Lemeshko, S.; Li, J.-F.; Luszcz, K.; Maleki, T.; Meixner, A. J.; Minne, S.; Raschke, M. B.; Ren, B.; Rogalski, J.; Roy, D.; Stephanidis, B.; Wang, X.; Zhang, D.; Zhong, J.-H.; Zenobi, R., Tip-enhanced Raman spectroscopy – An interlaboratory reproducibility and comparison study. *J. Raman Spectrosc.* 2014, *45*(1), 22–31.

27. Stadler, J.; Schmid, T.; Zenobi, R., Developments in and practical guidelines for tip-enhanced Raman spectroscopy. *Nanoscale* 2012, *4*(6), 1856–1870.

28. Yang, B.; Kazuma, E.; Yokota, Y.; Kim, Y., Fabrication of sharp gold tips by three-electrode electrochemical etching with high controllability and reproducibility. *J. Phys. Chem. C* 2018, *122*(29), 16950–16955.

29. Ren, B.; Picardi, G.; Pettinger, B., Preparation of gold tips suitable for tip-enhanced Raman spectroscopy and light emission by electrochemical etching. *Rev. Sci. Instrum.* 2004, *75*(4), 837–841.

30. Kharintsev, S. S.; Rogov, A. M.; Kazarian, S. G., Nanopatterning and tuning of optical taper antenna apex for tip-enhanced Raman scattering performance. *Rev. Sci. Instrum.* 2013, *84*(9), 093106.

31. Xu, G.; Liu, Z.; Xu, K.; Zhang, Y.; Zhong, H.; Fan, Y.; Huang, Z., Constant current etching of gold tips suitable for tip-enhanced Raman spectroscopy. *Rev. Sci. Instrum.* 2012, *83*(10), 103708.

32. Eligal, L.; Culfaz, F.; McCaughan, V.; Cade, N. I.; Richards, D., Etching gold tips suitable for tip-enhanced near-field optical microscopy. *Rev. Sci. Instrum.* 2009, *80*(3), 033701.

33. Boyle, M. G.; Feng, L.; Dawson, P., Safe fabrication of sharp gold tips for light emission in scanning tunnelling microscopy. *Ultramicroscopy* 2008, *108*(6), 558–566.

34. Zhang, C.; Gao, B.; Chen, L. G.; Meng, Q. S.; Yang, H.; Zhang, R.; Tao, X.; Gao, H. Y.; Liao, Y.; Dong, Z. C., Fabrication of silver tips for scanning tunneling microscope induced luminescence. *Rev. Sci. Instrum.* 2011, *82*(8), 083101.

35. Zhang, W.; Yeo, B. S.; Schmid, T.; Zenobi, R., Single molecule tip-enhanced Raman spectroscopy with silver tips. *J. Phys. Chem. C* 2007, *111*(4), 1733–1738.

36. Opilik, L.; Dogan, Ü.; Szczerbiński, J.; Zenobi, R., Degradation of silver near-field optical probes and its electrochemical reversal. *Appl. Phys. Lett.* 2015, *107*(9), 091109.

37. Iwami, M.; Uehara, Y.; Ushioda, S., Preparation of silver tips for scanning tunneling microscopy imaging. *Rev. Sci. Instrum.* 1998, *69*(11), 4010–4011.

38. Lloyd, J. S.; Williams, A.; Rickman, R. H.; McCowen, A.; Dunstan, P. R., Reproducible electrochemical etching of silver probes with a radius of curvature of 20 nm for tip-enhanced Raman applications. *Appl. Phys. Lett.* 2011, *99*(14), 143108.

39. Sasaki, S. S.; Perdue, S. M.; Perez, A. R.; Tallarida, N.; Majors, J. H.; Apkarian, V. A.; Lee, J., Note: Automated electrochemical etching and polishing of silver scanning tunneling microscope tips. *Rev. Sci. Instrum.* 2013, *84*(9), 096109.

40. Fujita, Y.; Chiba, R.; Lu, G.; Horimoto, N. N.; Kajimoto, S.; Fukumura, H.; Uji-i, H., A silver nanowire-based tip suitable for STM tip-enhanced Raman scattering. *Chem. Commun.* 2014, *50*(69), 9839–9841.

41. Huang, T.-X.; Li, C.-W.; Yang, L.-K.; Zhu, J.-F.; Yao, X.; Liu, C.; Lin, K.-Q.; Zeng, Z.-C.; Wu, S.-S.; Wang, X.; Yang, F.-Z.; Ren, B., Rational fabrication of silver-coated AFM TERS tips with a high enhancement and long lifetime. *Nanoscale* 2018, *10*(9), 4398–4405.

42. Tallarida, N.; Lee, J.; Apkarian, V. A., Tip-enhanced Raman spectromicroscopy on the Angstrom scale: Bare and CO-terminated Ag tips. *ACS Nano* 2017.

43. Berweger, S.; Atkin, J. M.; Olmon, R. L.; Raschke, M. B., Light on the tip of a needle: plasmonic nano-focusing for spectroscopy on the nanoscale. *J. Phys. Chem. Lett.* 2012, *3*(7), 945–952.

44. Mahapatra, S.; Li, L.; Schultz, J. F.; Jiang, N., Methods to fabricate and recycle plasmonic probes for ultrahigh vacuum scanning tunneling microscopy-based tip-enhanced Raman spectroscopy. *J. Raman Spectrosc.* 2021, *52*(2), 573–580.

45. Schmucker, S. W.; Kumar, N.; Abelson, J. R.; Daly, S. R.; Girolami, G. S.; Bischof, M. R.; Jaeger, D. L.; Reidy, R. F.; Gorman, B. P.; Alexander, J.; Ballard, J. B.; Randall, J. N.; Lyding, J. W., Field-directed sputter sharpening for tailored probe materials and atomic-scale lithography. *Nat. Commun.* 2012, *3*(1), 935.

46. Yang, Y.; Li, Z.-Y.; Nogami, M.; Tanemura, M.; Huang, Z., The controlled fabrication of "tip-on-tip" TERS probes. *RSC Adv.* 2014, *4*(9), 4718–4722.

47. Böckmann, H.; Liu, S.; Müller, M.; Hammud, A.; Wolf, M.; Kumagai, T., Near-field manipulation in a scanning tunneling microscope junction with plasmonic fabry-pérot tips. *Nano Lett.* 2019, *19*(6), 3597–3602.

48. Ropers, C.; Neacsu, C. C.; Elsaesser, T.; Albrecht, M.; Raschke, M. B.; Lienau, C., Grating-coupling of surface plasmons onto metallic tips: A nanoconfined light source. *Nano Lett.* 2007, *7*(9), 2784–2788.

49. Behr, N.; Raschke, M. B., Optical antenna properties of scanning probe tips: Plasmonic light scattering, tip–sample coupling, and near-field enhancement. *J. Phys. Chem. C* 2008, *112*(10), 3766–3773.

50. Berweger, S.; Atkin, J. M.; Olmon, R. L.; Raschke, M. B., Adiabatic tip-plasmon focusing for nano-Raman spectroscopy. *J. Phys. Chem. Lett.* 2010, *1*(24), 3427–3432.

51. Ma, X.; Zhu, Y.; Yu, N.; Kim, S.; Liu, Q.; Apontti, L.; Xu, D.; Yan, R.; Liu, M., Toward high-contrast atomic force microscopy-tip-enhanced Raman spectroscopy imaging: Nanoantenna-mediated remote-excitation on sharp-tip silver nanowire probes. *Nano Lett.* 2019, *19*(1), 100–107.

52. Bharadwaj, P.; Bouhelier, A.; Novotny, L., Electrical excitation of surface plasmons. *Phys. Rev. Lett.* 2011, *106*(22), 226802.

53. Zrimsek, A. B.; Chiang, N.; Mattei, M.; Zaleski, S.; McAnally, M. O.; Chapman, C. T.; Henry, A.-I.; Schatz, G. C.; Van Duyne, R. P., Single-molecule chemistry with surface- and tip-enhanced Raman spectroscopy. *Chem. Rev.* 2016.

54. Neacsu, C. C.; Dreyer, J.; Behr, N.; Raschke, M. B., Scanning-probe Raman spectroscopy with single-molecule sensitivity. *Phys. Rev. B* 2006, *73*(19), 193406.

55. Domke, K. F.; Pettinger, B., Comment on "scanning-probe Raman spectroscopy with single-molecule sensitivity". *Phys. Rev. B* 2007, *75*(23), 236401.

56. Neacsu, C. C.; Dreyer, J.; Behr, N.; Raschke, M. B., Reply to comment on "scanning-probe Raman spectroscopy with single-molecule sensitivity". *Phys. Rev. B* 2007, *75*(23), 236402.

57. Sonntag, M. D.; Klingsporn, J. M.; Garibay, L. K.; Roberts, J. M.; Dieringer, J. A.; Seideman, T.; Scheidt, K. A.; Jensen, L.; Schatz, G. C.; Van Duyne, R. P., Single-molecule tip-enhanced Raman spectroscopy. *J. Phys. Chem. C* 2012, *116*(1), 478–483.

58. Liu, Z.; Ding, S.-Y.; Chen, Z.-B.; Wang, X.; Tian, J.-H.; Anema, J. R.; Zhou, X.-S.; Wu, D.-Y.; Mao, B.-W.; Xu, X.; Ren, B.; Tian, Z.-Q., Revealing the molecular structure of single-molecule junctions in different conductance states by fishing-mode tip-enhanced Raman spectroscopy. *Nat. Commun.* 2011, 2(1), 305.

6.3.3 TERS STUDY IN VACUUM: HV AND UHV

Ultrahigh vacuum TERS measurements are technically and instrumentally challenging, requiring technical expertise that spans multiple areas, ranging from UHV technologies to surface science and optics. However, they have demonstrated the potential to study chemistry at the spatial limit and in a manner otherwise considered inaccessible. As a result, UHV-TERS has experienced dynamic growth with an increasing number of applications.

Initially, efforts to develop STM-TERS focused on improving the spatial resolution by relying on improving tip fabrication methods and increasing the stability of both the tip and sample. In an early STM-TERS study, Pettinger et al. studied CN⁻ ions and MG isothiocyanate on Au (111) with STM-TERS. However, they observed rapid photo-bleaching effects even with a relatively low laser power of 5 mW [1]. The rate of bleaching was used to understand electric field enhancement of the tip apex [2], but rapid sample photo-induced decomposition prevents spatially resolved measurements that require longer acquisition times. In an effort to avoid the effects of photo-bleaching observed in ambient conditions, Steidtner and Pettinger developed a UHV instrument capable to perform STM-TERS measurements in UHV conditions [3]. Their design required the inclusion of many optical elements within the vacuum chamber, with the alignments of nearly all of them adjusted prior to pumping down to vacuum, as they became inaccessible for manual adjustment once the UHV chamber was sealed. However, the final element, a parabolic mirror, for focusing and collection was mounted on a piezo-motor-driven alignment table for in-vacuo adjustments to optimize the alignment for each sample and tip. By sufficiently reducing the rate of photo-bleaching in UHV, they were able to acquire TERS spectra of BCB molecules on a gold substrate with a lateral spatial resolution of 15 nm [4]. Critically, they prepared their samples outside of the UHV chamber in ambient conditions. Later, other research groups would prepare their sample fully in UHV conditions, benefiting from the pristine samples typically used in UHV surface science.

In 2011, the Van Duyne group used UHV-STM-TERS to study copper phthalocyanine (CuPc) molecules on a Ag(111) surface. Jiang et al. performed all of the sample preparation within UHV conditions, including cleaning the Ag(111) single crystal and vapor depositing CuPc molecules on the substrate [5]. Optical coupling into the STM junction was accomplished via focusing and Raman collection lenses outside of the chamber aligned with large viewports located on opposite sides of the UHV-STM chamber as shown in Figure 6.34. In this design, all the optical elements could be aligned for each tip and sample without needing to break the vacuum of the UHV-STM. However, since the lenses were located outside of the viewport, and therefore a rather large distance from the tip-sample junction, the collection angle was rather small.

Since these measurements were acquired at RT, they could not observe isolated molecules on the surface with the STM due to on-surface molecular diffusion and thermal drift. Therefore, the authors had to rely on the formation of a full monolayer of CuPc molecules on the Ag(111) surface to stabilize the molecules on the surface for imaging. UHV-STM-TERS spectra of CuPc on Ag(111) acquired alongside STM images with molecular resolution, were supplemented with SERS spectra of CuPc on an AgFON substrate (Figure 6.35a and b) [5]. The UHV-SERS and UHV-STM-TERS spectra showed good agreement, but notably, the signal-to-noise ratio of the UHV-STM-TERS spectrum is significantly worse. This can be explained by the fact that the UHV-SERS spectrum samples countless hot spots due to the beam spot size, while the UHV-STM-TERS spectrum uses only a single hot spot located at the apex of the STM tip. Despite the overall less intense spectrum, TERS spectra with an enhancement factor of 7.1×10^5 still allowed the identification of eight vibrational modes through comparison with density functional theory (DFT) calculations, as visible in the

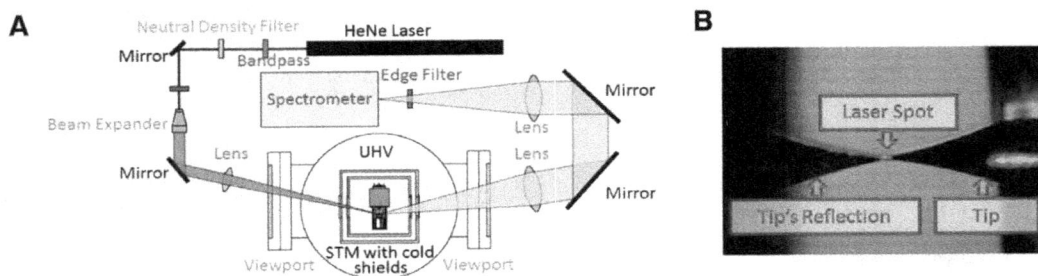

FIGURE 6.34 Images of one of the first UHV-STM-TERS setups. (a) Schematic of the experimental apparatus. (b) Camera image of the Ag tip – Ag(111) sample junction. (Figure and caption reprinted (adapted) with permission from Ref. [5], Copyright © 2012 American Chemical Society.)

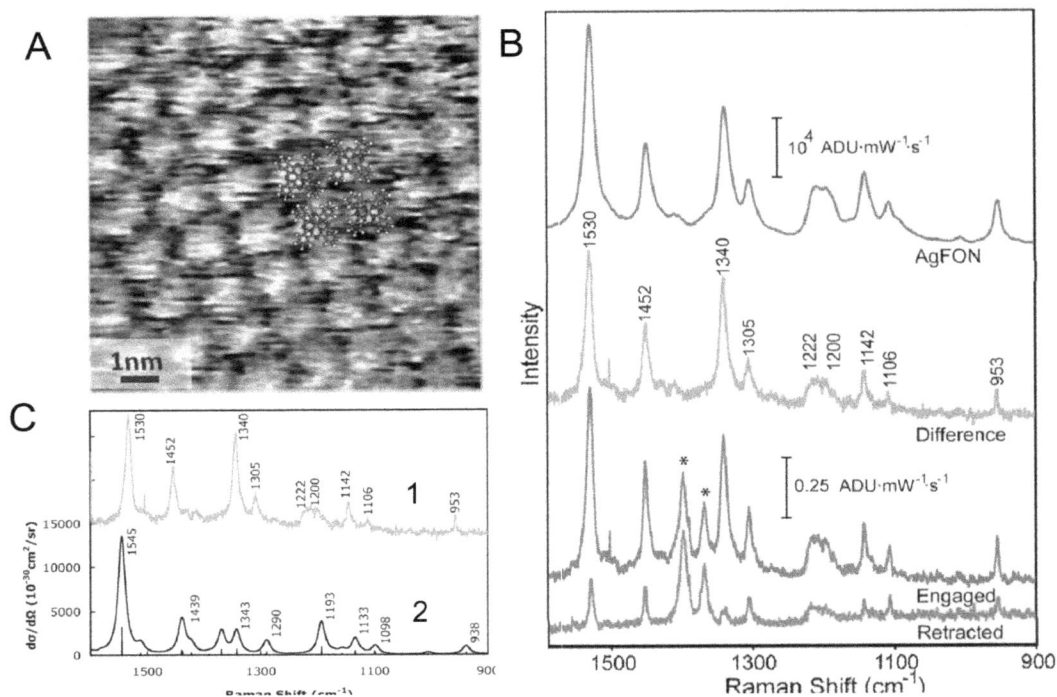

FIGURE 6.35 UHV-STM, UHV enhanced Raman spectra, and comparison with a simulated spectrum. (a) Molecular resolution topographic RT-UHV-STM image ($10 \times 10\,nm^2$) of a CuPc adlayer on Ag(111) with the tip illuminated by the continuous wave laser. (STM imaging condition: $U = +1.5\,V$, $I_t = 300\,pA$). (b) UHV-TERS spectra of a CuPc adlayer on Ag(111) and SERS spectrum of a CuPc adlayer on AgFON. The asterisks (*) indicate peaks at 1,374 and 1,403 cm^{-1} that result from the sapphire windows. (c) Simulated resonance Raman spectrum (black, 2) of CuPc at an incident wavelength of 631.48 nm. Peaks are broadened by a Lorentzian with a full width at half-maximum of 15 cm^{-1}. The scale is for the broadened spectrum. Experimental TERS spectrum (green, 1) is displayed for comparison. (Figure and caption reprinted (adapted) with permission from Ref. [5], Copyright © 2012 American Chemical Society.)

correlation between the experimental UHV-STM-TERS spectrum and DFT-calculated spectrum in Figure 6.35c. Significantly the samples were fully prepared in UHV conditions in this study, setting the stage for future UHV-STM-TERS studies of molecules on surfaces.

Following this initial realization, some groups quickly moved from RT studies to cryogenic UHV-STM-TERS studies, further improving the stability of the system. However, although the decrease in temperature can potentially improve the spatial resolution of UHV-STM-TERS, it can also prevent the observation of certain properties that require a certain degree of thermal energy. As a result, UHV-STM-TERS studies conducted at RT have been used to capture chemistry or molecular dynamics and conformations that require a certain degree of thermal or kinetic energy. Based on the promising first studies, a few research groups sought to implement and further develop STM-TERS into high vacuum (HV) and UHV at RT.

Although TERS is commonly considered to be a noninvasive technique, the tip plasmon can be used to induce chemical reactions. Initially, this effect was first observed in SERS, when para-aminothiophenol, an important probe molecule, was found to dimerize to form 4,4'-dimercaptoazobenzene (DMAB) when adsorbed on a roughened silver surface and illuminated with a 632.8 nm laser to perform SERS measurements [6] with a similar dimerization of para-nitrothiophenol molecules observed to form DMAB [7]. This type of reaction was studied at the nanoscale with AFM-TERS operated in ambient conditions relying on a dual-wavelength setup to trigger the reaction with one wavelength and probe the TERS response with a longer wavelength to monitor the reaction [8]. Contemporaneously, this reaction was investigated with HV-STM-TERS. However, in this case, the samples were prepared outside of vacuum, perhaps preventing the imaging of local surface structure and molecular coverage [9,10].

As previously mentioned in reference to the UHV-STM-TERS study by Jiang et al. of CuPc molecules on Ag(111), other studies that fully prepared the samples within the UHV environment have been able to achieve simultaneous high-resolution STM imaging in addition to TERS spectra. For example, it was possible to fabricate graphene nanoribbons (GNRs) on Au(111) via an on-surface polymerization technique that relies on two sequential reactions on the surface, Ullmann-like coupling followed by dehydrogenation as shown in Figure 6.36a [11]. The high-resolution STM imaging enabled the clear identification of intermediate polyethylene and final product 7-armchair GNRs as shown in Figure 6.36b–d, allowing the authors to combine their far-field (Figure 6.36e) and RT-UHV-STM-TERS spectra (Figure 6.36f), with an understanding of the surface-adsorbed species derived from the UHV-STM images. Importantly, they noted characteristic fluctuations in the Raman intensity, or blinking, in the near-field (tip-enhanced) spectra, that they attributed to thermal fluctuations in the effective radius of the Au STM tip apex. As a result, the authors suggested the need to develop more robust TERS probes that are resistant to thermally induced fluctuations.

Outside of the previously mentioned study of graphene nanoribbons with a gold UHV-STM-TERS tip, all the UHV-STM-TERS studies that will now be discussed relied upon a silver tip. This suggests that in contrast to ambient conditions where Au tips are primarily used to acquire TERS measurements, due to its inertness, UHV conditions permit the use of probe materials that may be slightly more reactive but may be slightly more mechanically robust.

Porphyrin-like molecules can adopt multiple conformations due to molecule–substrate interactions as the central macrocycle and peripheral functional groups distort. Due to its chemical sensitivity, TERS can provide a clear characterization of different binding conformations, yielding a view of the effects of molecule–substrate interactions [12,13]. The Van Duyne group used a UHV-STM-TERS line profile across meso-tetrakis (3,5-di-tertiarybutylphenyl)-porphyrin (H$_2$TBPP) molecules on Cu(111) in different conformations. As shown in Figure 6.37, they were able to distinguish adjacent "bowl-up" and "bowl-down" conformations with 2.6 Å resolution [14].

Continuing with their study of H$_2$TBPP molecules, Chiang et al. examined a SAM of H$_2$TBPP on Ag(111) with RT-UHV-STM-TERS [12]. The combination of STM, TERS, and time-dependent density functional theory (TDDFT) simulations suggested that the H$_2$TBPP molecules weakly couple with the Ag(111) substrate. As shown in Figures 6.38a and b, due to the C$_{2v}$ symmetry of the

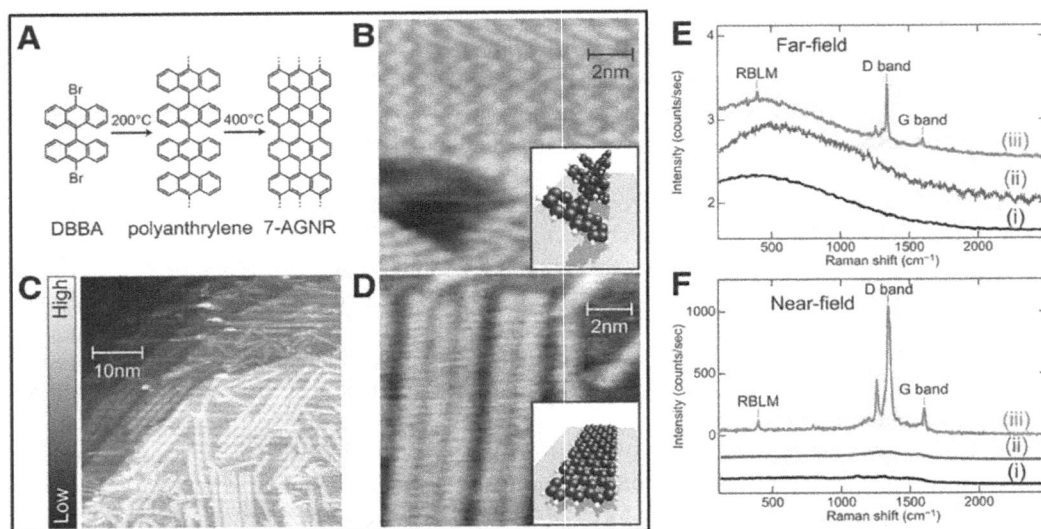

FIGURE 6.36 RT-UHV-STM-TERS of graphene nanoribbons on Au(111). (a) Fabrication scheme of 7-armchair graphene nanoribbons (7-AGNRs) from 10,10′-dibromo-9,9′-bianthryl (DBBA) molecules. STM images of Au(111) annealed at (b) ~200°C and (c) ~500°C after DBBA deposition. (d) Magnified image of (c). The STM images were acquired at $U = -0.5$ V, $I_t = 100$ pA. Insets in (b) and (d) show a schematic illustration of the molecular structure for polyethylene and 7-AGNRs, respectively. Far-field Raman spectra (e) and UHV-STM-TERS spectra (f) of (i) clean Au(111), (ii) polyethylene/Au(111), and (iii) 7-AGNR/Au(111). (Figure and caption reprinted (adapted) with permission from Ref. [11], Copyright © 2014 American Chemical Society.)

molecule, the in-plane skeletal porphyrin modes showed significant intensity in both the observed TERS and surface-enhanced resonance Raman spectra. In this manner, the authors developed a model of the DFT-optimized adsorption geometry of an H_2TBPP molecule on an atomically flat Ag(111) surface, finding that the bulky t-butylphenyl groups uplift the porphyrin ring by approximately 0.45 nm away from the Ag surface as shown in Figure 6.38c. As a result, they were able to acquire tip-enhanced fluorescence (TEF) spectra (Figure 6.38d), where this 0.45 nm distance prevents fluorescence quenching by the flat metal surface.

Beyond studies of the conformations of porphyrin-like molecules, RT-UHV-STM-TERS has also been used to study intermolecular and molecule–substrate interactions that can give rise to unique local environments. Since TERS spectra have a strong dependence on molecular orientation, it is possible to define the orientation of molecules on a surface even within nanoscale phases. According to the selection rules of SERS and TERS, vibrational modes with a substantial component perpendicular to the surface and parallel to the enhanced electric field of the tip are most strongly enhanced [15]. As a result, TERS spectra yield unique fingerprints for molecules depending on their adsorption orientation. In 2017, Jiang et al. investigated a molecular domain boundary of N-N'-bis(2,6-diisopropylphenyl)-1,7-(4′-t-butylphenoxy) perylene-3,4:9,10-bis(dicarboximide) (PPDI) molecules that formed on a Ag(100) surface using UHV-STM-TERS [16] (Figure 6.39). UHV-STM-TERS vibrational fingerprints were acquired for two different phases of PPDI, with diffusing PPDI exhibiting significantly more intense spectra as shown in Figure 6.39b. The boundary between a stable island of PPDI and a diffusing phase was imaged with the STM at RT as shown in Figure 6.39c and d. Molecular orientations within each phase were identified by interpreting TERS vibrational fingerprints with tilt angle-dependent Raman simulations. Ultimately, a planar orientation was found within stable molecular islands while the diffusing phase was found to consist of molecules in a perpendicular absorption orientation. By mapping the intensity of the most intense 1,350 cm^{-1} Raman peak, the authors were able to define the orientations of PPDI molecules at the

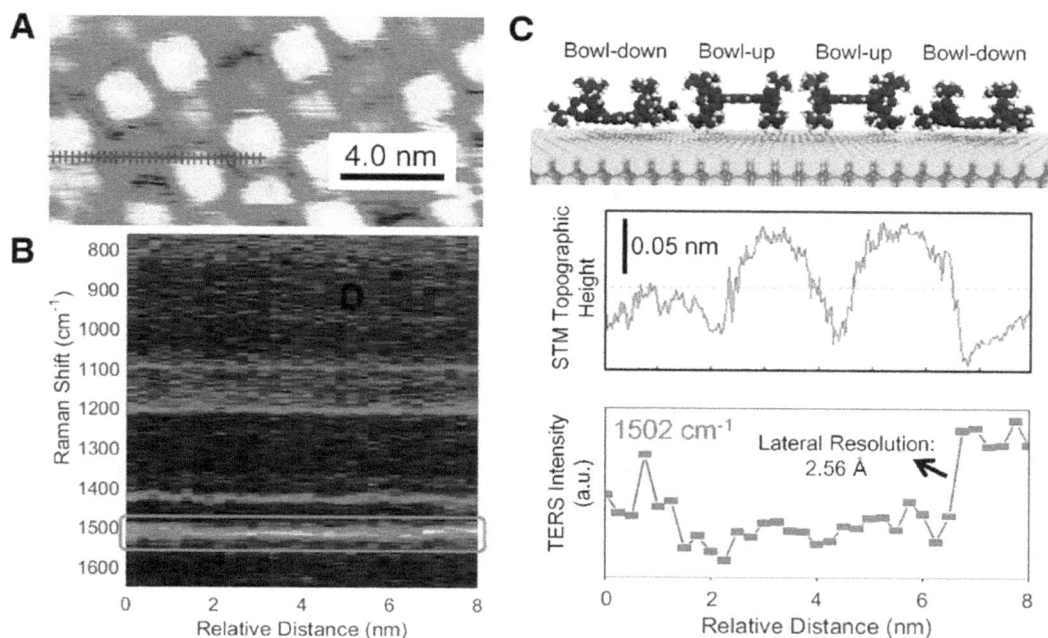

FIGURE 6.37 UHV-STM-TERS of H_2TBPP Conformations on Cu(111). (a) High-resolution STM image before the TERS imaging experiment. The dotted green line indicates the points where TERS spectra were obtained. (b) Vertical waterfall plot of H_2TBPP TERS obtained along the green dotted line in panel (a) (0.25 nm/step). (c) Top: Schematic of H_2TBPP porphyrin rings on Cu(111) buckled up/down, middle: simultaneously acquired STM line scan, bottom: TERS line scan of the integrated 1,502 cm^{-1} mode ($\lambda_{ex}=561$ nm). (Figure and caption reprinted (adapted) with permission from Ref. [14], Copyright © 2016 American Chemical Society.)

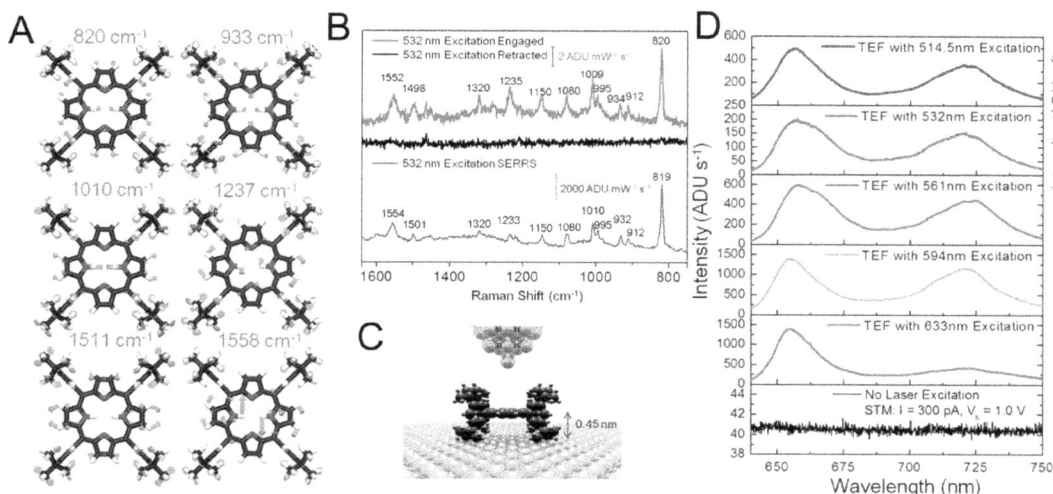

FIGURE 6.38 UHV-TERS and UHV-TEF of H_2TBPP molecules on Ag(111). (a) Selected TDDFT calculated Raman active normal modes of an H_2TBPP molecule. (b) RT-UHV-STM-TERS tip-engaged (red) and tip-retracted (black) spectra of H_2TBPP monolayer on Ag(111) with 532 nm excitation (0.1 V, 500 pA, 180 s) and SERS spectrum of an H_2TBPP adlayer on AgFON substrate with 532 nm excitation (10 s). (c) Schematic of an optimized H_2TBPP molecule adsorbed on the Ag(111) surface. (d) UHV-TEF spectra of a H_2TBPP adlayer on Ag(111) (tip-engaged spectra with tip-retracted spectra subtracted) with noted. (Figure and caption reprinted (adapted) with permission from Ref. [12], Copyright © 2015 American Chemical Society.)

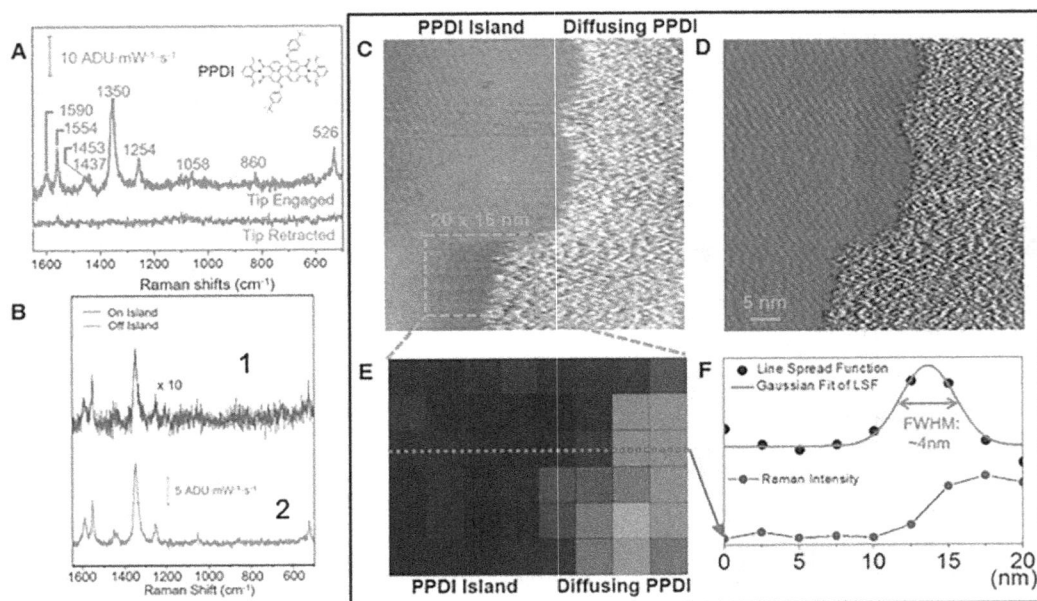

FIGURE 6.39 UHV-STM-TERS of a dynamic molecular boundary. (a) UHV-STM-TERS spectra of PPDI on Ag(111) (λ_{ex}=532 nm, P_{532}=1 mW, t_{acq}=30 s, 6 accumulations). (b) UHV-STM-TERS spectra (λ_{ex}=532 nm, P_{532}=1 mW, t_{acq}=15 s, 4 accumulations); blue (1), tip parked on top of the condensed phase; red (2), tip parked on top of the diffusing phase. (c) STM topographic image and (b) STM current image measured with a Ag tip illuminated by a 1 mW continuous wave (CW) laser. (c) TERS mapping of the 1,350 cm^{-1} Raman peak (8×6 pixels^2, ~2.5 nm/pixel, 10 s acquisition/pixel), processed from all individual TERS spectra acquired at each pixel. (d) TERS line scan shows ~4 nm spatial resolution. (Figure and caption reprinted (adapted) with permission from Ref. [16], Copyright © 2016 American Chemical Society.)

dynamic molecular domain boundary with ~4 nm spatial resolution. In a study of a similar molecular derivative on Ag(111) and Ag(100), Chiang et al. used RT-UHV-STM-TERS to discover the lifting of vibrational degeneracy of a mode of N-N'-bis(2,6-diisopropylphenyl)-perylene-3,4:9,10-bis(dicarboximide) (PDI) within SAMs on single-crystal silver substrates at RT [17]. Nanoscale spectroscopic characterizations of molecular interactions and binding conformations become possible with TERS, and the combination with the UHV-STM permits the ability to simultaneously image any surface species, free of potential contaminants, with a degree of spatial resolution that can remove many potential ambiguities.

Chemical reactions and catalytic systems lie at the extreme of the spectrum of potential molecular or atomic interactions. However, catalytic systems can quickly become complicated by the potential for multiple mechanisms that may become predominant based upon subnanoscale factors. Atomic-scale defects can serve as active sites or in the case of metal-ligand complex catalysis, the interactions between catalyst and reactant can give rise to chemical species that elude clear identification. One such system consists of cobalt (II) phthalocyanine (CoPc) molecules supported on a metal surface, where O_2 can coordinate as either molecular or atomic oxygen with the central cobalt atom, enabling the oxygen reduction reaction (ORR) [18]. Nguyen et al. considered the adsorption of molecular oxygen (O_2) on CoPc supported on a Ag(111) single-crystal surface with RT-UHV-STM-TERS [19]. STM imaging was used to define the initial self-assembly of CoPc molecules on the Ag(111) substrate as shown in Figure 6.40a. In the same manner, STM revealed the formation of multiple adsorption configurations following exposure to O_2. In combination with DFT-calculated adsorption geometries and STM images, STM imaging suggested the formation of multiple chemical species, specifically coordination between both atomic oxygen and molecular oxygen with CoPc atoms, i.e., O_2/CoPc and O/CoPc (Figure 6.40b to e). Furthermore, RT-UHV-TERS spectra of the distinct chemical species, O_2/CoPc (Figure 6.40f) and O/CoPc (Figure 6.40g) revealed significant differences in combination with isotopologue substitution. Specifically vibrational mode coupling of the O–O and CO–O vibrations with the Pc ring were observed, suggesting the ability to probe oxygen binding chemistry at the molecular scale.

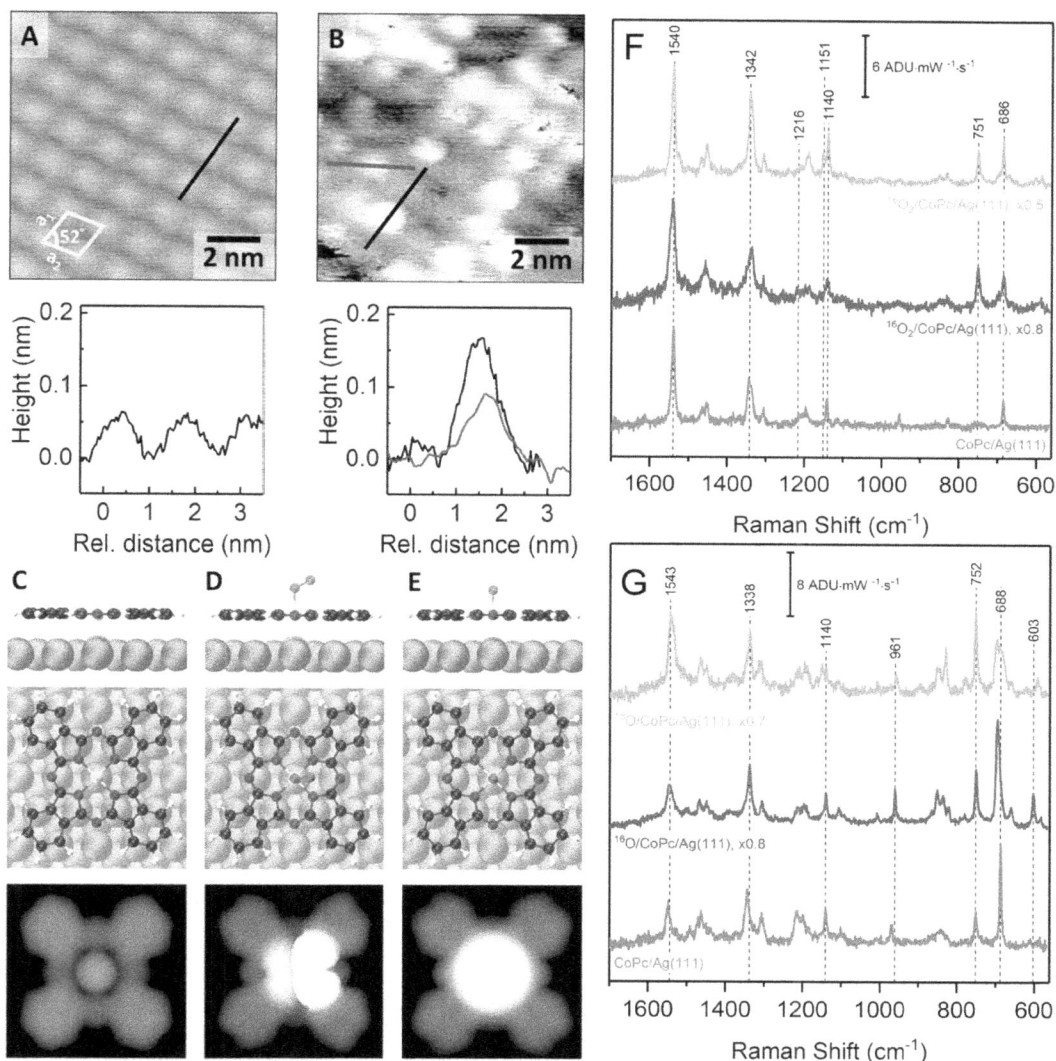

FIGURE 6.40 RT-UHV-STM-TERS of the interactions of isotopes of O_2 with CoPc on Ag(111). High-resolution STM images CoPc/Ag(111) (a) before and (b) after dosing with ~1,800L O_2, showing two different oxygen adsorption configurations. Height profiles are shown below. (a) CoPc/Ag(111) before O_2 dosing. The unit cell of the CoPc monolayer is highlighted in yellow with $a_1 \approx a_2 \approx 1.53$ nm. (b) O_2/CoPc/Ag(111) (brighter spot, black line profile) and O/CoPc/Ag(111) (dimmer spot, blue line profile) configurations. An unreacted CoPc/Ag(111) molecule is highlighted by the green dot. (c–e) Calculated STM images and adsorption geometries. Top row: side views, middle row: top views of adsorption geometries. Bottom row: calculated STM images. (c) CoPc/Ag(111). (d) O_2/CoPc/Ag(111). (e) O/CoPc/Ag(111). (f) UHV-TER spectra before (red spectrum, 3) and after $^{16}O_2$ (blue spectrum, 2) and $^{18}O_2$ (green spectrum, 1) dosing measured on the O_2/CoPc/Ag(111) molecules. (g) UHV-TER spectra before and after O_2 dosing measured on the O/CoPc/Ag(111) molecules. (Figure and caption reprinted (adapted) with permission from Ref. [19], Copyright © 2018 American Chemical Society.)

Despite the previously mentioned studies and the documented potential of RT-UHV-STM-TERS, by and large, most recent UHV-STM-TERS research has relied on low temperatures to further improve the stability of the system and therefore achieve the greatest degree of spatial resolution possible [20–22]. Notably, cryogenic temperatures enable the study of submonolayer coverages of molecules, as on-surface molecular diffusion can be effectively limited, in certain cases, this even enables the study of individual molecules with both STM and TERS. The next section will focus on the ongoing development and applications of low-temperature (LT) UHV-STM-TERS towards the study of both materials and molecules, ultimately concluding with recent LT-UHV-STM-TERS imaging measurements that have achieved submolecular resolution. These extreme levels of spatial resolution have challenged current theoretical understandings and necessitated new approaches and theoretical explanations resulting in an interesting synergy between experiments and theory.

REFERENCES

1. Pettinger, B.; Ren, B.; Picardi, G.; Schuster, R.; Ertl, G., Nanoscale probing of adsorbed species by tip-enhanced Raman spectroscopy. *Phys. Rev. Lett.* 2004, *92*(9), 096101.
2. Pettinger, B.; Ren, B.; Picardi, G.; Schuster, R.; Ertl, G., Tip-enhanced Raman spectroscopy (TERS) of malachite green isothiocyanate at Au(111): Bleaching behavior under the influence of high electromagnetic fields. *J. Raman Spectrosc.* 2005, *36*(6–7), 541–550.
3. Steidtner, J.; Pettinger, B., High-resolution microscope for tip-enhanced optical processes in ultrahigh vacuum. *Rev. Sci. Instrum.* 2007, *78*(10), 103104.
4. Steidtner, J.; Pettinger, B., Tip-enhanced Raman spectroscopy and microscopy on single dye molecules with 15 nm resolution. *Phys. Rev. Lett.* 2008, *100*(23), 236101.
5. Jiang, N.; Foley, E. T.; Klingsporn, J. M.; Sonntag, M. D.; Valley, N. A.; Dieringer, J. A.; Seideman, T.; Schatz, G. C.; Hersam, M. C.; Van Duyne, R. P., Observation of multiple vibrational modes in ultrahigh vacuum tip-enhanced Raman spectroscopy combined with molecular-resolution scanning tunneling microscopy. *Nano Lett.* 2012, *12*(10), 5061–7.
6. Huang, Y.-F.; Zhu, H.-P.; Liu, G.-K.; Wu, D.-Y.; Ren, B.; Tian, Z.-Q., When the signal is not from the original molecule to be detected: Chemical transformation of para-aminothiophenol on Ag during the SERS measurement. *J. Am. Chem. Soc.* 2010, *132*(27), 9244–9246.
7. Pozzi, E. A.; Goubert, G.; Chiang, N.; Jiang, N.; Chapman, C. T.; McAnally, M. O.; Henry, A.-I.; Seideman, T.; Schatz, G. C.; Hersam, M. C.; Duyne, R. P. V., Ultrahigh-vacuum tip-enhanced Raman spectroscopy. *Chem. Rev.* 2017, *117*(7), 4961–4982.
8. van Schrojenstein Lantman, E.; Deckert-Gaudig, T.; Mank, A. J. G.; Deckert, V.; Weckhuysen, B. M., Catalytic processes monitored at the nanoscale with tip-enhanced Raman spectroscopy. *Nat. Nanotechnol.* 2012, *7*, 583.
9. Zhang, Z.; Chen, L.; Sun, M.; Ruan, P.; Zheng, H.; Xu, H., Insights into the nature of plasmon-driven catalytic reactions revealed by HV-TERS. *Nanoscale* 2013, *5*(8), 3249–3252.
10. Sun, M.; Zhang, Z.; Zheng, H.; Xu, H., In-situ plasmon-driven chemical reactions revealed by high vacuum tip-enhanced Raman spectroscopy. *Sci. Rep.* 2012, *2*, 647.
11. Shiotari, A.; Kumagai, T.; Wolf, M., Tip-enhanced Raman spectroscopy of graphene nanoribbons on Au(111). *J. Phys. Chem. C* 2014, *118*(22), 11806–11812.
12. Chiang, N.; Jiang, N.; Chulhai, D. V.; Pozzi, E. A.; Hersam, M. C.; Jensen, L.; Seideman, T.; Van Duyne, R. P., Molecular-resolution interrogation of a porphyrin monolayer by ultrahigh vacuum tip-enhanced Raman and fluorescence spectroscopy. *Nano Lett.* 2015, *15*(6), 4114–4120.
13. Park, K.-D.; Muller, E. A.; Kravtsov, V.; Sass, P. M.; Dreyer, J.; Atkin, J. M.; Raschke, M. B., Variable-temperature tip-enhanced Raman spectroscopy of single-molecule fluctuations and dynamics. *Nano Lett.* 2016, *16*(1), 479–487.
14. Chiang, N.; Chen, X.; Goubert, G.; Chulhai, D. V.; Chen, X.; Pozzi, E. A.; Jiang, N.; Hersam, M. C.; Seideman, T.; Jensen, L.; Van Duyne, R. P., Conformational contrast of surface-mediated molecular switches yields Angstrom-scale spatial resolution in ultrahigh vacuum tip-enhanced Raman spectroscopy. *Nano Lett.* 2016, *16*(12), 7774–7778.
15. Gao, X.; Davies, J. P.; Weaver, M. J., Test of surface selection rules for surface-enhanced Raman scattering: The orientation of adsorbed benzene and monosubstituted benzenes on gold. *J. Phys. Chem. C* 1990, *94*(17), 6858–6864.

16. Jiang, N.; Chiang, N.; Madison, L. R.; Pozzi, E. A.; Wasielewski, M. R.; Seideman, T.; Ratner, M. A.; Hersam, M. C.; Schatz, G. C.; Van Duyne, R. P., Nanoscale chemical imaging of a dynamic molecular phase boundary with ultrahigh vacuum tip-enhanced Raman spectroscopy. *Nano Lett.* 2016, *16*(6), 3898–3904.

17. Chiang, N.; Jiang, N.; Madison, L. R.; Pozzi, E. A.; Wasielewski, M. R.; Ratner, M. A.; Hersam, M. C.; Seideman, T.; Schatz, G. C.; Van Duyne, R. P., Probing intermolecular vibrational symmetry breaking in self-assembled monolayers with ultrahigh vacuum tip-enhanced Raman spectroscopy. *J. Am. Chem. Soc.* 2017, *139*(51), 18664–18669.

18. Nguyen, D.; Kang, G.; Hersam, M. C.; Schatz, G. C.; Van Duyne, R. P., Molecular-scale mechanistic investigation of oxygen dissociation and adsorption on metal surface-supported cobalt phthalocyanine. *J. Phys. Chem. Lett.* 2019, *10*(14), 3966–3971.

19. Nguyen, D.; Kang, G.; Chiang, N.; Chen, X.; Seideman, T.; Hersam, M. C.; Schatz, G. C.; Van Duyne, R. P., Probing molecular-scale catalytic interactions between oxygen and cobalt phthalocyanine using tip-enhanced Raman spectroscopy. *J. Am. Chem. Soc.* 2018, *140*(18), 5948–5954.

20. Schultz, J. F.; Mahapatra, S.; Li, L.; Jiang, N., The expanding frontiers of tip-enhanced Raman spectroscopy. *Appl. Spectrosc.* 2020, *74*(11), 1313–1340.

21. Schultz, J. F.; Li, S.; Jiang, S.; Jiang, N., Optical scanning tunneling microscopy based chemical imaging and spectroscopy. *J. Phys.: Condens. Matter* 2020, *32*(46), 463001.

22. Mahapatra, S.; Li, L.; Schultz, J. F.; Jiang, N., Tip-enhanced Raman spectroscopy: Chemical analysis with nanoscale to Angstrom scale resolution. *J. Chem. Phys.* 2020, *153*(1), 010902.

6.3.4 Low-Temperature UHV-TERS for Submolecular Resolution

As an analytical tool, Raman spectroscopy provides rich chemical information and vibrational fingerprint identification of molecules and materials [1–3]. In UHV, TERS has been proven to provide highly sensitive vibrational fingerprints of chemical species below the diffraction limit of light, even reaching the subnanoscale [4–6]. The sample and tip stability inherent to room temperature (RT) UHV-STM-TERS experiments has yielded exceptional levels of spatial resolution. However, by decreasing the temperature it becomes possible to begin to unravel the intricacies of adsorbate–substrate interactions and other such subtle or highly localized chemical effects that are inaccessible by other means. Cryogenic temperatures simultaneously limit the diffusion of any surface adsorbates, while also improving the thermal stability of the tip-sample junction even under focused illumination. Ultimately, this has resulted in the ability to obtain spectroscopic information at the nanometer and even Angstrom-scale. This section will be dedicated to an overview of various low-temperature (LT) UHV-STM-TERS studies of both molecules and materials, ultimately concluding with a focus on studies where it was possible to resolve individual vibrational modes with submolecular resolution. LT-UHV-STM topographic images of a molecule or nanoscale area can be directly compared with a TERS map, where the spectral features of a Raman mode are plotted against the positions of the tip in real space. This results in a two-dimensional image where the pixel color represents the intensity or other spectral property a Raman mode at that tip position. Mapping measurements and well as line profiles provide a method to discover and understand the extreme spatial resolution of LT-UHV-STM-TERS and at the same time provide detailed properties of the system since a full spectrum is acquired at each pixel, essentially hyperspectroscopy.

However, beyond simple measurements of spatial resolution, cryogenic temperatures also result in a significant narrowing of observed line widths that correspond to vibrational modes in TERS spectra. In combination with a high degree of spatial resolution resulting in the ability to observe a sub-ensemble or even single-molecule, LT-UHV-STM-TERS spectra become essentially limited only by the resolution of the experimental setup. For example, the first UHV-STM-TERS experiment performed at liquid helium temperature revealed the adsorbate–substrate interactions of rhodamine 6G (R6G) on Ag(111). The Van Duyne group observed that spectral peaks narrowed significantly due to low temperature, as shown in Figure 6.41a. This enabled them to identify that molecule–substrate interactions were localized to the xanthene moiety and its ethylamine substituents [7]. In a similar manner, the Raschke Group used variable-temperature HV-AFM-TERS (Figure 6.41b) to uncover temperature-dependent line narrowing and splitting in TERS spectra of MG molecules adsorbed on a Au substrate [8] (Figure 6.41c). Based on these observations, they were able to quantify ultrafast vibrational dephasing, intramolecular coupling, and conformational heterogeneity of MG molecules adsorbed on a Au substrate. While these studies illustrate the power to use TERS at cryogenic temperatures to resolve single-molecule phenomena, the LT-UHV-STM-TERS studies that will now be discussed use the high-resolution imaging capabilities of the STM to develop understandings of molecules and materials with respect to atomic landscapes.

At perhaps the simplest level, Raman spectroscopy can be used to acquire a vibrational fingerprint of a chemical species that can be used to identify a molecule or material. In LT-UHV-STM-TERS, a vibrational fingerprint can be acquired of a single molecule on a surface, providing unambiguous chemical identification of individual molecules closely packed on a surface. In a study of free-base meso-tetrakis (3,5-di-tertiarybutyl-phenyl)-porphyrin (H_2TBPP), Dong and co-workers were able to use LT-UHV-STM-TERS (Figure 6.42a) measurements, performed at ~80K, to identify H_2TBPP (Figure 6.42b) and zinc-5,10,15,20-tetraphenyl-porphyrin (ZnTPP) (Figure 6.42c) on a Ag(111) surface [9]. The authors first established reference TERS fingerprints by depositing each molecule independently onto the silver surface, observing self-assembled molecular islands with STM (Figure 6.42d and e) and unique vibrational fingerprints for each species that corresponded well with DFT-calculated TERS spectra (Figure 6.42f). Then, as shown in Figure 6.42g, they acquired sequential TERS spectra across adjacent islands of H_2TBPP and ZnTPP molecules. The spectral evolution, represented in Figure 6.42h, provided the ability to identify ZnTPP (position 6) and H_2TBPP (position 8) molecules even within Van der Waals contact. The authors later returned to this system of co-deposited H_2TBPP and ZnTPP on Ag(111) and argued that by considering their subnanometer-resolved TERS imaging with advanced multivariate analysis, they were able to achieve high-quality chemical images despite limited signal-to-noise ratios in individual pixel-level spectra [10]. In this work, they stated that they could not only unambiguously distinguish adjacent molecules with a resolution of ~0.4 nm, but that they could also resolve submolecular features and differences in molecular adsorption conformations.

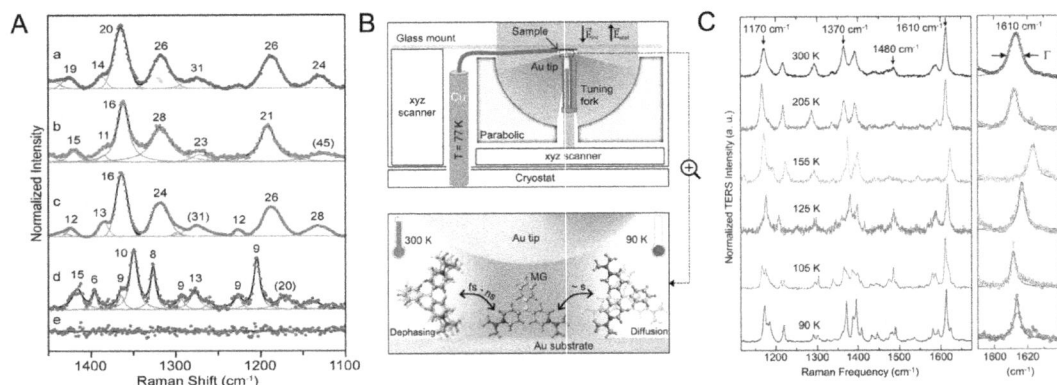

FIGURE 6.41 LT-TERS demonstrating spectral peak narrowing and the observation of new vibrational modes. (a) Enhanced Raman spectra of R6G (a) RT-SERS, (b) RT-TERS, (c) LT-SERS, and (d) LT-TERS spectra of R6G. The displayed values are full widths at half-maximum (FWHMs) for the corresponding peaks. Spectra were background-subtracted and fit with Gaussians (SERS) or Lorentzians (TERS). Data are represented by colored circles, individual peak fits by gray lines, and composite fits by black lines. (e) Residuals after fitting for LT-TERS. (Reprinted (adapted) with permission from Ref. [7], Copyright © 2014 American Chemical Society.) (b) Schematic of the cryogenic HV-AFM-TERS experiment of MG on a Au substrate. Incident laser (E_{inc}) is focused onto the apex of an etched Au AFM tip and Raman signal (E_{scat}) is collected by a parabolic mirror in a back-scattering geometry. Illustration of the tip-substrate junction showing molecular motion of malachite green (MG) with characteristic vibrational signatures at different temperatures (below). (c) Temperature-dependent TERS spectra of MG showing observed narrowing. (Reprinted (adapted) with permission from Ref. [8], Copyright © 2015 American Chemical Society.)

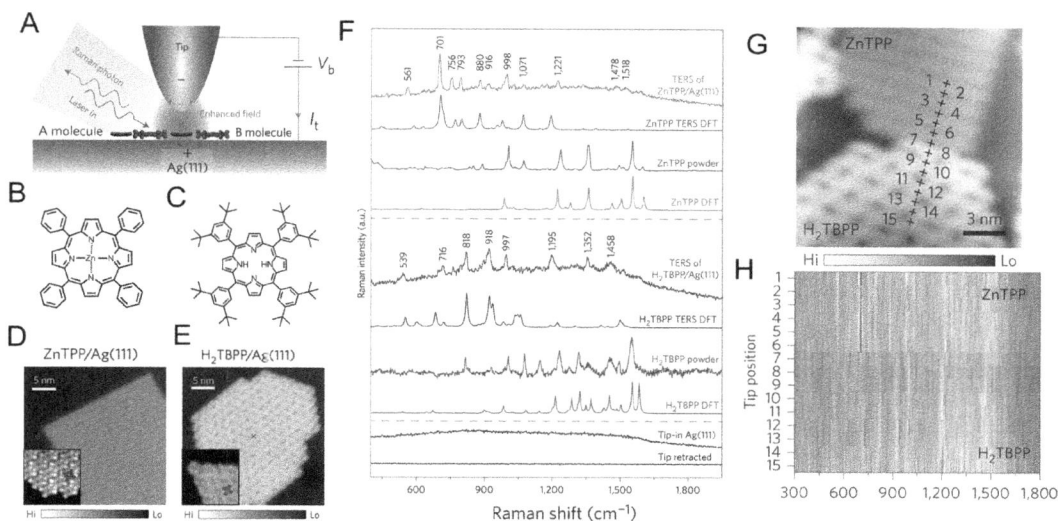

FIGURE 6.42 Distinguishing adjacent molecules with LT-UHV-STM-TERS. (a) Schematic of STM-controlled TERS in a confocal-type side-illumination configuration on Ag(111) using a Ag tip. [(a) and (b)] Molecular structures of ZnTPP (b) and H_2TBPP (c). Insets are high-resolution images acquired with a modified tip. [(d) and (e)] STM images of self-assembled molecular islands on Ag(111) for ZnTPP (d) and H_2TBPP (e). (f) LT-UHV-STM-TERS spectra acquired above ZnTPP or H_2TBPP molecular islands. For comparison, power Raman spectra and Raman spectra calculated via DFT are also shown. The brown spectrum was taken on bare Ag(111) and the black spectrum was measured on top of a molecular island but with the tip retracted 5 nm from the surface. (Figure and caption reprinted by permission from Springer Nature: *Nature*, Copyright © 2015 [9].)

In more recent work, TERS has been used to identify adjacent regioisomers that have very similar structures based on their vibrational fingerprints. The Jiang group used LT-UHV-STM-TERS at ~77K to spectroscopically define and image regioisomeric porphodilactone molecules [11]. STM images of the co-deposition of trans- and cis-tetrakis pentafluorophenyl porphodilactone (H_2F_{20}TPPDL) molecules onto a Ag(100) substrate revealed the formation of two distinct self-assembled superstructures as noted in Figure 6.43a. The regioisomers differ only in the relative positions of the lactone moieties. However, this small structural difference results in distinct vibrational fingerprints. Mahapatra et al. used a TERS line profile measurement to identify a spatial resolution of 8 Å across the boundary of regioisomeric assemblies as shown in Figure 6.43b, finding that they were able to demonstrate single molecule sensitivity. They also used large-scale TERS mapping to characterize a two-component organic heterojunction consisting of the regioisomers (Figure 6.43c to e). TERS imaging resulted in the unambiguous identification of the positions of trans- and cis-porphodilactones based on their distinct vibrational fingerprints (Figure 6.43f). The spectral assignments were supported by good agreement with time-dependent DFT simulated spectra (Figure 6.43g). Ultimately, the authors were able to identify distinct regioisomers on the Ag(100) surface with Angstrom-scale spatial resolution. Mahapatra et al. later investigated the molecule–substrate interactions of these H_2F_{20}TPPDLregioisomers on different substrates, using LT-UHV-STM-TERS measurements to consider trans- and cis-H_2F_{20}TPPDL molecules on (100) surfaces of Ag, Cu and Au [12]. In this later work, they identified that the regioisomers adopt distinct adsorption conformations with a strong dependence on the substrate identity.

In contrast to the previous studies where the co-deposition of organic molecules resulted in distinct self-assemblies, it is also possible for two distinct chemical species to form mixed lateral interfaces or even mixed phases that result in bicomponent supramolecular self-assemblies. The formation of mixed networks significantly complicates the identification of individual molecules. This

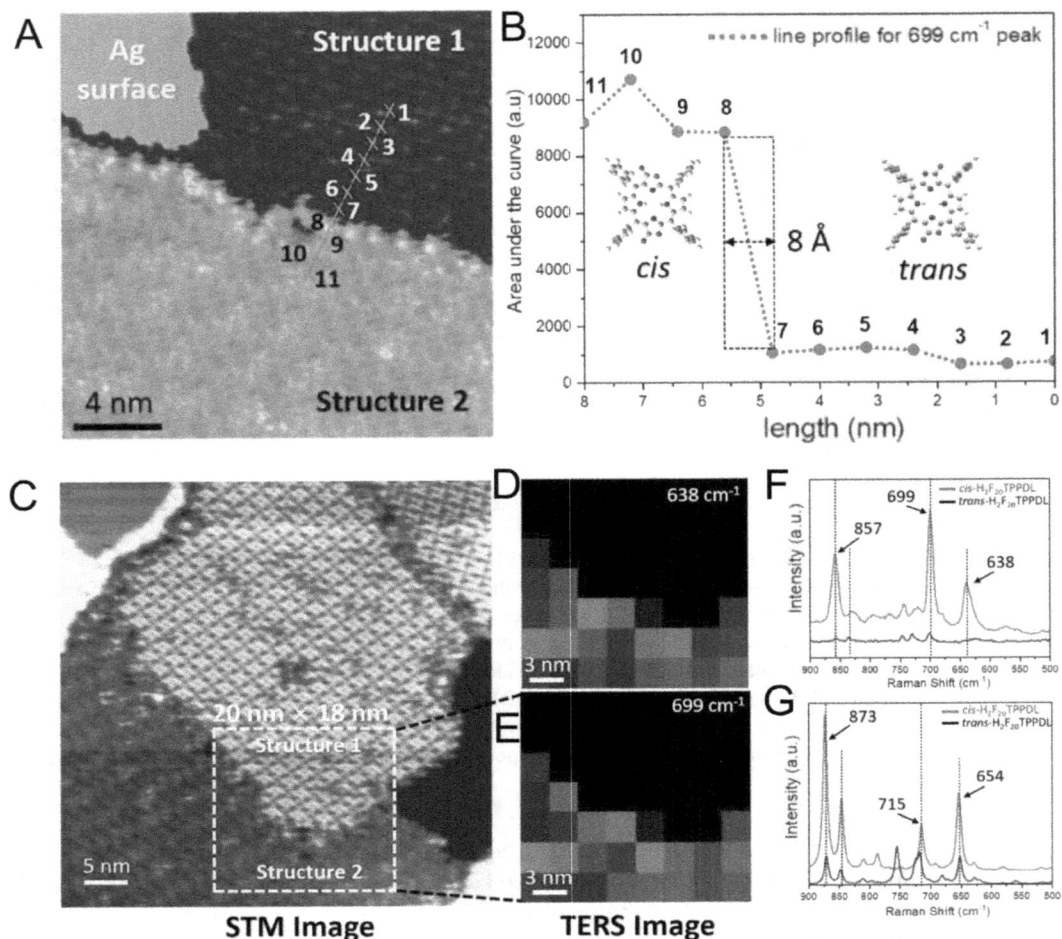

FIGURE 6.43 LT-UHV-STM-TERS study of self-assemblies of regioisomers. (a) STM image of co-deposited trans- and cis-porphodilactone molecules on Ag(100) ($V=-1.50$ V, $I=150$ pA). The location and direction of a TERS line profile measurement is marked. Structure 1 (blue), Structure 2 (orange), and the bare Ag surface (green) are marked. (c) TERS line profile measured with a Ag tip for the 699 cm^{-1} peak (area under the curve) showing a spatial resolution of 8 Å. $\lambda_{ex}=633$ nm, $P_{acq}=0.5$ mW, $t_{acq}=10$ s at each point. (c) STM image of coadsorbed trans- and cis-isomers ($V=-1.46$ V, $I=150$ pA). A white dotted rectangle marks the area where TERS images were used to identify Structures 1 and 2. [(d) and (e)] TERS images obtained by tracking the intensity of the 638 and 699 cm^{-1} peaks, respectively (8×7 pixels^2). $\lambda_{ex}=633$ nm, $P_{acq}=0.5$ mW, 2.5 nm step size/pixel, $t_{acq}=10$ s at each pixel. (f) Comparisons of experimental TERS spectra for cis- and trans-isomers. (g) TDDFT-simulated spectra for cis- and trans-isomers. (Figure and caption reprinted (adapted) with permission from Ref. [11], Copyright © 2019 American Chemical Society.)

can prevent a clear understanding of the underlying intermolecular or molecule–substrate interactions. As a result, the ability to recognize individual molecules in real space becomes essential. Zhang et al., used LT-UHV-STM-TERS at 80K to distinguish individual DNA bases in a hydrogen-bonded network [13]. As shown in the STM image in Figure 6.44a, the co-adsorption of adenine (A) and thymine (T) molecules on Ag(111) surface resulted in adenine networks that were surrounded by molecules arranged in repeating trimer structures. The spectral evolution of a TERS line profile measurement across this area (Figure 6.44b) resulted in the ability to identify these repeating trimer structures as hydrogen-bonded thymine molecules that surround the adenine network as

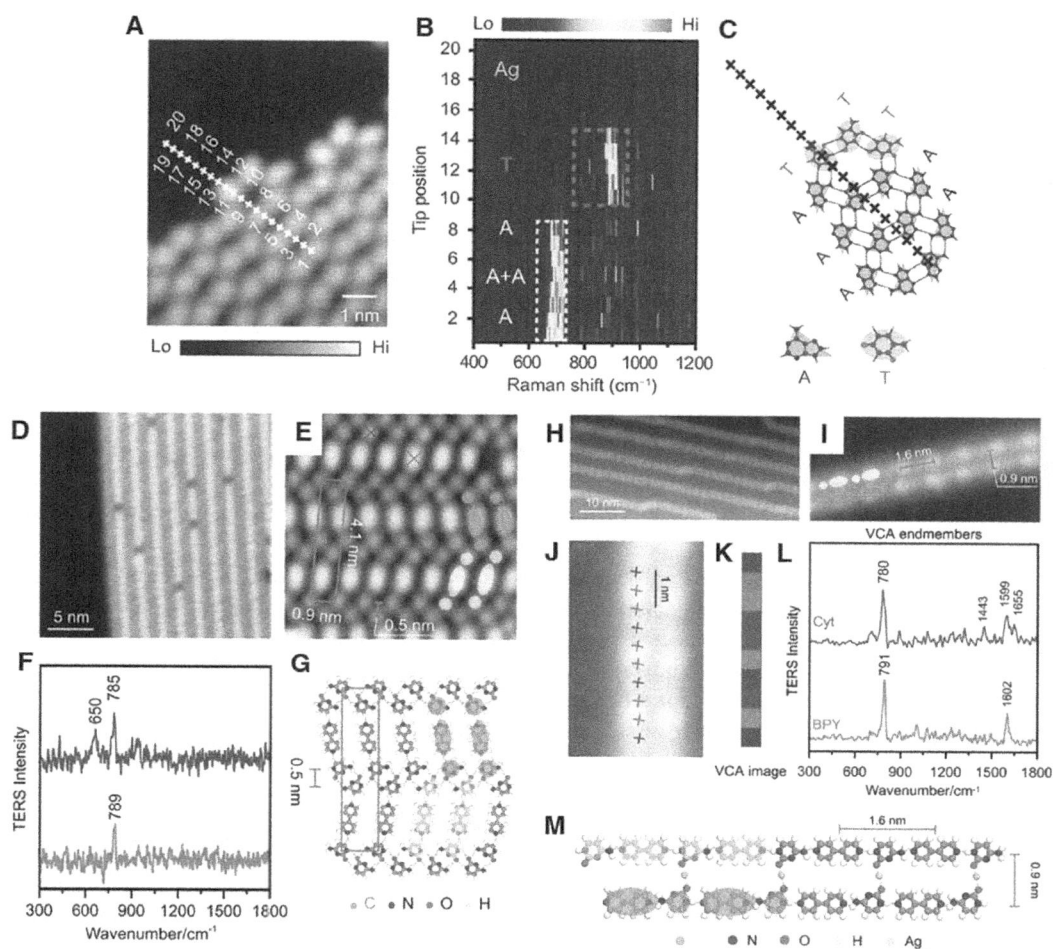

FIGURE 6.44 LT-UHV-STM-TERS characterizations of mixed supramolecular self-assemblies. (a) High-resolution STM imaging (−1 V, 5 pA) around the co-adsorption boundary of thymine molecules and an adenine molecular network. (b) Corresponding spectral evolution for 20 sequential TERS measurements along the line trace marked by crosses in (a) (0.3 mW, −0.1 V, 1 nA, 30 s) with a step size of 0.27 nm. (c) A proposed molecular arrangement along the line trace in (a). (Figure and caption reprinted (adapted) with permission from Angewandte Chemie International Edition [13], Copyright © 2017 John Wiley and Sons.) [(d) and (e)] Large-scale (d) and high-resolution (e) STM images of bicomponent monolayer of Cyt and BPY. STM imaging conditions: −1 V, 10 pA. (f) TERS spectra acquired from the positions marked in (e). TERS measurement conditions: −0.1 V, 1 nA, 30 s. (g) Possible packing model of the bicomponent molecular monolayer. [(h) and (i)] Large-scale (h) and high-resolution (i) STM images of bicomponent chains following thermal annealing of Cyt and BPY on Ag(111). (j) STM image showing the positions for the TERS line profile measurement with a step size of 0.51 nm. STM imaging conditions: −0.2 V, 10 pA. (k) Vertex component analysis abundance image for the line profile in (j). (l) Corresponding endmember spectra corresponding to Cyt and BPY identification. TERS measurement condition: −0.2 V, 500 pA, 5 s. (m) Possible packing model of the bicomponent molecular chain. (Figure and caption reprinted (adapted) with permission from Journal of Raman Spectroscopy [14], Copyright © 2020 John Wiley and Sons.)

illustrated in Figure 6.44c. More recently, Jiang et al. used a similar method to study bicomponent supramolecular self-assemblies of cytosine (Cyt) and 4,4′-bipyridine (BPY) on Ag(111) [14]. In this work, they used LT-UHV-STM-TERS to image and spectroscopically characterize the well-ordered hydrogen-bonded monolayers that these molecules form on Ag(111), which were found to transform into well-aligned double-stranded chains after thermal annealing. As shown in Figure 6.44d and e, they used STM and TERS vibrational fingerprints to identify Cyt and BPY molecules within the initial hydrogen-bonded networks. As can be seen in the STM image in Figure 6.44h, thermal annealing at ~370K for 3 min, resulted in the formation of double-stranded chains. By considering TERS spectra with multivariate analysis, specifically vertex component analysis, the authors were able to identify individual molecules within the bicomponent chains (Figure 6.44j–m). Additionally, based on spectral analysis, the double-stranded chains were found to result from the formation of metal–organic coordination bonds between strands, providing a tool to probe and define local molecular interactions.

Raman spectroscopy provides more than just chemical identification. Vibrational fingerprints are sensitive to chemical environments and intermolecular forces [15]. Additionally, the surface selection rules for enhanced Raman spectroscopies (SERS and TERS) result in spectra that are highly sensitive to the orientation of molecules on a surface [16]. Due to its high degree of spatial resolution and sensitivity, LT-UHV-STM-TERS can identify the intrinsic properties and molecule–substrate interactions of individual molecules even when they lie directly adjacent to each other on a surface. Simultaneous SPM imaging with atomic-scale spatial resolution provides the ability to correlate vibrational and topographic information to understand the effects of highly localized chemical environments.

Recently LT-UHV-STM-TERS has been used to compare the orientation of different molecules on surfaces, even when they are adjacent to each other on a surface. As shown in Figure 6.45a and b, the Dong group were able to spectroscopically identify multiple orientations of 4,4′-bipyridine (BPY) molecules along atomic step edges of a Ag(111) substrate. Furthermore, they found that along an island edge BPY molecules adopt a perpendicular orientation in contrast to the flat-lying molecules that occur within the interior of self-assembled molecular islands [17]. In a similar manner, the Jiang group studied the binding conformations of rubrene that result in multiple distinct supermolecular structures on a Ag(100) surface. By using TERS spectra with interpretation supported by DFT calculations, they identified rubrene molecules in different binding orientations on a Ag(100) surface. 1D chains and the boundaries of molecular islands of planar rubrene molecules were found to consist of rubrene molecules in a perpendicular binding orientation, as shown in Figure 6.45c–f [18]. They reported 5 Å spatial resolution of molecular binding orientations. Through these studies and others, LT-UHV-STM-TERS provides a view of molecule–substrate interactions [19] which are essential to catalytic processes and the realization of devices through methods of bottom-up assembly. The spatial resolution of TERS alongside its sensitivity to molecular orientation results in a tool that provides the unambiguous identification of binding orientations and the effects of highly localized chemical environments.

Although LT-UHV-STM-TERS studies have resulted in impressive chemical imaging at the Angstrom-scale, in certain cases cryogenic temperatures have been found to prevent the ability to acquire TERS spectra of surface-adsorbed molecules. The current prevailing explanation for this phenomenon relies on a few different effects. Perhaps first and foremost, as previously mentioned, is that TERS is governed by so-called selection rules, where out-of-plane vibrational modes are most strongly enhanced due to the directionality of the enhanced electric field, it is oriented parallel to the tip and therefore perpendicular to the surface [6,20]. As a result, a flat molecule that lies flat or parallel to the surface has a poor enhanced Raman scattering cross section. This enables the ability to easily differentiate molecules with different surface orientations [21,22]. However, low temperatures can also lead to molecules freezing in a planar adsorption state on a surface, reducing the possibility of vibrational modes with a strong out-of-plane component and therefore yielding substantially less intense enhanced Raman spectra. The second effect is attributed to strong electronic interactions between a molecule and the supporting substrate that can result in coupling of the molecule with the surface state, therefore screening it from the enhanced electric field of the tip [23] or resulting in strong hybridization of the molecular orbitals with the metal substrate that prevents

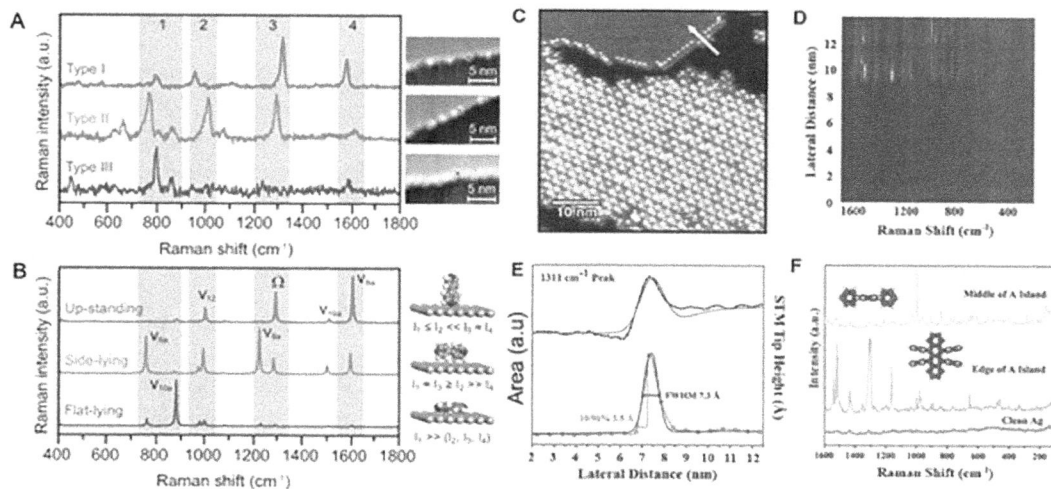

FIGURE 6.45 Resolving molecular orientations with LT-UHV-STM-TERS. (a) Representative TERS spectra ($\lambda_{ex} = 532$ nm) for three types of BPY molecules adsorbed at the step edge with corresponding STM shown on the right. The characteristic peaks are categorized into four regions. (b) Calculated Raman spectra for single BPY molecules with three different orientations: up-standing, side-lying, and flat-lying. The schematics of molecules with different adsorption orientations. (Figure and caption reprinted (adapted) with permission from Ref. [17], *ChemPhysChem* Copyright © 2018 John Wiley and Sons.) (c) STM image of two self-assemblies of rubrene molecules. A bright boundary of molecules surrounding one island is apparent. The arrow denotes the direction and location of the TERS line profile that was acquired from the bare Ag substrate to the island. (d) Waterfall plot representing the spectra acquired for the line profile. (e) STM and TERS line profiles across the island boundary. (f) Representative spectra from the TERS line profile of the clean Ag substrate where no signal was observed (red), the fingerprint from the edge of the island (green), and a spectrum from the middle of the island (yellow) with corresponding molecular orientations ($\lambda_{ex} = 515$ nm). (Figure and caption reprinted (adapted) with permission from Ref. [18], Copyright © 2019 American Chemical Society.)

resonance Raman enhancement [24]. In either case, LT-UHV-STM-TERS measurements of certain molecule–substrate systems require certain considerations or unique methods.

The primary method relies on decoupling the molecule from the metal surface to obtain LT-UHV-STM-TERS spectra. This can be accomplishment by purposely transferring a single molecule from the surface to the tip, a technique that has found widespread use in LT-UHV-STM and LT-UHV non-contact AFM (NC-AFM). The supreme stability enables the functionalization of the probe tip with a single atom or molecule via a voltage pulse leading to exceptionally high-resolution images of molecular structures, including bonding within and between molecules [25]. Lee et al. investigated azobenzyl thiol (ABT) molecules adsorbed on a Au(111) substrate with LT-UHV-STM-TERS at 80K Although the enhanced Raman spectrum of azobenzene thiol molecules was found to be suppressed at the tip–sample junction, they found that by transferring a single molecule to the tip, the enhanced Raman spectrum increased by a factor of ~5 as the tip was retracted away from the substrate across a range of 1 nm to 161 nm [26]. This provided the means to identify a single adsorbed cis-azobenzene thiol molecule. In other work, the Apkarian group explored the photoisomerization of ABT [27]. They found that ABT molecules lie flat on a Au(111) surface, preventing photoisomerization or detection with TERS. However, when an ABT molecule was chemisorbed onto the apex of the silver STM-TERS tip they were able to track cis-trans photoisomerization at the single-molecule level based upon the observation of two distinct spectra with two discrete, on- and off-levels. These two distinct spectra were attributed to the cis and trans forms of the molecule, as it flip-flopped between isomers. Overall, these studies focused on the behavior of the molecule adsorbed onto the tip. However, to capture the intrinsic properties of the molecule or focus on molecules on the surface an alternative approach is required.

Instead of removing the molecule from the surface, it is possible to instead add something to the surface to disrupt the molecule–substrate interactions that would otherwise result in silent LT-UHV-STM-TERS spectra. Lee et al. found that at RT cobalt(II) tetraphenylporphyrin (CoTPP) forms an ordered monolayer on Au(111) that can be observed with both STM and hyperspectral TERS images. However, upon being cooled to 80K, the molecular TERS signal disappeared. By introducing CO molecules, which eventually desorb, but could serve as local field spoilers or change the conformation of CoTPP molecules through the formation of coordination bonds, they were able to recover the ability to record STM (Figure 6.46a) and TERS (Figure 6.46b) images of a molecular self-assembly with an observed spectrum appearing in Figure 6.46c that is in good agreement with a simulated spectrum (Figure 6.46d). Interestingly, the contrast between an STM image (Figure 6.46e) and a map of the TERS intensity of the $1,270\,cm^{-1}$ (Figure 6.46f) is inverted which was attributed to the conductivity of the junction. However, the authors stated that a map of the frequency shift of this vibrational mode (Figure 6.46g) shows significantly more spatial detail than the STM image with line profiles shown in Figure 6.46h.

While CO adsorption can sufficiently disrupt molecule–substrate interactions, it can also change the identity or intrinsic properties of the molecule. Alternatively, the molecule can be placed onto a relatively inert insulating layer. In this manner, Jaculbia et al. used a thin film of NaCl to screen effects from the metal surface on the electronic properties of the molecule under investigation, specifically copper naphthalocyanine (CuNc) [28]. By first growing 2 and 3 monolayer-thick NaCl films on a Ag(111) surface and then vapor depositing CuNc molecules onto the modified substrate at low temperatures (<4.8K), they found that varying the NaCl film thickness changes the degree of screening effects from the surface, leading to variations in molecular resonance phenomena. They later expanded on this work to consider resonance Raman effects at the single-molecule level with respect

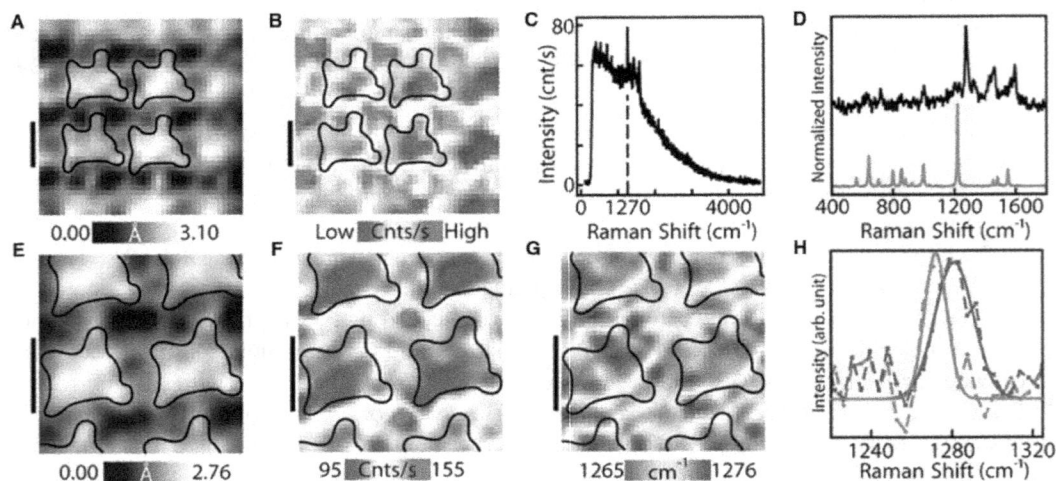

FIGURE 6.46 Simultaneous LT-UHV-STM-TERS of self-assembled CoTPP molecules on a Au(111) substrate. (a) STM topography of CoTPP lattice at 0.1 nA, 9.8 mV. (b) Raman image anticorrelated with topography produced by integrating $1,270\,cm^{-1}$ mode intensity (634 nm laser at 5 μW/μm², 1 s accumulation/pixel). (c) Degraded TERS spectrum. (d) Background-subtracted experimental spectrum and simulation. (e) Zoomed-in topography. [(f)–(h)] Super-resolution TERS. The $1,270\,cm^{-1}$ mode is fitted to a Gaussian (h), and the extracted intensity and frequency shift are mapped in (f) and (g), respectively. The high pyrroles of saddled CoTPPs are resolved in the frequency map. (h) Line profiles from two pixels in (g). Scale bars are 1 nm long. Images are filtered due to invasive imaging conditions. (Figure and caption reprinted (adapted) under the terms of the ACS AuthorChoice License from Ref. [23], Copyright © 2017 American Chemical Society.)

to molecular symmetry with an investigation of a CuNc molecule adsorbed on a three-monolayer-thick insulating NaCl film supported on a Ag(111) substrate, as shown in Figure 6.47a [29]. A CuNc molecule was found to adsorb aligned 45° with respect to the [010] direction of the NaCl thin film, as confirmed with STM imaging (Figure 6.47b and c). As shown in Figure 6.47d, they were able to obtain TERS spectra of a single molecule. The excitation wavelength was tuned to match an electronic transition of the molecule and achieve resonance Raman conditions. With resonance conditions, the different vibrational modes for a single CuNc molecule on NaCl were investigated, identifying vibrational modes with three types of symmetries. As shown in Figure 6.47e, Raman active modes of the D_{4h} molecule exhibited A_{1g}, B_{2g}, and B_{1g} symmetry in TERS images (Figure 6.47f), which was confirmed by simulations of the corresponding vibrational modes (Figure 6.47g). Although the use of insulating layers is a necessity in single-molecule studies of scanning tunneling microscope-induced luminescence [20], this direction remains relatively unexplored in relation to LT-UHV-STM-TERS, perhaps due to instrumental demands, but appears to be an exciting direction to consider the intrinsic properties of a molecule decoupled from the metal substrate.

Historically, LT-UHV-STM-TERS studies have pushed the spatial resolution of chemical imaging to its limits. Initially, TERS was the eventual realization of a theoretical proposal to confine light. However, technique developments have resulted in unprecedented Angstrom-scale spatial resolution surpassing the expected spatial resolution and requiring new theoretical approaches.

TERS imaging with Angstrom-scale spatial resolution has necessitated the development of new theoretical explanations. Early work relied on a classical approach, approximating the tip as a sphere. The tip radius at the microscale was considered to be the primary factor that determines the confinement of the near-field enhancement [30]. However, the first demonstration of subnanometer lateral resolution for TERS made it necessary to reconsider the underlying theory. In an optomechanical-based method, the dynamical backaction amplification of molecular vibrations leads to the observed nonlinear enhancement [31,32]. In a similar manner, the near-field self-interaction of a molecule in the plasmonic nanogap

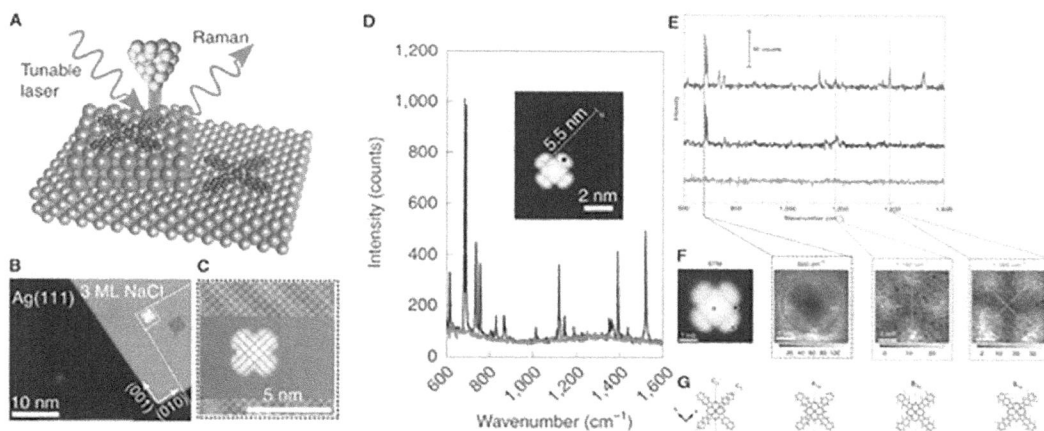

FIGURE 6.47 Resonance STM-TERS of a molecule on an insulating layer. (a) Experimental configuration for performing STM-TERS under resonance ($\lambda_{ex} = 738$ nm). (b) STM scan showing an individual CuNc molecule on three-monolayer-thick NaCl and on Ag(111). (c) High-resolution STM image of the molecule on NaCl. (d) STM-TERS data collected when the tip is placed at the molecular lobe (black) and when the tip is more than 5.5 nm away from the molecular center (red). Inset shows the tip locations during data collection. [(e) and (f)] Three different STM-TERS spectra (e) at positions indicated in the STM topography of (f). TERS images obtained simultaneously under constant current mode ($V_b = 1$ V, $I_t = 50$ pA; TERS maps: 40×40 pixels^2, 1 Å/pixel, 1 s exposure time/pixel). The TERS images show three distinct patterns: circle (680 cm^{-1}), plus (1,182 cm^{-1}), and cross (1,389 cm^{-1}). (g) Diagrams of the calculated vibrational modes corresponding to the TERS images above in (f) showing A_{1g}, B_{2g}, and B_{1g} symmetry. (Figure and caption reprinted by permission from Springer Nature: *Nature Nanotechnology* Ref. [29], Copyright © 2020.)

through multiple elastic scattering events was proposed to results in the tremendous enhancement and spatial confinement found in TERS [33–35]. More recently, atomic-scale protrusions that lead to the confinement of light at the atomic scale in "picocavities" have come to be considered essential [36–40]. Expanding from the atomic-scale confinement of light, another group has suggested that such an extreme level of confinement leads to significant field-gradient effects that result in submolecular TERS images and the breakdown of selection rules [41–44]. New experimental demonstrations of increasing spatial resolution have resulted in the ongoing development of new theoretical explanations in a synergistic manner.

To further improve the stability of UHV-STM-TERS experiments, the methods were brought into cryogenic temperatures. LT-UHV-STM-TERS performed at liquid nitrogen temperature (~77K) by the Dong group yielded truly impressive results (Figure 6.48a). As shown in Figure 6.48b, meso-tetrakis (3,5-di-tertiarybutylphenyl)-porphyrin (H_2TBPP) molecules occur as self-assembled molecular islands as well as isolated single molecules on a Ag(111) surface [45]. By matching the excitation wavelength to the absorbance of the molecule, they made use of resonance Raman effects to enhance Raman scattering beyond that accomplished solely by TERS. TERS maps of a single H_2TBPP molecule revealed the localization of multiple vibrational modes within the structure of a single molecule (Figure 6.48c and d), allowing the central cavity of the porphyrin core to be resolved. The porphyrin core offered a powerful means to validate the subnanometer resolution of the TERS spectral map, with the TERS image showing notable contrast across the central cavity of the molecule, matching the appearance visible in STM imaging (Figure 6.48e and f). With single-molecule sensitivity, they were also able to consider the adsorption orientation of individual molecules. TERS spectra were acquired for a single molecule, a molecule within a molecular island, and a molecule tilted across an atomic step edge. By comparing with a series of calculated spectra for a molecule at different tilt angles they were able to identify the orientation of the molecule adsorbed on the step edge. Ultimately, they spectroscopically visualized the conformation and distinct vibrational modes of a single molecule with subnanometer resolution. More recently, the Dong group continued their study of (H_2TBPP) molecules but on a Ag(100) surface and at a lower temperature (7K). The further increased stability resulted in TERS mapping images where the

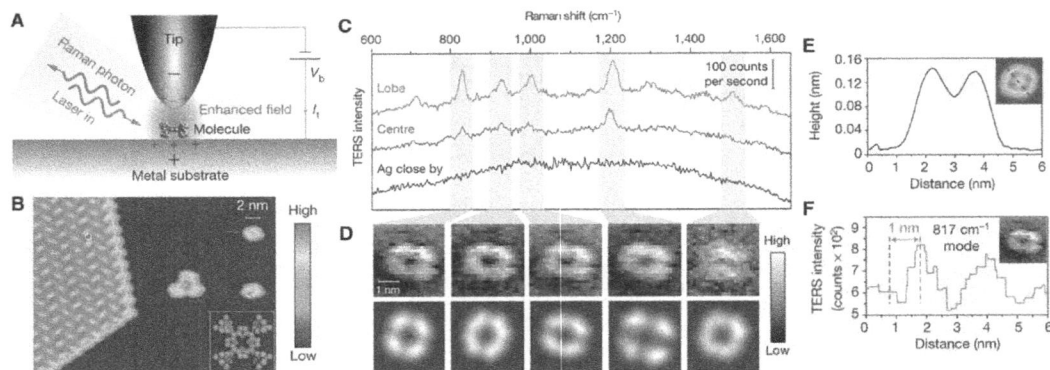

FIGURE 6.48 First demonstration of LT-UHV-STM-TERS with submolecular resolution. (a) Schematic tunneling-controlled TERS in a confocal-type side-illumination configuration, in which V_b is the sample bias and I_t is the tunneling current. (b) STM topograph of sub-monolayered H_2TBPP molecules on Ag(111) (1.5 V, 30 pA, 35×27 nm²). The inset shows the chemical structure of H_2TBPP. (c) Representative single-molecule TERS spectra (λ_{ex}=532 nm) on a lobe (red, 1) and center (blue, 2) of a flat-lying porphyrin-type molecule on Ag(111). The TERS spectrum on the bare Ag about 1 nm away from the molecule is also shown in black, 3. (d) The panels show experimental TERS mapping of a single molecule for different Raman peaks (~0.16 nm/pixel), processed from all individual TERS spectra acquired at each pixel. (e) Height profile of a line trace in the inset STM topograph (1 V, 20 pA). (d) TERS intensity profile of the same line trace for the inset Raman map associated with the 817 cm⁻¹ Raman peak, integrated over 800–852 cm⁻¹. (Figure and caption reprinted/adapted by permission from Springer Nature: *Nature* Ref. [45].)

location and orientation of the N–H bonds within the porphyrin core were resolved with subnanometer spatial resolution [46].

Indeed, nearly all the LT-UHV-STM-TERS studies that have achieved submolecular resolution rely upon liquid helium cooling to <10K. Under these conditions, the tip–sample junction becomes exceptionally stable, enabling studies where the STM feedback loop can be used to maintain sub-atomic separation between the tip and molecule adsorbed on the surface. Maintaining the absolute minimum tip–sample distance and therefore optimizing the enhancement and confinement of light at the tip apex [47,48], results in what is referred to as atomistic near-field tip-enhanced Raman spectroscopy where light is confined to essentially a picocavity [49]. In this manner, it becomes possible to image the localization of individual vibrational modes within the structure of a single molecule.

Two of the first demonstrations of atomistic near-field TERS considered porphyrin molecules. The Apkarian group studied a single Co(II)-tetraphenylporphyrin (CoTPP) molecule on a Cu(100) substrate at a temperature of 6K [47]. First, the authors scanned with the STM tip kept exceptionally close with the feedback loop to flatten the CoTPP molecule on the surface. This sufficiently increased the stability of the molecule on the surface resulting in the ability to maintain the tip within the atomistic contact limit and therefore image the vibrational modes of the planar molecule (Figure 6.49a). TERS images of the various vibrational modes are shown in Figure 6.49b. Significantly, the strong molecule–substrate interactions of the flattened molecule with the copper substrate and low temperature are essential to maintaining subatomic separation between the tip and

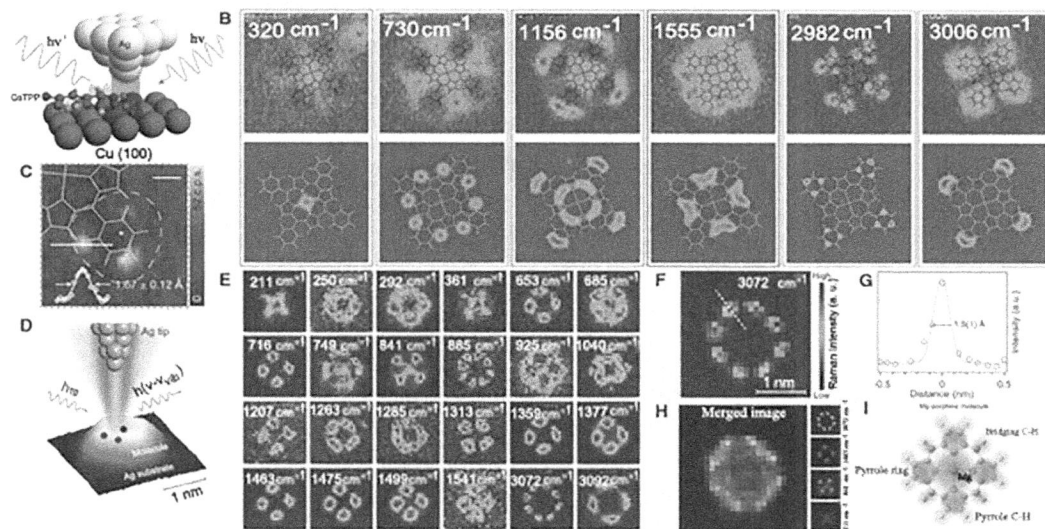

FIGURE 6.49 First demonstrations of atomistic near-field TERS for submolecular imaging. (a) Schematic of TERS of the CoTPP/Cu(100) system (λ_{ex}=634 nm). (b) Top, experimental maps of vibrational normal modes (29×29 Å^2, 64×64 pixels^2) with overlaid molecular frames. The vibrational frequencies are indicated. Bottom, simulations according to Hirshfeld partitioning. (Figure and caption reprinted by permission from Springer Nature: *Nature* Ref. [47], Copyright © 2019). (d) Schematic diagram of scanning Raman picoscopy (SRP) technique (λ_{ex}=532 nm). The STM topograph of a single target molecule adsorbed on Ag(100) is shown below. (e) SRP spatial mapping images (−0.02 V, 8 nA, 2.5×2.5 nm^2, 25×25 pixels^2, 2 s/pixel) corresponding to the Raman peaks noted on the respective images, revealing different spatial distribution patterns for different Raman modes. (f) SRP image at 3,072 cm^{-1} used for the estimation of spatial resolution. (g) Line profile of Raman signal intensities corresponding to the dashed line in (f), exhibiting a lateral spatial resolution down to 1.5(1) Å. (h) Merged SRP image from overlaying four different image patterns shown on the right corresponding to vibrational modes. (i) Mg-porphine molecule with the different substructures color-coded to match their characteristic vibrational modes shown in (h). (Figure and caption reprinted (adapted) from Ref. [48] under the terms of the Creative Commons Attribution (CC-BY) License, Copyright © 2019 The Authors.)

surface, as any instability would lead to undesirable tip–sample interactions. Ultimately, with this method they were able to visualize individual vibrational modes with submolecular resolution, finding the vibrational signatures of C–H bonds on a flattened phenyl group to be laterally localized to ~1.67 (Figure 6.49c). In a similar manner, the Dong group considered a single magnesium-porphine molecule on a Ag(100) substrate with liquid helium cooling (~7K) as shown in Figure 6.49d [48]. They were able to resolve distinct vibrational modes for a single molecule as represented by images of observed Raman peaks in Figure 6.49e with a reported lateral resolution of ~1.5 based on the localization of C–H stretching modes (Figure 6.49f and g). The authors reported the ability to localize specific vibrational modes that correspond to specific chemical groups in real space. They were able to use these TERS images to construct the chemical structure of a single molecule. They likened this process to assembling LEGO. As seen in Figure 6.49h, when combined into a single image, and color coded, the vibrational modes corresponding to the different parts of the molecule, such as the C-H groups, pyrroles, pyridines, and Mg–N bonds, resulted in a merged image where the complete molecular structure of a Mg-porphine molecule could be defined (Figure 6.49i).

Expanding beyond studies of individual porphyrin molecules on a surface, LT-UHV-STM-TERS has been used to investigate adsorption-induced distortion in [12] Cycloparaphenylene ([12]CPP) molecules [50]. The hoop-shaped π-conjugated molecules were considered on a Cu(100) surface at ~80K (Figure 6.50a) and a Ag(110) surface at ~70K. As shown in Figure 6.50b, a [12]CPP molecule adopts a conformation where the benzene units in the molecular hoop are alternately tilted, resulting in a geometry that is also reproduced in an STM image of [12]CPP on a Cu(100) surface (Figure 6.50c). LT-UHV-STM-TERS maps (Figure 6.50d) of such a molecule reveal that the notable vibrational modes (Figure 6.50e) exhibit a blurring spot feature across the entire molecule. This

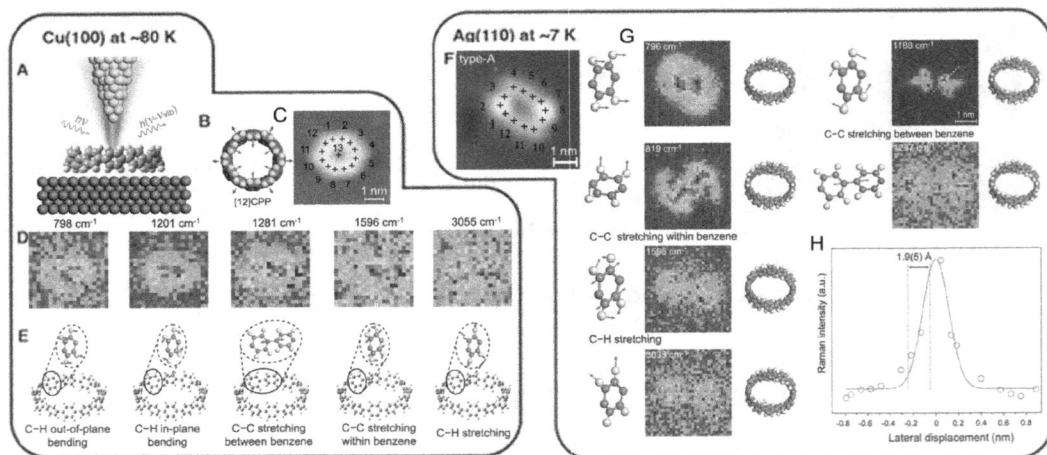

FIGURE 6.50 LT-UHV-STM-TERS of ([12]CPP) on Cu(100) and Ag(110) at two different cryogenic temperatures. (a) Schematic of TERS measurement. (b) Top view of the chemical structure of [12]CPP in free space, where the blue arrows indicate the benzene rings tilted outward and the red arrows indicate the benzene rings tilted inward. (c) STM image of a single [12]CPP molecule adsorbed on Cu(100) (−0.1 V, 2 pA). (d) TERS mapping images (−0.1 V, 400 pA, 4×4 nm², 20×20 pixels², 4 s/pixel) for specific vibrational modes. (e) Schematic of the corresponding vibrational modes showing the atomic displacements involved in the benzene ring units, highlighted by the red arrows for each mode. (f) STM image of a type A [12]CPP molecule on Ag(11). (−0.1 V, 2 pA). (g) Middle: TERS mapping images (−0.1 V, 1 nA, 4×4 nm², 32×32 pixels², 1 s/pixel) of type A [12]CPP corresponding to the six noted vibrational modes. Right: molecular structures highlighting the atoms involved in vibrational motions in pink. Left: the schematic of vibrations for the involved benzene units for each mode. (h) Line profile of Raman intensities corresponding to the white dashed line in (b), exhibiting a lateral spatial resolution of about 1.9(5) Å. (Figure and caption reprinted from Ref. [50] with the permission of AIP Publishing.)

suggests that the five modes can be observed either on the hoop or at the center. In contrast, when they considered a [12]CPP molecule on Ag(110) at ~7K, STM imaging revealed that the molecule adopts two types of saddle-like conformations. Only the type A conformation (Figure 6.50f) was stable enough to map with LT-UHV-STM-TERS. Here higher-resolution TERS maps revealed difference in symmetries and spatial distributions for the observed vibrational modes (Figure 6.50g) with a lateral spatial resolution estimated to be ~1.9(5) Å (Figure 6.50h). Although the molecule–substrate interactions are substantially different between the Cu(100) and Ag(110) surfaces, this study serves as one of the few instances where the spatial resolution of LT-UHV-STM-TERS performed at two different cryogenic temperatures can be compared for the study of the same single molecule. The increased stability found at <10K enables a new regime of submolecular resolution through atomistic near-field TERS.

With the ability to study vibrational modes in real space that correspond to signatures of individual chemical bonds established, two recent studies have used LT-UHV-STM-TERS to study the formation or dissociation of a single chemical bond. Xu et al. used pentacene on a Ag(110) surface as a model system to consider chemical transformation with LT-UHV-STM-TERS and non-contact (NC) AFM [51]. They found that by applying voltage pulses with the STM tip located over the molecule they could transform intact pentacene (α) into β species, with a slightly distorted dumbbell-like shape, and γ species, with a spindle-like shape. Furthermore, with TERS they could identify the dissociation of C–H bonds of the central benzene by tracking the C–H stretching mode (~2,800 cm^{-1}), finding that the transformation from α through β to γ results in the disappearance of the telltale vibrational signature. By combining single-molecule STM imaging with submolecular TERS maps, they were able to identify that the β and γ species arise from the dissociation of central C–H bonds, as confirmed with simulated Raman maps. Exceptional lateral spatial resolution revealed a shift in the real-space position of the observed Raman intensity of the C–H stretching of 0.89 Å for the β compared to the intact α species. Finally, they used the same method that they used previously for Mg-porphine, to create a merged image of a γ species from the combination of distinct vibrational modes resolved with submolecular resolution. Ultimately, they were able to determine the structural and chemical heterogeneity of a surface species at the single-bond limit.

While the previously mentioned study used submolecular LT-UHV-STM-TERS to consider the dissociation of a single chemical bond, Kim and co-workers found that it was possible to identify the nature of a chemical bond that connects molecules on a surface. Expanding from an initial study where they observed a nanoscale dehydrogenation reaction with LT-UHV-STM-TERS [53], they considered π-skeletons generated by the coupling of terminal alkynes at the single-chemical-level with topographic imaging and chemical imaging as shown in the schematic in Figure 6.51a [52]. Nondehydrogenative coupling of 4,4′-diethynyl-1,1′-biphenyl (DEBP) molecular precursors was found to result in sp–sp^2-carbon skeletons (enyne and cumulene), with the charge densities of the possible dimers illustrated in Figure 6.51b and c. By acquiring LT-UHV-STM-TERS point spectra over different connecting bonds, they were able to identify enyne C≡C and cumulene C=C carbon bonds (Figure 6.51d) based on the characteristic CC stretching modes of sp-carbon. Although there have only been a few studies that have used LT-UHV-STM-TERS to study for dissociation or formation of a single chemical bond, they have demonstrated the powerful ability to probe chemistry at the most fundamental limit even from this nascent stage.

Despite the potential to develop LT-UHV-STM-TERS as a "Chemiscope" to track individual molecular events in real time and with atomic resolution, most single-molecule TERS studies with submolecular resolution of vibrational modes have thus far focused largely on demonstrations of exceptional spatial resolution as opposed to studying dynamic chemistry. Atomic-scale factors become paramount, as everything from substrate and molecule selection to atomic asperities on the tip plays a role in confining light to a picocavity requiring exceptional levels of stability. Submolecular TERS images are truly impressive and yield new insight into chemical bonds and molecular vibrations. By focusing on the intrinsic vibrational features of individual molecules the spatial resolution of TERS has pushed beyond the initial theoretical expectations. At the same

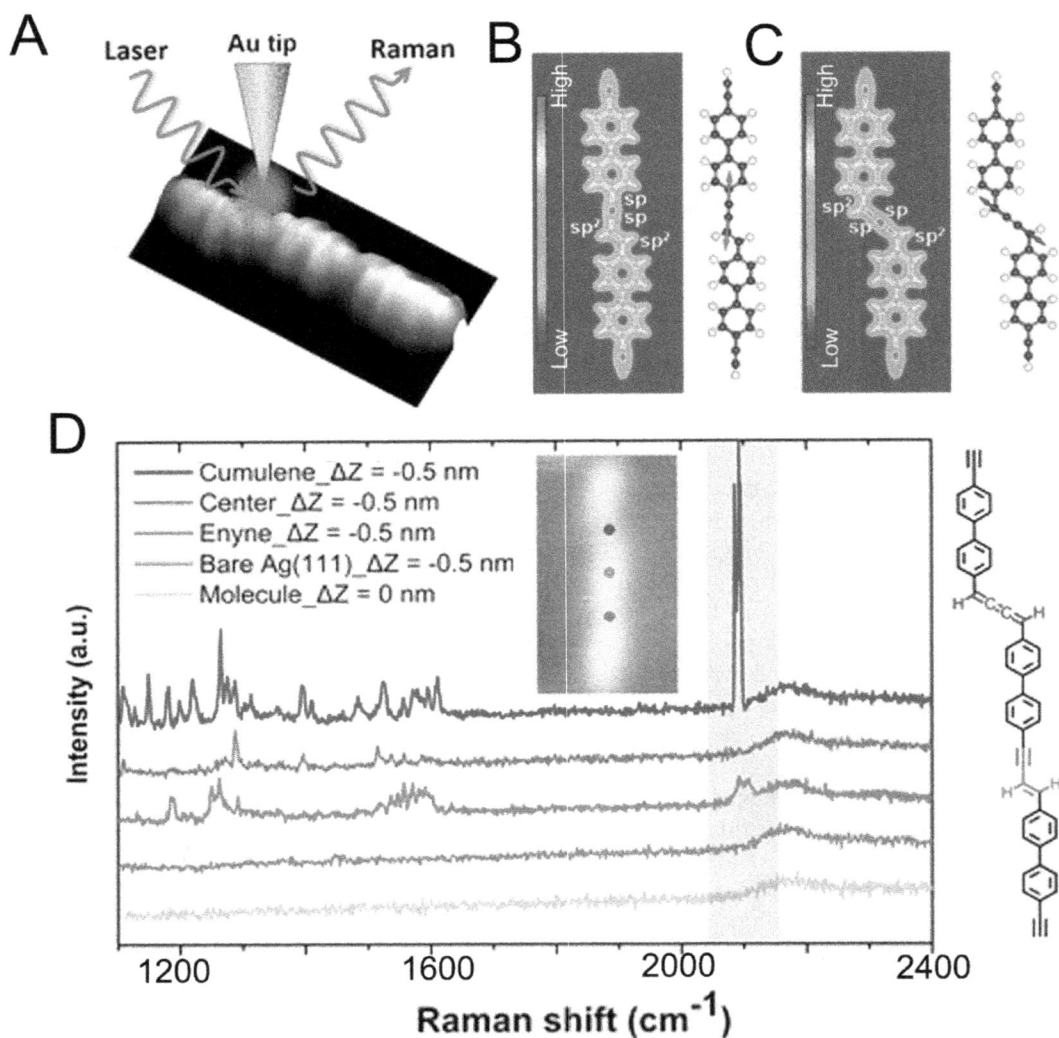

FIGURE 6.51 Submolecular resolution LT-UHV-STM-TERS of a chemical reaction. (a) Schematic illustration of the STM-TERS measurement to identify chemical bonds. [(b) and (c)] Charge density distributions of (b) enyne and (c) cumulene on Ag(111) (left panels), and the corresponding CC stretching modes of sp-carbon involved (right panels), respectively. (d) STM-TERS spectra (left) collected when the tip was placed at positions marked in the inset and the corresponding chemical structure of the trimer (right). ΔZ is the tip height offset with respect to the set point of $V_s = 0.1$ V, $I_t = 5$ pA above the molecule (with the feedback loop off). (Figure and caption reprinted (adapted) with permission from Ref. [52], Copyright © 2021 American Chemical Society.)

time, at stable cryogenic temperatures, it becomes possible to use the STM tip to manipulate single molecules to probe and manipulate chemistry or other properties that would otherwise elude characterization at the single-molecule level.

The manipulation of a single molecule provides the ability to isolate and study molecular interactions at the fundamental level. Recently, Li et al. used LT-UHV-STM-TERS to study interfacial strain in a vertical heterostructure of tetraphenyldibenzoperiflanthene (DBP) and borophene on a Ag(111) surface [54] (Figure 6.52a). TERS line profile measurements before and after the removal of molecules were used to verify the molecular origin of interfacial strain (Figure 6.52b and c).

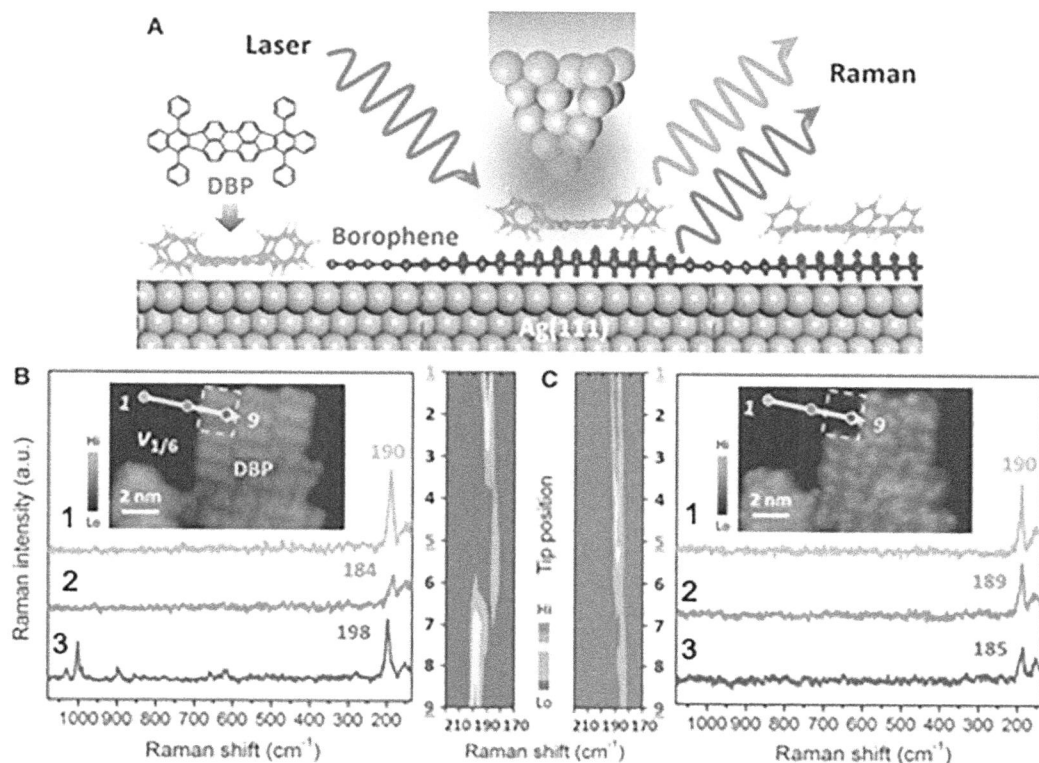

FIGURE 6.52 LT-UHV-STM-TERS study of interfacial strain in a vertical heterostructure with molecular manipulation. (a) Experimental schematic including the molecular structure of DBP. The blue arrows denote boron atomic displacements in the borophene monolayer due to molecular adsorption. [(b) and (c)] The verification of molecular-adsorption-induced localized interfacial strain with TERS line profiles and molecular manipulation. (b) The position and direction of a TERS line profile measurement are noted on the inset STM image, the characteristic spectra correspond to pristine (green, 1), borophene laterally adjacent to vertical DBP–borophene (red, 2), and DBP–borophene (blue, 3). The spectral evolution is adjacent. (c) Following the removal of two molecules marked with a white dashed box in the STM image inset (b), the TERS line profile measurement was repeated. Molecular removal was found to release highly localized strains. (Figure and caption reprinted (adapted) with permission from Ref. [54], Copyright © 2021 American Chemical Society.)

Furthermore, due to the spatial resolution of TERS, in this case, the lateral resolution was found to be 5 Å, the interfacial strain interaction was found to propagate ~1 nm beyond the molecular region into the adjacent borophene lattice. The high-energy resolution resulted in the ability to identify and therefore control local strain at the atomic scale with magnitudes as small as ~0.6%. Overall, through the combination of LT-UHV-STM-TERS with molecular manipulations Li et al. were able to spectroscopically investigate buried and highly localized interfacial interactions in an unprecedented manner [54].

Aside from laterally displacing a molecule on a surface, the STM tip can also be used to pick up a single molecule. In this manner, the probe tip can be functionalized to use a molecule adsorbed at the apex of the tip as a reporter that enables unique studies. There are many LT-UHV-STM and LT-UHV-NC-AFM studies that have used a probe tip functionalized by a single carbon monoxide (CO) molecule to image the topography of molecules and materials, resolving individual chemical bonds [25,55,56]. By functionalizing the LT-UHV-STM-TERS tip with a single CO molecule, the Schatz and Apkarian group found that the vibrational frequency of the CO stretching mode is sensitive to the bias applied with the STM function [57]. Furthermore, the Apkarian group found that it was possible to use the large Stark tuning rate of the CO stretch to map mechanical forces and electrostatic potentials with remarkable spatial resolution. As shown in Figure 6.53a, by monitoring the intensity, peak position, and full-width half max of the spectral peak corresponding to the CO stretching mode across a single CoTPP molecule on a Au(111) substrate they were able to visualize electrostatic interactions between the molecule and the CO functionalized probe [58]. They suggested that the apparent distribution between these images was the result of hydrogen bonding between CO and the positively charge H atoms found in the CoTPP molecule (Figure 6.53b). In another study, a CO-terminated tip was used to identify a single Ag atom and CO molecule on a Au(111) substrate [59]. The CO terminus of the scanning tip was used to map out intermolecular forces, based on spectral shifts of the CO vibration (Figure 6.53c). The STM topography suggested that a CO molecule was bound to a single Ag atom and locked into one orientation (Figure 6.53d and e). As can be seen in Figure 6.53f, the Ag atom and surface CO molecule resulted in opposite shifts in the frequency of the CO probe molecule revealing the different electrostatic forces between the probe and modified substrate. In later work, it is found that this method can be adapted to functionalize the tip with other polyatomic molecules besides for the traditional CO molecule. They were able to selectively adsorb a single CoTPP molecule through its cobalt center onto a Ag tip apex. And then use shifts in its observed vibrational modes to obtain ion-selective, atomically resolved images of an insulating Cu_2N on a Cu(100) substrate. Although these

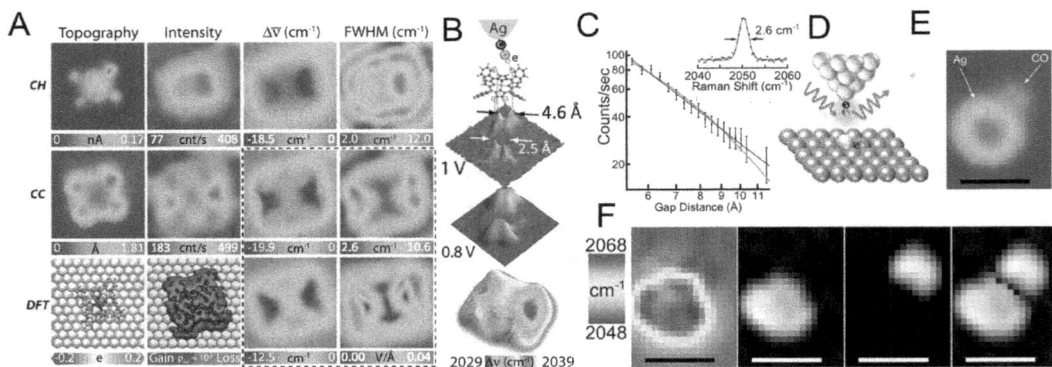

FIGURE 6.53 LT-UHV-STM-TERS studies with a CO-terminated tip. (a) STM topography and TERS-based images of a CoTPP molecule on Au(111). The $\Delta\bar{v}$ maps in CH and CC mode are referenced to 2,048.1 and 2,046.9 cm^{-1}, respectively. (b) Atomically resolved forces due to hydrogen bonding between CO and the indicated H atoms. The indicated voltages are the potential differences relative to the gold substrate, as measured by the Stark shift of CO. Electrostatic field mapped on the isosurface of local density of states (LDOS) (bottom). The common image size is 27×27 Å^2. The set point is 0.1 nA, 1.2 V. (Figure and caption reprinted (adapted) under a Creative Commons Attribution (CC-BY) License from Ref. [58], Copyright © 2018 The Authors). [(c)–(f)] Chemical speciation of a single Ag atom and CO molecule using a CO-terminated tip. (c) Decay of the integrated intensity of the CO vibrational line (inset) as a function of gap distance. Inset: A CO spectrum recorded at 1.2 V. (d) Schematic of the measurement. (e) Topographic image (1.2 V, 0.1 nA) of a Ag atom and CO molecule. (f) TERS relayed molecular force microscopy (0.37 V, 1.0 nA), from left to right: Map of CO stretch frequency obtained via a Gaussian fit. The frequency red shifts over the Ag atom and blueshifts over the CO molecule with respect to the Au(111) surface. When integrating the Raman intensity, the silver atom appears in the spectral window (e) 2,042–2,055 cm^{-1} while the CO molecule is best mapped in the window 2,072–2,077 cm^{-1}. A composite image. Scale bar = 1 nm. (Figure and caption reprinted (adapted) under the terms of the ACS AuthorChoice License from Ref. [59], Copyright © 2017 American Chemical Society.)

methods are highly technically and instrumentally demanding, these early demonstrations of using LT-UHV-STM-TERS with a functionalized probe tip have provided new insight into the complicated interactions that occur at the atomic and molecular scale. Future applications of methods such as these offer the possibility to investigate chemistry and physical properties at an unprecedented scale.

In conclusion, even though LT-UHV-STM-TERS is still a novel technique its potential has been established with an increasing number of applications. Improvements in experimental conditions, including a move towards use of liquid helium cooling to achieve <10K, and theories have led to a new regime of spatial resolution. Although most studies with submolecular spatial resolution rely on porphyrin molecules to benefit from resonance Raman, more recent research has shown that this is not inherently necessary. And as a result, research is expanding from proofs of exceptional spatial resolution to instead addressing interesting and unique chemistry and materials questions, including heterogeneous catalysis and the fabrication of new low-dimensional materials. As integrated instruments capable of LT-UHV-STM-TERS become commercially available, the barrier to entry becomes significantly lower. However, although UHV conditions enable the relatively long-term use of a probe tip, the initial fabrication and maintenance of a reliable tip remains a challenge. Optical access and alignment into the LT-UHV-STM tip–sample junction will be a constant challenge, but the studies previously discussed highlight the incredible potential for this method to consider and understand chemistry at the spatial limit.

REFERENCES

1. Lyon, L. A.; Keating, C. D.; Fox, A. P.; Baker, B. E.; He, L.; Nicewarner, S. R.; Mulvaney, S. P.; Natan, M. J., Raman spectroscopy. *Anal. Chem.* 1998, *70*(12), 341–362.
2. Tackman, E. C.; Trujillo, M. J.; Lockwood, T.-L. E.; Merga, G.; Lieberman, M.; Camden, J. P., Identification of substandard and falsified antimalarial pharmaceuticals chloroquine, doxycycline, and primaquine using surface-enhanced Raman scattering. *Analytical Methods* 2018, *10*(38), 4718–4722.
3. Gu, X.; Trujillo, M. J.; Olson, J. E.; Camden, J. P., SERS sensors: Recent developments and a generalized classification scheme based on the signal origin. *Annu. Rev. Anal. Chem.* 2018, *11*(1), 147–169.
4. Domke, K. F.; Zhang, D.; Pettinger, B., Toward Raman fingerprints of single dye molecules at atomically smooth Au(111). *J. Am. Chem. Soc.* 2006, *128*(45), 14721–14727.
5. Hayazawa, N.; Watanabe, H.; Saito, Y.; Kawata, S., Towards atomic site-selective sensitivity in tip-enhanced Raman spectroscopy. *J. Chem. Phys.* 2006, *125*(24), 244706.
6. Schultz, J. F.; Mahapatra, S.; Li, L.; Jiang, N., The expanding frontiers of tip-enhanced Raman spectroscopy. *Appl. Spectrosc.* 2020, *74*(11), 1313–1340.
7. Klingsporn, J. M.; Jiang, N.; Pozzi, E. A.; Sonntag, M. D.; Chulhai, D.; Seideman, T.; Jensen, L.; Hersam, M. C.; Duyne, R. P. V., Intramolecular insight into adsorbate–substrate interactions via low-temperature, ultrahigh-vacuum tip-enhanced Raman spectroscopy. *J. Am. Chem. Soc.* 2014, *136*(10), 3881–3887.
8. Park, K.-D.; Muller, E. A.; Kravtsov, V.; Sass, P. M.; Dreyer, J.; Atkin, J. M.; Raschke, M. B., Variable-temperature tip-enhanced Raman spectroscopy of single-molecule fluctuations and dynamics. *Nano Lett.* 2016, *16*(1), 479–487.
9. Jiang, S.; Zhang, Y.; Zhang, R.; Hu, C.; Liao, M.; Luo, Y.; Yang, J.; Dong, Z.; Hou, J. G., Distinguishing adjacent molecules on a surface using plasmon-enhanced Raman scattering. *Nat. Nanotechnol.* 2015, *10*, 865.
10. Jiang, S.; Zhang, X.; Zhang, Y.; Hu, C.; Zhang, R.; Zhang, Y.; Liao, Y.; Smith, Z. J.; Dong, Z.; Hou, J. G., Subnanometer-resolved chemical imaging via multivariate analysis of tip-enhanced Raman maps. *Light: Science & Applications* 2017, *6*, e17098.
11. Mahapatra, S.; Ning, Y.; Schultz, J. F.; Li, L.; Zhang, J.-L.; Jiang, N., Angstrom scale chemical analysis of metal supported trans- and cis-regioisomers by ultrahigh vacuum tip-enhanced Raman mapping. *Nano Lett.* 2019, *19*(5), 3267–3272.
12. Mahapatra, S.; Schultz, J. F.; Ning, Y.; Zhang, J.-L.; Jiang, N., Probing surface mediated configurations of nonplanar regioisomeric adsorbates using ultrahigh vacuum tip-enhanced Raman spectroscopy. *Nanoscale* 2019, *11*(42), 19877–19883.

13. Zhang, R.; Zhang, X.; Wang, H.; Zhang, Y.; Jiang, S.; Hu, C.; Zhang, Y.; Luo, Y.; Dong, Z., Distinguishing individual DNA bases in a network by non-resonant tip-enhanced Raman scattering. *Angew. Chem. Int. Ed.* 2017, *56*(20), 5561–5564.

14. Jiang, S.; Zhang, R.; Zhang, X.-B.; Zhang, Y.; Zhang, Y.; Dong, Z.-C., Bicomponent supramolecular self-assemblies studied with tip-enhanced Raman spectroscopy. *J. Raman Spectrosc.* 2021, *52*(2), 366–374.

15. Srivastava, R. P.; Zaidi, H. R., Intermolecular forces revealed by Raman scattering. In *Raman Spectroscopy of Gases and Liquids*, Weber, A., Ed. Berlin, Heidelberg: Springer, 1979, pp. 167–201.

16. Moskovits, M., Surface selection rules. *J. Chem. Phys.* 1982, *77*(9), 4408–4416.

17. Zhang, Y.; Zhang, R.; Jiang, S.; Zhang, Y.; Dong, Z.-C., Probing adsorption configurations of small molecules on surfaces by single-molecule tip-enhanced Raman spectroscopy. *ChemPhysChem* 2018, *0*(0).

18. Schultz, J. F.; Li, L.; Mahapatra, S.; Shaw, C.; Zhang, X.; Jiang, N., Defining multiple configurations of rubrene on a Ag(100) surface with 5 å spatial resolution via ultrahigh vacuum tip-enhanced Raman spectroscopy. *J. Phys. Chem. C* 2020, *124*(4), 2420–2426.

19. Whiteman, P. J.; Schultz, J. F.; Porach, Z. D.; Chen, H.; Jiang, N., Dual binding configurations of sub-phthalocyanine on Ag(100) substrate characterized by scanning tunneling microscopy, tip-enhanced Raman spectroscopy, and density functional theory. *J. Phys. Chem. C* 2018, *122*(10), 5489–5495.

20. Schultz, J. F.; Li, S.; Jiang, S.; Jiang, N., Optical scanning tunneling microscopy based chemical imaging and spectroscopy. *J. Phys.: Condens. Matter* 2020, *32*(46), 463001.

21. Jiang, N.; Chiang, N.; Madison, L. R.; Pozzi, E. A.; Wasielewski, M. R.; Seideman, T.; Ratner, M. A.; Hersam, M. C.; Schatz, G. C.; Van Duyne, R. P., Nanoscale chemical imaging of a dynamic molecular phase boundary with ultrahigh vacuum tip-enhanced Raman spectroscopy. *Nano Lett.* 2016, *16*(6), 3898–3904.

22. Schultz, J. F.; Li, L.; Mahapatra, S.; Shaw, C.; Zhang, X.; Jiang, N., Defining multiple configurations of rubrene on a Ag(100) surface with 5 å spatial resolution via ultrahigh vacuum tip-enhanced Raman spectroscopy. *J. Phys. Chem. C* 2020, *124*(4), 2420–2426.

23. Lee, J.; Tallarida, N.; Chen, X.; Liu, P.; Jensen, L.; Apkarian, V. A., Tip-enhanced Raman spectro-microscopy of Co(II)-tetraphenylporphyrin on Au(111): Toward the chemists' microscope. *ACS Nano* 2017, *11*(11), 11466–11474.

24. Jaculbia, R. B.; Imada, H.; Miwa, K.; Iwasa, T.; Takenaka, M.; Yang, B.; Kazuma, E.; Hayazawa, N.; Taketsugu, T.; Kim, Y., Single-molecule resonance Raman effect in a plasmonic nanocavity. *Nat. Nanotechnol.* 2020.

25. Gross, L., Recent advances in submolecular resolution with scanning probe microscopy. *Nat. Chem.* 2011, *3*(4), 273–278.

26. Lee, J.; Tallarida, N.; Rios, L.; Ara Apkarian, V., The Raman spectrum of a single molecule on an electrochemically etched silver tip. *Applied Spectroscopy* 2020, *74*(11), 1414–1422.

27. Tallarida, N.; Rios, L.; Apkarian, V. A.; Lee, J., Isomerization of one molecule observed through tip-enhanced Raman spectroscopy. *Nano Lett.* 2015, *15*(10), 6386–6394.

28. Jaculbia, R.; Hayazawa, N.; Imada, H.; Kim, Y., Controlling the resonance Raman effect in tip-enhanced Raman spectroscopy using a thin insulating film. *Applied Spectroscopy* 2020, *74*(11), 1391–1397.

29. Jaculbia, R. B.; Imada, H.; Miwa, K.; Iwasa, T.; Takenaka, M.; Yang, B.; Kazuma, E.; Hayazawa, N.; Taketsugu, T.; Kim, Y., Single-molecule resonance Raman effect in a plasmonic nanocavity. *Nat. Nanotechnol.* 2020, *15*, 105–110.

30. Pettinger, B.; Domke, K. F.; Zhang, D.; Picardi, G.; Schuster, R., Tip-enhanced Raman scattering: Influence of the tip-surface geometry on optical resonance and enhancement. *Surf. Sci.* 2009, *603*(10–12), 1335–1341.

31. Roelli, P.; Galland, C.; Piro, N.; Kippenberg, T. J., Molecular cavity optomechanics as a theory of plasmon-enhanced Raman scattering. *Nat. Nanotechnol.* 2016, *11*(2), 164–169.

32. Aspelmeyer, M.; Kippenberg, T. J.; Marquardt, F., Cavity optomechanics. *Rev. Mod. Phys.* 2014, *86*(4), 1391–1452.

33. Zhang, C.; Chen, B.-Q.; Li, Z.-Y., Optical origin of subnanometer resolution in tip-enhanced Raman mapping. *J. Phys. Chem. C* 2015, *119*(21), 11858–11871.

34. Chen, B.-Q.; Zhang, C.; Li, J.; Li, Z.-Y.; Xia, Y., On the critical role of Rayleigh scattering in single-molecule surface-enhanced Raman scattering via a plasmonic nanogap. *Nanoscale* 2016, *8*(34), 15730–15736.

35. Schmidt, M. K.; Esteban, R.; González-Tudela, A.; Giedke, G.; Aizpurua, J., Quantum mechanical description of Raman scattering from molecules in plasmonic cavities. *ACS Nano* 2016, *10*(6), 6291–6298.

36. Benz, F.; Schmidt, M. K.; Dreismann, A.; Chikkaraddy, R.; Zhang, Y.; Demetriadou, A.; Carnegie, C.; Ohadi, H.; de Nijs, B.; Esteban, R.; Aizpurua, J.; Baumberg, J. J., Single-molecule optomechanics in "picocavities". *Science* 2016, *354*(6313), 726–729.

37. Baumberg, J. J.; Aizpurua, J.; Mikkelsen, M. H.; Smith, D. R., Extreme nanophotonics from ultrathin metallic gaps. *Nat. Mater.* 2019, *18*(7), 668–678.

38. Carnegie, C.; Griffiths, J.; de Nijs, B.; Readman, C.; Chikkaraddy, R.; Deacon, W. M.; Zhang, Y.; Szabó, I.; Rosta, E.; Aizpurua, J.; Baumberg, J. J., Room-temperature optical picocavities below 1 nm^3 accessing single-atom geometries. *The Journal of Physical Chemistry Letters* 2018, *9*(24), 7146–7151.

39. Alvarez-Puebla, R.; Liz-Marzán, L. M.; García de Abajo, F. J., Light concentration at the nanometer scale. *The Journal of Physical Chemistry Letters* 2010, *1*(16), 2428–2434.

40. Barbry, M.; Koval, P.; Marchesin, F.; Esteban, R.; Borisov, A. G.; Aizpurua, J.; Sánchez-Portal, D., Atomistic near-field nanoplasmonics: Reaching atomic-scale resolution in nanooptics. *Nano Lett.* 2015, *15*(5), 3410–3419.

41. Liu, P.; Chulhai, D. V.; Jensen, L., Single-molecule imaging using atomistic near-field tip-enhanced Raman spectroscopy. *ACS Nano* 2017, *11*(5), 5094–5102.

42. Chen, X.; Jensen, L., Morphology dependent near-field response in atomistic plasmonic nanocavities. *Nanoscale* 2018, *10*(24), 11410–11417.

43. Chen, X.; Liu, P.; Hu, Z.; Jensen, L., High-resolution tip-enhanced Raman scattering probes submolecular density changes. *Nat. Commun.* 2019, *10*(1), 2567.

44. Liu, P.; Chen, X.; Ye, H.; Jensen, L., Resolving molecular structures with high-resolution tip-enhanced Raman scattering images. *ACS Nano* 2019, *13*(8), 9342–9351.

45. Zhang, R.; Zhang, Y.; Dong, Z. C.; Jiang, S.; Zhang, C.; Chen, L. G.; Zhang, L.; Liao, Y.; Aizpurua, J.; Luo, Y.; Yang, J. L.; Hou, J. G., Chemical mapping of a single molecule by plasmon-enhanced Raman scattering. *Nature* 2013, *498*(7452), 82–6.

46. Ghafoor, A.; Yang, B.; Yu, Y.-J.; Zhang, Y.-F.; Zhang, X.-B.; Chen, G.; Zhang, Y.; Zhang, Y.; Dong, Z.-C., Site-dependent TERS study of a porphyrin molecule on Ag(100) at 7 k. *Chin. J. Chem. Phys.* 2019, *32*(3), 287–291.

47. Lee, J.; Crampton, K. T.; Tallarida, N.; Apkarian, V. A., Visualizing vibrational normal modes of a single molecule with atomically confined light. *Nature* 2019, *568*(7750), 78–82.

48. Zhang, Y.; Yang, B.; Ghafoor, A.; Zhang, Y.; Zhang, Y.-F.; Wang, R.-P.; Yang, J.-L.; Luo, Y.; Dong, Z.-C.; Hou, J. G., Visually constructing the chemical structure of a single molecule by scanning Raman picoscopy. *National Science Review* 2019, *6*(6), 1169–1175.

49. Mahapatra, S.; Li, L.; Schultz, J. F.; Jiang, N., Tip-enhanced Raman spectroscopy: Chemical analysis with nanoscale to Angstrom scale resolution. *J. Chem. Phys.* 2020, *153*(1), 010902.

50. Li, H.; Zhang, Y.-F.; Zhang, X.-B.; Farrukh, A.; Zhang, Y.; Zhang, Y.; Dong, Z.-C., Probing the deformation of [12]cycloparaphenylene molecular nanohoops adsorbed on metal surfaces by tip-enhanced Raman spectroscopy. *J. Chem. Phys.* 2020, *153*(24), 244201.

51. Xu, J.; Zhu, X.; Tan, S.; Zhang, Y.; Li, B.; Tian, Y.; Shan, H.; Cui, X.; Zhao, A.; Dong, Z.; Yang, J.; Luo, Y.; Wang, B.; Hou, J. G., Determining structural and chemical heterogeneities of surface species at the single-bond limit. *Science* 2021, *371*(6531), 818–822.

52. Zhang, C.; Jaculbia, R. B.; Tanaka, Y.; Kazuma, E.; Imada, H.; Hayazawa, N.; Muranaka, A.; Uchiyama, M.; Kim, Y., Chemical identification and bond control of π-skeletons in a coupling reaction. *J. Am. Chem. Soc.* 2021, *143*(25), 9461–9467.

53. Chaunchaiyakul, S.; Setiadi, A.; Krukowski, P.; Catalan, F. C. I.; Akai-Kasaya, M.; Saito, A.; Hayazawa, N.; Kim, Y.; Osuga, H.; Kuwahara, Y., Nanoscale dehydrogenation observed by tip-enhanced Raman spectroscopy. *J. Phys. Chem. C* 2017, *121*(33), 18162–18168.

54. Li, L.; Schultz, J. F.; Mahapatra, S.; Liu, X.; Shaw, C.; Zhang, X.; Hersam, M. C.; Jiang, N., Angstrom-scale spectroscopic visualization of interfacial interactions in an organic/borophene vertical heterostructure. *J. Am. Chem. Soc.* 2021, *143*(38), 15624–15634.

55. Gross, L.; Mohn, F.; Moll, N.; Liljeroth, P.; Meyer, G., The chemical structure of a molecule resolved by atomic force microscopy. *Science* 2009, *325*(5944), 1110–1114.

56. Liu, X.; Wang, L.; Li, S.; Rahn, M. S.; Yakobson, B. I.; Hersam, M. C., Geometric imaging of borophene polymorphs with functionalized probes. *Nat. Commun.* 2019, *10*(1), 1642.

57. Gieseking, R. L. M.; Lee, J.; Tallarida, N.; Apkarian, V. A.; Schatz, G. C., Bias-dependent chemical enhancement and nonclassical stark effect in tip-enhanced Raman spectromicroscopy of CO-terminated Ag tips. *J. Phys. Chem. Lett.* 2018, *9*(11), 3074–3080.

58. Lee, J.; Tallarida, N.; Chen, X.; Jensen, L.; Apkarian, V. A., Microscopy with a single-molecule scanning electrometer. *Sci. Adv.* 2018, *4*(6).

59. Tallarida, N.; Lee, J.; Apkarian, V. A., Tip-enhanced Raman spectromicroscopy on the Angstrom scale: Bare and CO-terminated Ag tips. *ACS Nano* 2017, *11*(11), 11393–11401.

6.4 TERS STUDY

6.4.1 TERS CHARACTERIZATION OF INORGANIC MATERIALS

Inorganic materials, in particular various 2D materials benefited the most from advances in TERS instrumentation and methodology. We've already discussed TERS imaging in CNTs and graphene in prior chapters and will cover here some inorganic materials that have been successfully characterized using TERS or TEPL:

- Strained silicon
- MoO_3
- MXenes
- Quantum dots
- Vertical and lateral heterostructures of transition metal dichalcogenides (TMD)

This list is by far not exhausting, many more materials have been successfully imaged with TERS/TEPL – SnS/SnS_2, $TaSe_3$, InSe, WTe_2, BiOCl, alloys of various TMD, etc. This list is growing every year as many more scientifically relevant samples are being probed by TERS and TEPL imaging.

6.4.1.1 Strained Silicon

Strained silicon has been broadly used in the semiconductor industry to improve the mobility of the charge carriers in semiconductor devices. Precise control of the degree of strain was obviously of high importance. In 2007, the group of Alexei Sokolov [1] demonstrated that the side-access TERS excitation–collection scheme combined with precise polarization control in the excitation and collection channels can lead to high-contrast TERS sensing of thin (30 nm) strained silicon grown on silicon oxide layer (Figure 6.54).

Unfortunately, this early success did not lead to establishing TERS as the methodology of choice for the strain control in the semiconductor industry due to the lack at that time of commercially available instrumentation and highly enhancing TERS probes with consistent performance.

6.4.1.2 MoO_3

α-MoO_3 belongs to the class of Van der Waals materials with peculiar structure of the "monolayer" which consists of two offset lattices of MoO_6 octahedra (Figure 6.55). Crystals that were directly grown on or exfoliated to the substrate show reach Raman response with over a dozen sharp Raman

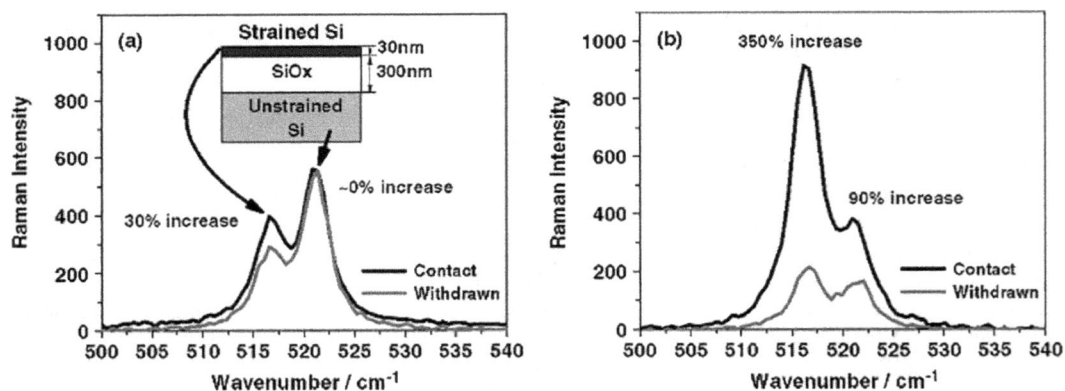

FIGURE 6.54 TERS (black lines) and the far-field (gray line) spectra from the strained silicon sample sketched in the insert. (a) Non-optimized polarization, only the signal from strained silicon is enhanced by about 30% of the far-field signal level. (b) Optimized (70°) polarization angle in the excitation and collection lines leads to much-improved contrast. Adopted with permission from Ref. [1].

lines, relative intensities of which depend on the mutual orientation of the polarizers in the excitation and collection lines [2].

Tip-enhanced Raman spectroscopy imaging of thin, 15–25 nm crystals of α-MoO$_3$ proved to be successful [3] with 638 or 785 nm excitation. Somewhat surprisingly, TERS spectra were dominated by the 995 cm^{-1} peak which corresponds to the out-of-plane vibration of the only non-shared oxygen atom at the tips of octahedral lattices (Figure 6.56).

The degree of this preferential enhancement was rather unexpected, especially taking into account the fact that the thickness of the crystals (15–25 nm) excluded the classic gap-mode conditions. Another interesting aspect of TERS response in MoO$_3$ is the fact that this material is translucent, while most other inorganic materials on which successful TERS imaging was demonstrated absorb strongly in visible range and in the bulk form look dark.

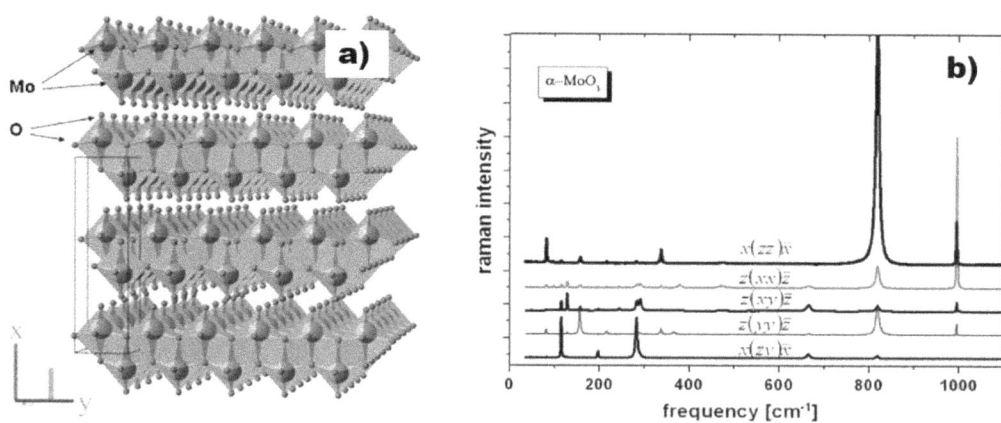

FIGURE 6.55 (a) Structure of α-MoO$_3$ crystal showing "monolayers" consisting of layers of two offset octahedral lattices. (b) Polarized Raman spectra of α-MoO$_3$ with different orientations of the polarization in excitation and collection channels. Adopted with permission from Ref. [2].

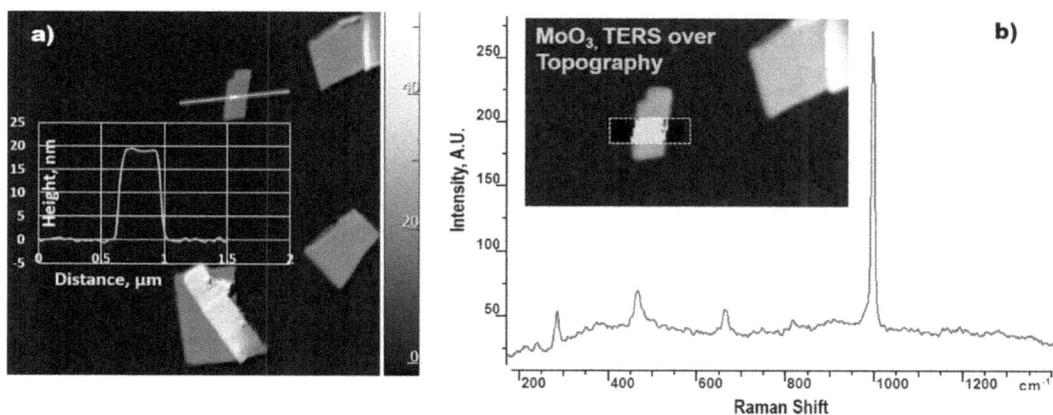

FIGURE 6.56 (a) Topography and corresponding section graph of a MoO$_3$ crystal from the same batch as reported in Ref. [3], showing that the height of the crystal to be probed with TERS was about 20 nm. (b) Averaged TERS spectrum of MoO$_3$ crystal, in the insert – a TERS map (intensity of the 995 cm^{-1} band) overlaid over topography. Images courtesy of HORIBA Scientific.

6.4.1.3 MXenes

MXenes are the members of the diverse class of novel 2-dimensional materials consisting of alternating layers of transition metal atoms such as Ti, V, Mo, Nb, Ta, etc., and carbon, or nitrogen [4]. Monolayers of MXenes can consist of three, five or seven atomic layers in which correspondingly one, two, or three layers of transition metal atoms are sandwiched between the layers of carbon or nitrogen. MXenes are obtained by selective etching of so-called MAX phases where "A" atoms (usually Al) are covalently bound to the transition metal carbide/nitride layers (Figure 6.57).

This possibility to vary the number of sublayers along with the possibility to use atoms of several transition metals within the same MXene, either in the form of stochastic alloy or with alternating continuous layers of atoms of different transition metals creates immense diversity of MXenes and allows fine tuning of their properties.

Raman spectroscopy proved to be extremely helpful for characterization of MXenes [5], though Raman *imaging* of this novel class of materials remained extremely challenging due to their fairly weak Raman response and consequently – necessity to integrate signal for 10 or even 100 of seconds in order to achieve reasonable signal-to-noise ratio, which is de facto a prohibitive condition for the reasonable pixel density Raman imaging. Increasing the laser power in Raman experiments has its limitation as excessive heating of the sample in the laser focus can alter or even completely destroy MXene flakes.

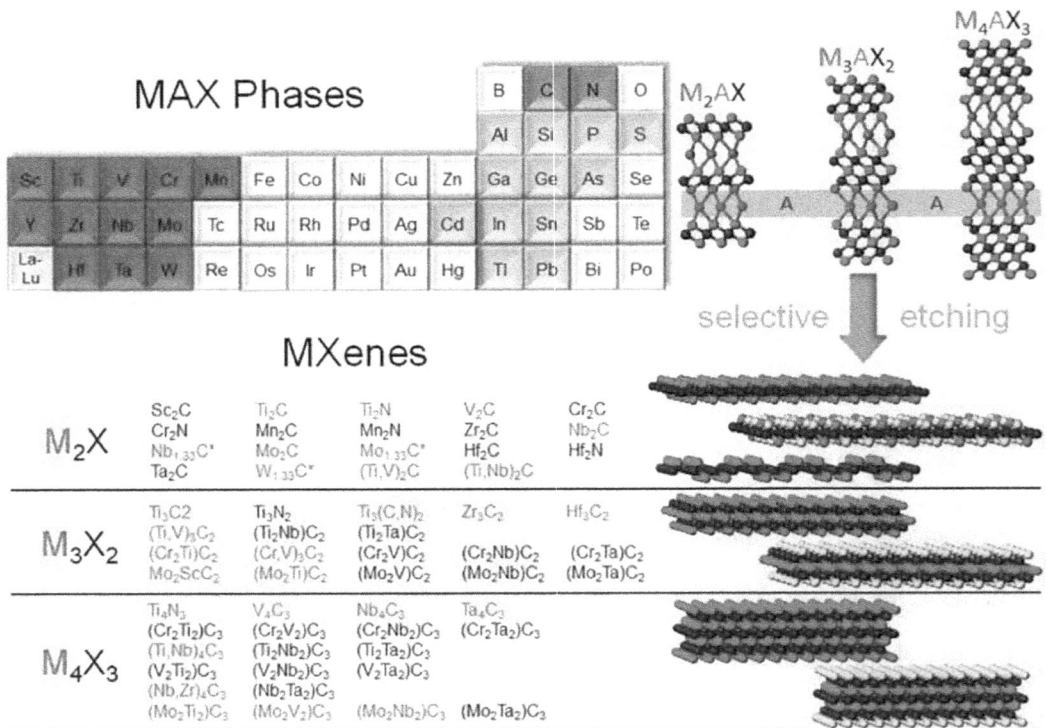

FIGURE 6.57 MAX and MXene genomes: With combinations of 12 transition metals (red), 12 group A elements (blue), and 2X elements (black), close to 100 MAX phases that have either M2AX, M3AX2, or M4AX3 structures have been reported to date. Moreover, transition metals in the M layer can form a solid solution and/or double-M-ordered phases leading to numerous complex multielemental phases. By selective etching of A layer from MAX phases, close to 30 MXenes have been experimentally synthesized (marked in green) and many more theoretically predicted. The asterisks indicate MXenes with ordered divacancies. Used with permission from Ref. [4].

In late 2020 it was demonstrated that Ti_3C_2, by far the most popular MXene with the most pol-ished synthetic protocol, produces a strong TERS signal with 785 nm excitation when deposited on gold or silver substrate. It enabled first time ever efficient Raman imaging of mono- to few-layer-thick crystals of Ti_3C_2 with the laser power density on the sample several times lower compared to earlier confocal Raman measurements. In the course of TERS imaging study of this MXene, it was found that the intensity of the major TERS band in Ti_3C_2 at 201 cm^{-1} decreases considerably faster with the increase of the MXene crystal thickness compared to the intensity of two other sharp bands at 126 and 725 cm^{-1} which allows assessment of the number of layers in MXene flakes based on their TERS spectra.

In addition to the thickness dependence of TERS spectra of Ti_3C_2, it turned out that TERS response from the wrinkles which inevitably form during the deposition of MXene flakes from aqueous suspensions is different compared to the adjacent flat portion of the flake (Figure 6.58) – the absolute intensity of the side 126 and 725 cm^{-1} bands was increasing over the wrinkles.

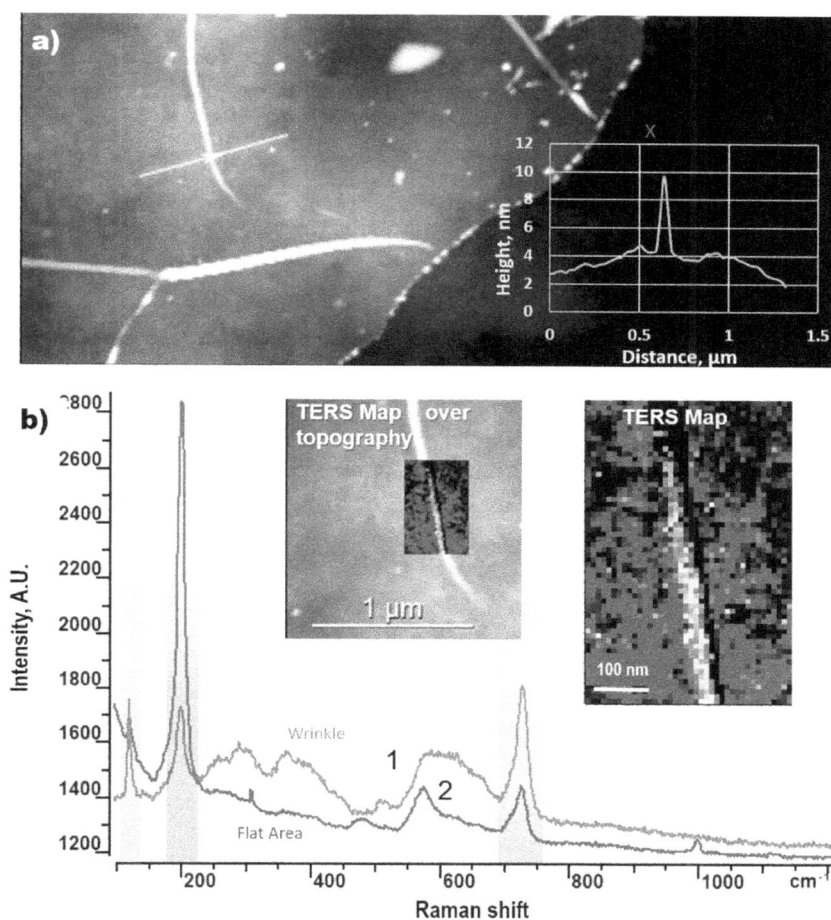

FIGURE 6.58 (a) Topography image of a monolayer Ti_3C_2 flake deposited on template-stripped gold sub-strate. In the insert section, graph of an approximately 5–6 nm high wrinkle. (b) Averaged TERS spectra from the flat area (blue spectrum, 2) and from the wrinkle (red spectrum, 1). In the insert – combined TERS map showing the distribution of the intensity of correspondingly colored bands overlaid over the topography and expanded view of the same TERS map. Sample courtesy of Asia Sarycheva and Yury Gogotsi, Drexel University. Images courtesy of HORIBA Scientific.

What's particularly interesting about this fact is that these two bands are of different nature – the 126 cm⁻¹ is an in-plane vibration, while the 725 cm⁻¹ is out-of-plane one [5], which indicates that the mechanism of the increase of the TERS signal over Ti_3C_2 wrinkles is different compared to the one observed in wrinkles of graphene and graphene oxide [6], which was assigned to the mutual orientation of the in-plane D, G and 2D modes with the optical electric field in the tip-substrate cavity.

The first manuscript on TERS imaging of MXenes has been submitted at the time of writing of this chapter. Many more exciting discoveries in this class of materials revealed by TERS imaging are expected as the first proof-of-principle measurements demonstrated the possibility of TERS imaging not only of several pristine MXenes, but also their vertical heterostructures with TMD and graphene oxide.

6.4.1.4 Quantum Dots

Quantum dots are just another important class of nanomaterials that are actively explored/used in such important applications as light-emitting diodes (LEDs) [7] and quantum information processing [8]. First, TEPL experiments on CdSe/ZnS QDs were reported by Protasenko et al. [9], where the authors used evanescent excitation scheme with perturbation caused by the apex of Si AFM probe. The early and relatively recent TEPL measurements on QD are nicely summarized in a fresh review by Lee et al. [10]. Advances in TERS/TEPL methodology benefited QD applications as can be seen from a cutting-edge report by Wang, Zamkov, and El-Khoury [11], in which the authors demonstrated about 5 nm spatial resolution (Figure 6.59) in ambient TEPL imaging of ZnS/CdSe QD using the same TERS/TEPL imaging methodology as was successfully applied earlier for characterization of 1D, 2D materials and molecular layers in which the maximum enhancement of the TERS/TEPL signal is achieved when the TERS probe is in direct gentle contact with the sample, while the TERS/TEPL enhancement largely disappears when the tip is oscillating with 20–40 nm

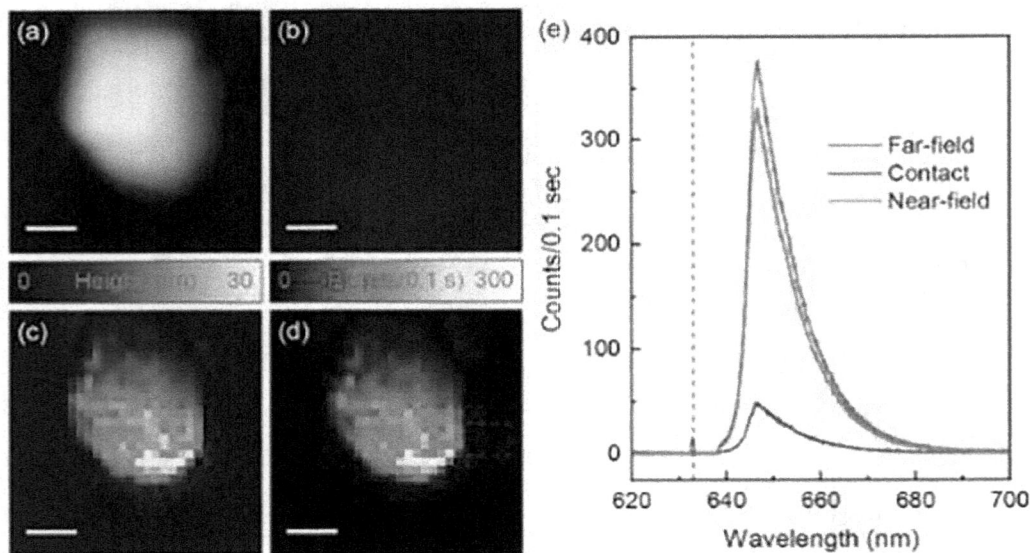

FIGURE 6.59 AFM (a), far-field (b), and contact mode (c) images. These spectra were recorded using intermittent AFM feedback. The near-field image (d) is obtained by subtracting the far-field (part b) from the contact mode (part c) image. Spatially averaged spectra taken from a region tracing the entire nanodot island are shown on the same plot (e). Note that the vertical dashed line in this panel indicated the excitation wavelength used. The optical spectra were recorded using a 633 nm laser (12 μW/μm²) and were time-integrated for 0.1 s. The scale bar indicates 30 nm in panels (a–d), and the lateral/vertical step sizes were both 5 nm. The PL images were frequency-integrated in the 640–680 nm range. Used with permission from Ref. [11].

amplitude above the surface, a condition corresponding to the far-field signal that can be subtracted from the combined TERS/TEPL signal if necessary.

Interestingly, similar spatial resolution in QD TEPL imaging was achieved with 633 nm and 532 nm excitation with the same gold-coated AFM probe, though the TEPL enhancement was significantly higher for the case of red laser. Authors attribute this rather non-trivial fact to the significant influence of the nanoscale asperities at the apex of a TERS probe on overall enhancement in the near-field imaging experiments.

Another exciting development in TEPL characterization of QDs was the achievement of a strong coupling regime between the tip-plasmonic substrate surface plasmon polaritons and the QD's excitons. Precise control over the plasmonic properties of the tip-substrate cavity achieved by controlling the tip-sample distance allowed the authors observation of Rabi splitting in the emission spectra of QD in the cavity [12]. Somewhat similar results were obtained by another group through the use of a broad-band slit plasmonic nanoresonator [13].

6.4.1.5 Vertical and Lateral Heterostructures of TMDs

Unique advantages of the TERS/TEPL imaging are particularly well suited for nanoscale characterization of vertical and lateral heterostructures of TMDs.

Lateral heterolayers of TMDs consist of two or more different materials within the continuous layer of practically the same thickness (Figure 6.60).

For the two-component heterolayer, several options are available: material A and material B may feature the same chalcogen atoms, but different transition metals (MoS_2–WS_2 for example), the same transition metal but different chalcogen atoms (WS_2–WSe_2) or finally both the transition metal and the chalcogen atoms may be different (WS_2–$MoSe_2$) as an example). The width and chemical composition of the junction between the constituents of the heterolayer depends on many factors and may scale from the atomically sharp borderline to the mixed composition of two materials that can extend far away from the formal junction line. Electric and optoelectronic properties of heterolayers may be obviously strongly affected by the electronic properties of its components and the spatial scale and chemical composition of the transition area and therefore its characterization is of great importance for understanding and future applications of such heterostructures.

Van der Waals nature of TMDs allows another type of heterostructures when for example the monolayers of two different materials are stacked vertically, ideally with controlled twist

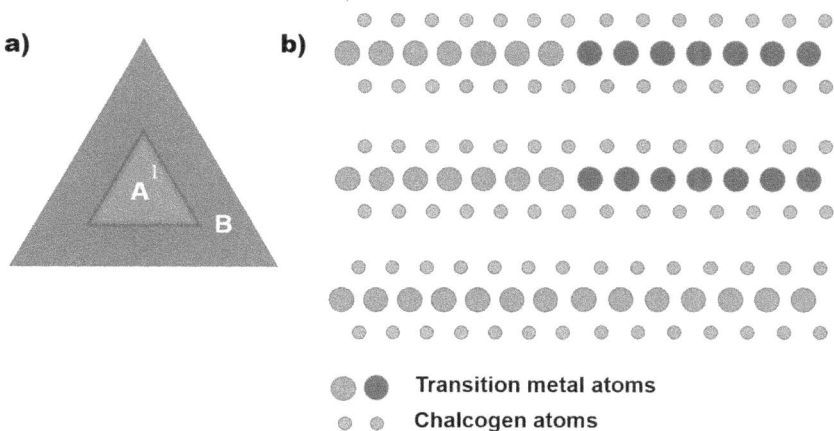

FIGURE 6.60 (a) A cartoon of a two-component lateral heterolayer. Red line marks (1) the transition area between components A and B. (b) Side view on the junction showing several types of the heterojunction in TMDs; bottom: when the transition atom is the same and the chalcogen atoms are different, middle: the chalcogen atoms are the same and the transition metals are different, top: when both the chalcogen and the transition metal atoms are different.

angle. Such heterostructures may be obtained by either consecutive exfoliation (by far the most popular method at the moment) or by consecutive growth, a method which at least at this stage has lower control over the mutual orientation of the crystals in vertical heterostructure, but may result, as we'll see further in this chapter in formation of interesting vertical–lateral heterostructures.

A very special type of vertical heterostructures in TMDs can be achieved when the top chalcogen layer is replaced with different chalcogen atoms which result in the formation of so-called Janus materials [14–18].

Tip-enhanced photoluminescence has been successfully applied for sub-diffraction imaging of as-grown $MoSe_2$–WSe_2 lateral heteromonolayers and heteromonolayer lattices [19,20], the latter benefited particularly from the much-improved spatial resolution provided by the tip-enhanced spectroscopic imaging which allowed unambiguous identification of the chemical composition of the alternating stripes in the multijunction lateral heteromonolayers.

In a very recent work, Garg et al. [21] performed the gap-mode TERS measurements on MoS_2–WS_2 heteromonolayers that were transferred from the growth sapphire substrate using a dry gold-assisted procedure published earlier [22].

Tip-enhanced Raman spectroscopy spectra of the junction area in a MoS_2–WS_2 crystal that was characterized by the confocal PL imaging prior to the transfer, unambiguously showed that the junction area whose width greatly varied along the junction line in the same crystal within 25–500 nm range (Figure 6.61), featured formation of the $Mo_xW_{1-x}S_2$ alloy, exact composition of which was changing across the junction line, which was independently confirmed by the presence of the alloy-specific Raman bands and the shift of the spectral position of the composition-specific line. It's interesting to note that the spatial resolution achieved in the gap-mode TERS maps of the MoS_2–WS_2 heteromonolayer was considerably better compared to the spatial resolution of the Kelvin probe imaging, which probably was related to extended charge depletion areas next to the junction line which affected the sharpness of the surface potential image.

Somewhat similar results have been obtained in TERS and TEPL imaging of the MoS_2–WS_2 lateral heterobilayers [23]. In this work, two distinctly different heterobilayer crystals have been identified – one (Type I) featuring sharp, pixel-limited transition between the WS_2 and MoS_2 bilayer components, while the other (Type II) crystal clearly showed extended transition region. Careful analysis of the TERS and TEPL spectra in the Type I junction showed the presence of W atoms on the MoS_2 side of the junction, while the WS_2 side looked pristine. This was in clear contrast with the

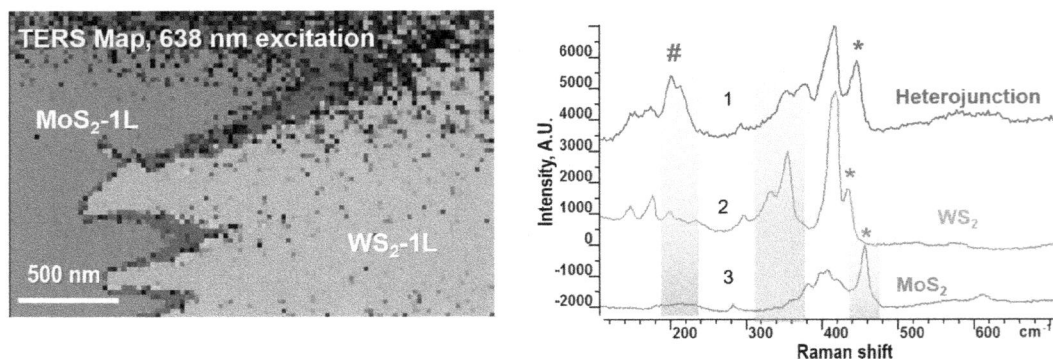

FIGURE 6.61 (a) Combined TERS map of the gold-transferred MoS_2–WS_2 heteromonolayer collected with 638 nm excitation showing the distribution of the intensity of the correspondingly colored bands in panel (b). The width of the alloyed transition area varied greatly along the junction line. (b) Averaged TERS spectra from the monolayer MoS_2 core (red, 3), WS_2 shell (green, 2), and the transition region (blue, 1). Spectra were offset vertically for clarity. Raman band marked with # is characteristic for the $Mo_xW_{1-x}S_2$ alloys, the band whose spectral position was composition-sensitive is marked with *. Adopted with permission from Ref. [21].

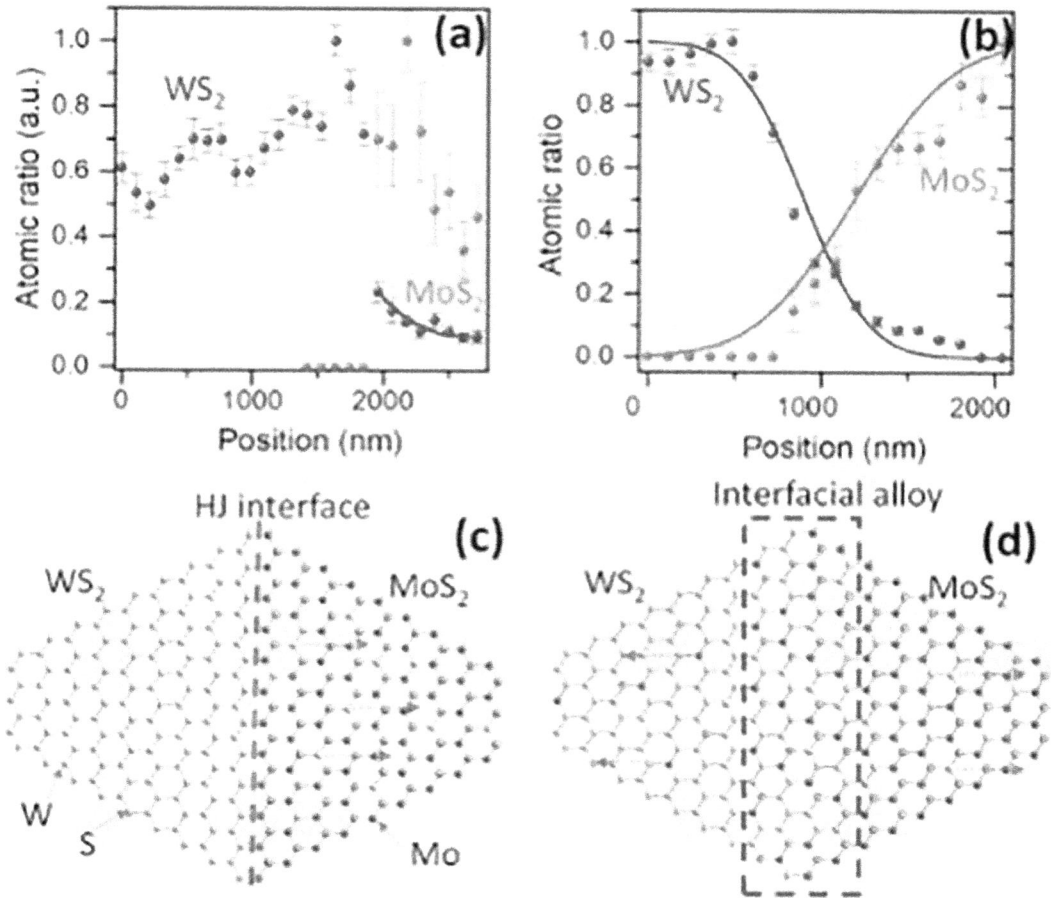

FIGURE 6.62 Atomic ratio of W (blue balls, 1) and Mo (pink balls, 2) across the Type I (a) and Type II (b) heterojunctions. Schematic of unidirectional diffusion in Type I junction (c) and more symmetric diffusion in Type II heterojunction (d), respectively. In (a) and (b), diffusion equations were used to fit the atomic ratio curves and plotted in solid lines. Used with permission from Ref. [23].

Type II heterojunction where more symmetric alloying was observed (Figure 6.62). Such dramatic difference in the composition in the junction area was attributed to different seeding conditions for the MoS_2 growth at the edges of WS_2 crystals that were grown first.

It's important to note that TERS and TEPL imaging, though lacking Angstrom-scale spatial resolution of STEM, allows obtaining chemical information from the heterojunction at much-improved sensitivity and spatial resolution as compared to the conventional confocal microscopy, and cross-correlating this nanoscale spectroscopic information with other observables provided by SPM preserving the sample for multiple cross-correlated studies, while STEM imaging of a sample is extremely damaging and direct cross-correlation of the atomic-resolution image with other properties from the same location is barely possible.

Tip-enhanced Raman spectroscopy imaging was successfully applied to characterization of the vertical heterostructures of TMDs. In just another very recent work [24], it was shown that TERS imaging cross-correlated with SKM can successfully identify various types of vertical WS_2–WSe_2 heterobi – and heterotrilayers, though it's very important to perform TERS imaging of these vertical heterostructures with several different excitation wavelengths, as the use of only 633(638) nm laser, which is the most popular choice for TERS/TEPL measurements, may, somewhat surprisingly, lead to incorrect conclusions.

Kelvin probe imaging of the area containing the monolayer WS_2 and several WSe_2–WS_2 vertical heteromultilayers shown in Figure 6.63a and c suggests the presence of three different components within imaged area based on three distinctly different levels of the CPD signal. TERS map of a portion of that area together with representative averaged TERS spectra collected with 785 nm laser unambiguously identifies the area corresponding to the lowest level of the CPD as the monolayer WS_2, while the areas with higher CPD were the WSe_2–WS_2 heterobilayer and $2LWSe_2$–WS_2 heterotrilayer, based on simultaneous presence of the peak at around ~250 cm^{-1} characteristic for WSe_2 and A'/A_{1g} peak of WS_2 at ~415 cm^{-1}. Areas corresponding to the highest value of the CPD featured the highest intensity of the ~250 cm^{-1} peak, which identified them as the $2LWSe_2$–WS_2 heterotrilayers.

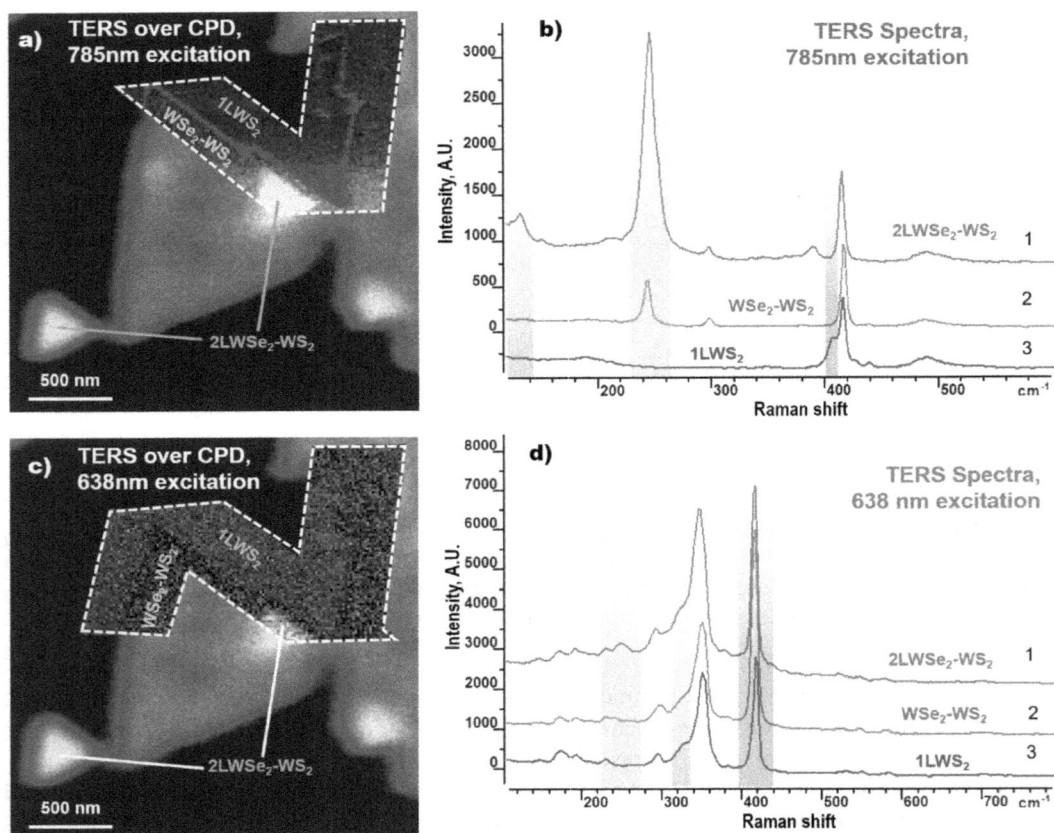

FIGURE 6.63 (a) TERS map showing the intensities of the correspondingly colored bands in (b) of WS_2–WSe_2 vertical heterostructures on gold collected with 785 nm excitation overlaid on the CPD image. (b) Averaged TERS spectra from the monolayer WS_2 (blue spectrum, 3), WSe_2–WS_2 heterobilayer (green spectrum, 2), and $2LWSe_2$–WS_2 heterotrilayer (red spectrum, 1). Spectra were offset vertically for clarity. (c) TERS map showing the intensities of the red and green colored bands in (d) and the spectral position of the band colored blue of the same WS_2–WSe_2 vertical heterostructures on gold collected with 638 nm excitation overlaid on the CPD image. (d) Averaged TERS spectra from the monolayer WS_2 (blue spectrum, 3), WSe_2–WS_2 heterobilayer (green spectrum, 2), and $2LWSe_2$–WS_2 heterotrilayer (red spectrum, 1). Spectra were offset vertically for clarity. Surprisingly, TERS spectrum of the WS_2–WSe_2 heterobilayer collected with the 638 nm laser did not show noticeable intensity of the ~250 cm^{-1} band. Used with permission from Ref. [24].

Quite surprisingly, TERS map collected with 638 nm laser (Figure 6.63c) with the same tip over practically the same area, produced the TERS spectra that were strikingly different from the TERS spectra collected with 785 nm laser (Figure 6.63b). TERS spectra from the area that was already identified as a heterobilayer, did not show any noticeable intensity of the ~250 cm^{-1} peak and the heterotrilayer featured only very weak peak at this frequency, nowhere close to the intense peak observed with 785 nm excitation. Were the TERS measurements performed only with 638 nm laser, the conclusions made based on the TERS spectra from the heterobilayer would be plain wrong.

The results obtained with 638 nm excitation on WSe$_2$–WS$_2$ heterotri – and, especially, heterobilayer were rather surprising, taking into account that strong TERS signal with significant intensity of the A'/2LA(M) peak at ~250 cm^{-1} from the mono- and few layer-thick crystals of WSe$_2$, both exfoliated and CVD grown, was demonstrated in several publications [22,25]. The exact nature of such a strong difference in TERS response of the vertical WSe$_2$–WS$_2$ heterostructures still remains to be understood, but it does not change the overall utility of TERS/TEPL for the characterization of both vertical and lateral heterostructures of TMDs and other 2D materials.

Progress with production of twisted bi- and tri-layers of TMDs and twisted heterobilayers that would have full control over the twist angle and the type of Moire patterns appearing as a result will open a huge new field of applications for TERS and TEPL. Development of the next-generation commercial TERS probes with 10–100 times improved enhancement factors and the operational spectral range extended into green or even blue part of visible spectrum may further facilitate nanoscale spectroscopic characterization of advanced 2D materials and their heterostructures.

REFERENCES

1. Lee, N.; Hartschuh, R. D.; Mehtani, D.; Kisliuk, A.; Maguire, J. F.; Green, M.; Foster, M. D.; Sokolov, A. P. High contrast scanning nano-Raman spectroscopy of silicon. *Journal of Raman Spectroscopy*, 2007, *38*(6). https://doi.org/10.1002/jrs.1698.

2. Atuchin, V. V.; Gavrilova, T. A.; Grigorieva, T. I.; Kuratieva, N. V.; Okotrub, K. A.; Pervukhina, N. V.; Surovtsev, N. V. Sublimation growth and vibrational microspectrometry of α-MoO$_3$ single crystals. *Journal of Crystal Growth*, 2011, *318*(1). https://doi.org/10.1016/j.jcrysgro.2010.10.149.

3. Smithe, K. K. H.; Krayev, A. V.; Bailey, C. S.; Lee, H. R.; Yalon, E.; Aslan, Ö. B.; Muñoz Rojo, M.; Krylyuk, S.; Taheri, P.; Davydov, A. V.; Heinz, T. F.; Pop, E. Nanoscale heterogeneities in monolayer MoSe$_2$ revealed by correlated scanning probe microscopy and tip-enhanced Raman spectroscopy. *ACS Applied Nano Materials*, 2018, *1*(2). https://doi.org/10.1021/acsanm.7b00083.

4. Hantanasirisakul, K.; Gogotsi, Y. Electronic and optical properties of 2D transition metal carbides and nitrides (MXenes). *Advanced Materials*, 2018, *30*(52). https://doi.org/10.1002/adma.201804779.

5. Sarycheva, A.; Gogotsi, Y. Raman spectroscopy analysis of the structure and surface chemistry of Ti$_3$C$_2$T$_x$MXene. *Chemistry of Materials*, 2020, *32*(8). https://doi.org/10.1021/acs.chemmater.0c00359.

6. Bhattarai, A.; Krayev, A.; Temiryazev, A.; Evplov, D.; Crampton, K. T.; Hess, W. P.; El-Khoury, P. Z. Tip-enhanced Raman scattering from nanopatterned graphene and graphene oxide. *Nano Letters*, 2018, *18*(6). https://doi.org/10.1021/acs.nanolett.8b01690.

7. Dai, X.; Zhang, Z.; Jin, Y.; Niu, Y.; Cao, H.; Liang, X.; Chen, L.; Wang, J.; Peng, X. Solution-processed, high-performance light-emitting diodes based on quantum dots. *Nature*, 2014, *515*(7525). https://doi.org/10.1038/nature13329.

8. Michler, P.; Kiraz, A.; Becher, C.; Schoenfeld, W. V.; Petroff, P. M.; Zhang, L.; Hu, E.; Imamoğlu, A. A quantum dot single photon source. In *Advances in Solid State Physics*. Berlin, Heidelberg: Springer. https://doi.org/10.1007/3-540-44946-9_1.

9. Protasenko, V. V; Kuno, M.; Gallagher, A.; Nesbitt, D. J. Fluorescence of single ZnS overcoated CdSe quantum dots studied by apertureless near-field scanning optical microscopy. *Optics Communications*, 2002, *210*(1–2). https://doi.org/10.1016/S0030-4018(02)01759-5.

10. Lee, H.; Lee, D. Y.; Kang, M. G.; Koo, Y.; Kim, T.; Park, K.-D. Tip-enhanced photoluminescence nano-spectroscopy and nano-imaging. *Nanophotonics*, 2020, *9*(10). https://doi.org/10.1515/nanoph-2020-0079.

11. Wang, C.-F.; Zamkov, M.; El-Khoury, P. Z. Ambient tip-enhanced photoluminescence with 5 nm spatial resolution. *The Journal of Physical Chemistry C*, 2021, *125*(22). https://doi.org/10.1021/acs.jpcc.1c04012.

12. Park, K.-D.; May, M. A.; Leng, H.; Wang, J.; Kropp, J. A.; Gougousi, T.; Pelton, M.; Raschke, M. B. Tip-enhanced strong coupling spectroscopy, imaging, and control of a single quantum emitter. *Science Advances*, 2019, *5*(7). https://doi.org/10.1126/sciadv.aav5931.

13. Groß, H.; Hamm, J. M.; Tufarelli, T.; Hess, O.; Hecht, B. Near-field strong coupling of single quantum dots. *Science Advances*, 2018, *4*(3). https://doi.org/10.1126/sciadv.aar4906.

14. Zhang, J.; Jia, S.; Kholmanov, I.; Dong, L.; Er, D.; Chen, W.; Guo, H.; Jin, Z.; Shenoy, V. B.; Shi, L.; Lou, J. Janus monolayer transition-metal dichalcogenides. *ACS Nano*, 2017, *11*(8). https://doi.org/10.1021/acsnano.7b03186.

15. Ju, L.; Bie, M.; Shang, J.; Tang, X.; Kou, L. Janus transition metal dichalcogenides: A superior platform for photocatalytic water splitting. *Journal of Physics: Materials*, 2020, *3*(2). https://doi.org/10.1088/2515-7639/ab7c57.

16. Zhang, K.; Guo, Y.; Ji, Q.; Lu, A.-Y.; Su, C.; Wang, H.; Puretzky, A. A.; Geohegan, D. B.; Qian, X.; Fang, S.; Kaxiras, E.; Kong, J.; Huang, S. Enhancement of van der Waals interlayer coupling through polar janus MoSSe. *Journal of American Chemical Society*, 2020, *142*(41). https://doi.org/10.1021/jacs.0c07051.

17. Petrić, M. M.; Kremser, M.; Barbone, M.; Qin, Y.; Sayyad, Y.; Shen, Y.; Tongay, S.; Finley, J. J.; Botello-Méndez, A. R.; Müller, K. Raman spectrum of janus transition metal dichalcogenide monolayers WSSe and MoSSe. *Physical Review B*, 2021, *103*(3). https://doi.org/10.1103/PhysRevB.103.035414.

18. Lu, A.-Y.; Zhu, H.; Xiao, J.; Chuu, C.-P.; Han, Y.; Chiu, M.-H.; Cheng, C.-C.; Yang, C.-W.; Wei, K.-H.; Yang, Y.; Wang, Y.; Sokaras, D.; Nordlund, D.; Yang, P.; Muller, D. A.; Chou, M.-Y.; Zhang, X.; Li, L.-J. Janus monolayers of transition metal dichalcogenides. *Nature Nanotechnology*, 2017, *12*(8). https://doi.org/10.1038/nnano.2017.100.

19. Sahoo, P. K.; Zong, H.; Liu, J.; Xue, W.; Lai, X.; Gutiérrez, H. R.; Voronine, D. V. Probing nano-heterogeneity and aging effects in lateral 2D heterostructures using tip-enhanced photoluminescence. *Optical Materials Express*, 2019, *9*(4). https://doi.org/10.1364/OME.9.001620.

20. Tang, C.; He, Z.; Chen, W.; Jia, S.; Lou, J.; Voronine, D. V. Quantum plasmonic hot-electron injection in lateral WSe$_2$–MoSe$_2$ heterostructures. *Physical Review B*, 2018, *98*(4). https://doi.org/10.1103/PhysRevB.98.041402.

21. Garg, S.; Fix, J. P.; Krayev, A. V.; Flanery, C.; Colgrove, M.; Sulkanen, A. R.; Wang, M.; Liu, G.-Y.; Borys, N. J.; Kung, P. Nanoscale Raman characterization of a 2D semiconductor lateral heterostructure interface. *ACS Nano*, 2022, *16*(1), 340–350. https://doi.org/10.1021/acsnano.1c06595.

22. Krayev, A.; Bailey, C. S.; Jo, K.; Wang, S.; Singh, A.; Darlington, T.; Liu, G.-Y.; Gradecak, S.; Schuck, P. J.; Pop, E.; Jariwala, D. Dry transfer of van der Waals crystals to noble metal surfaces to enable characterization of buried interfaces. *ACS Applied Materials & Interfaces*, 2019, *11*(41). https://doi.org/10.1021/acsami.9b09798.

23. Shao, J.; Chen, F.; Su, W.; Zeng, Y.; Lu, H.-W. Multimodal nanoscopic study of atomic diffusion and related localized optoelectronic response of WS$_2$/MoS$_2$ lateral heterojunctions. *ACS Applied Materials & Interfaces*, 2021, *13*(17). https://doi.org/10.1021/acsami.1c03061.

24. Krayev, A.; Chen, P.; Terrones, H.; Duan, X.; Zhang, Z.; Duan, X. Importance of multiple excitation wavelengths for TERS characterization of TMDCs and their vertical heterostructures. *The Journal of Physical Chemistry C*, 2022, *126*(11), 5218–5223. https://doi.org/10.1021/acs.jpcc.1c10469.

25. Jariwala, D.; Krayev, A.; Wong, J.; Robinson, A. E.; Sherrott, M. C.; Wang, S.; Liu, G.-Y.; Terrones, M.; Atwater, H. A. Nanoscale doping heterogeneity in few-layer WSe$_2$ exfoliated onto noble metals revealed by correlated SPM and TERS imaging. *2D Materials*, 2018, *5*(3). https://doi.org/10.1088/2053-1583/aab7bc.

6.4.2 TERS CHARACTERIZATION OF MOLECULAR SWITCHING IN ORGANIC DEVICES

The word memristor is the contraction of memory and resistance; a memristor is an electrical two-terminal device whose resistance depends on the history of the applied voltage. Memristors are non-volatile memories predicted to be the component of the future for high-density data storage and brain-inspired ultralow energy computing. Of different material systems used to realize memristive devices, oxide-based devices are close to a commercial deployment. Among other genres, a lot of research effort has also been put into organic memristors because of their nonstochastic and more uniform switching as well as their cheap fabrication cost. However, the emergence of organic memristors has been hindered by poor reproducibility, endurance stability scalability, and low switching speed. Knowing the primary driving mechanism at the molecular

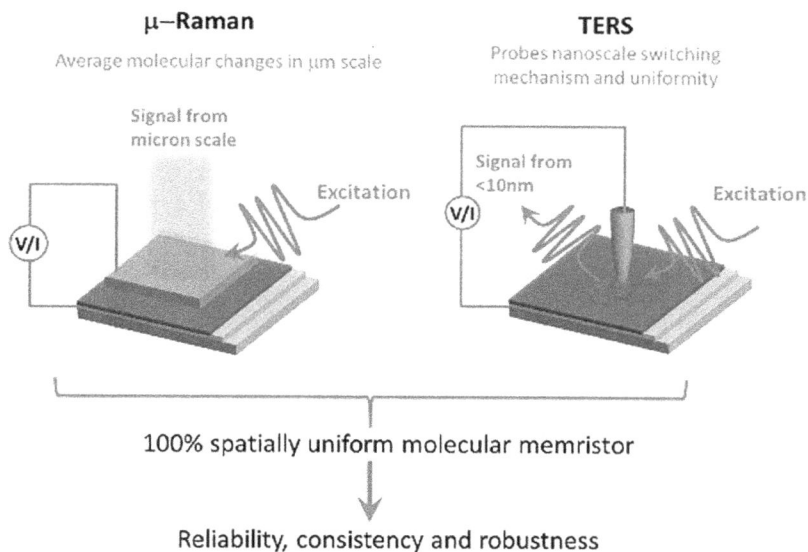

μ−Raman
Average molecular changes in μm scale

TERS
Probes nanoscale switching mechanism and uniformity

Signal from micron scale

Excitation

V/I

Signal from <10nm

Excitation

V/I

100% spatially uniform molecular memristor

Reliability, consistency and robustness

FIGURE 6.64 From microscale to molecular scale. (Reprinted with permission from Ref. [1], Copyright © 2020 John Wiley & Sons, Ltd.)

scale will be the key to improve the robustness and reliability of such organic-based devices. Deterministic tracking of molecular mechanisms necessitates a nanoscale in situ spectroscopy in tandem with nanoscopic current measurement that can correlate molecular changes with the current response. Such a measurement remains long-awaited yet elusive. The article by Goswami et al., in Advanced Materials, achieves this using correlated AFM and TERS measurement [1] (Figure 6.64).

6.4.2.1 Potential/Input from the TERS Technique

Tip-enhanced Raman spectroscopy has emerged as a powerful analytical technique providing high chemical sensitivity for surface mapping of molecules with nanoscale spatial resolution [2–7]. In-operando TERS and concurrent conductive AFM will allow to characterize the chemical fingerprints of the molecular switching mechanism as voltage across the memristive layer between the AFM tip and a bottom conductive layer is applied while Raman spectra are acquired and plasmonic enhancement permits nanoresolution at the contact point of the tip. In addition, μRaman comes into play to confirm the switching mechanism up to the microscale (Figure 6.65).

A device based on a Ru complex with an azo-aromatic ligand has been reported as giving great performance: stable (tested over hundreds of devices), enduring ($\approx 10^{12}$ write/erase cycles), fast (<30 ns), and ultralow energy (≈ 1.5 fJ) memristive switching properties [8–11]. Its switching mechanism has been studied using in situ μRaman and UV-vis spectroscopies. The role of the ligand in the molecular redox transition has been determined and confirmed by quantum chemical calculations. As schematically depicted in Figure 6.64, to establish areal switching uniformity, chemical fingerprints obtained in μRaman will be compared with the molecular changes probed by in-operando TERS [1].

6.4.2.2 Description of the Molecular Device and TERS Measurement

The memristive organic layers are prepared from solutions of the two precursor azo-aromatic complexes in acetonitrile (Figure 6.65): (i) system A $[Ru(L_1)_3]$ $(PF_6)_2$ $(L_1 = 2$(phenylazo)pyridine) and

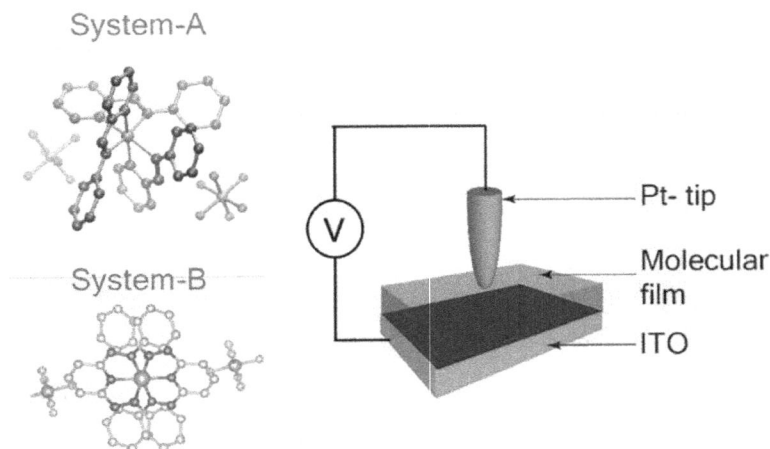

FIGURE 6.65 Left: Azo-aromatic molecules of system A: $[Ru(L_1)_3]$ $(PF_6)_2$ $(L_1=2(phenylazo)pyridine)$ and system B: $[Ru(L_2)_2]$ $(PF_6)_2$ $(L_2=2,6\text{-bis(phenylazo)pyridine})$. Right: schematic diagram of the test device. (Reprinted with permission from Ref. [1], Copyright © 2020 John Wiley & Sons, Ltd.)

(ii) system B $[Ru(L_2)_2]$ $(PF_6)_2$ $(L_2=2,6\text{-bis(phenylazo)pyridine})$. The test devices comprise a layer of these materials off-centered spin-coated (thickness ≈15–70 nm) on an epitaxial indium tin oxide (ITO) film of ≈60 nm grown via pulsed laser deposition (PLD) on an annealed yttria-stabilized zirconia (YSZ) substrate. The ITO thin film serves as the bottom electrode and the conductive c-AFM/TERS tip as the top electrode. The use of an atomically flat ITO bottom electrode (RMS roughness of ≈0.5 nm over $25\times25\,\mu m^2$) facilitates the formation of an ultrasmooth spin-coated film with RMS roughness ≈1.5 nm over $25\times25\,\mu m^2$. After deposition, the samples were stored inside a vacuum chamber with a pressure of about 10^{-8} Torr for 12 h.

Tip-enhanced Raman spectroscopy measurements were performed using a NanoRaman system from HORIBA Scientific integrating an atomic force microscope (OmegaScope, based on SmartSPM) and a Raman microscope (XploRA) with a 100× WD objective tilted by 60° with respect to the sample plane. A 638 nm p-polarized laser (80 μW) was focused onto the cantilever-based Au-coated AFM-TERS tip (OMNI TERS-SNC-Au, App Nano). This probe was conductive and thereby suitable to apply a voltage and acquire Raman spectra at any voltage. After recording an AFM topography image in AC mode ($2\times2\,\mu m^2$), the tip was positioned on a spot of interest, the potential voltage was applied on the tip, and a Raman spectrum was acquired for 50 s with the tip in contact with sample surface with a typical interaction of 2–10 nN. It had been checked that the spectrum acquired with the tip few nm away from the sample, which is the far-field or μRaman contribution, was background spectrum without Raman signature. As a result, the "in contact" spectrum was a pure near-field contribution generated from the nanoregion under the tip. Figure 6.66 compares the Raman spectra obtained in μRaman and in TERS for both systems A and B in their ON and OFF states and also in the intermediate state for system B. In both scale Raman spectroscopies, the changes in the azo-stretching modes are identical. The molecular redox process induces spectral weight transfer, from a dominant E_{0A} (unreduced) mode in the ON state to a dominant E_{1A} (singly reduced ligand) in the OFF state in the reduction process for system A (Figure 6.66b). For system B, the Raman spectrum features a higher E_{2B} (triply reduced ligand state) mode in the ON state, higher E_{1B} (singly reduced ligand state) in the intermediate state, and dominant E_{0B} (unreduced ligand state) in the OFF state (Figure 6.66c). The plots featuring color-coded intensity μ Raman and TERS spectra collected for voltages from −0.3 to +0.3 V (forward sweep) and from +0.3 to −0.3 V also indicate the very close voltage dependence

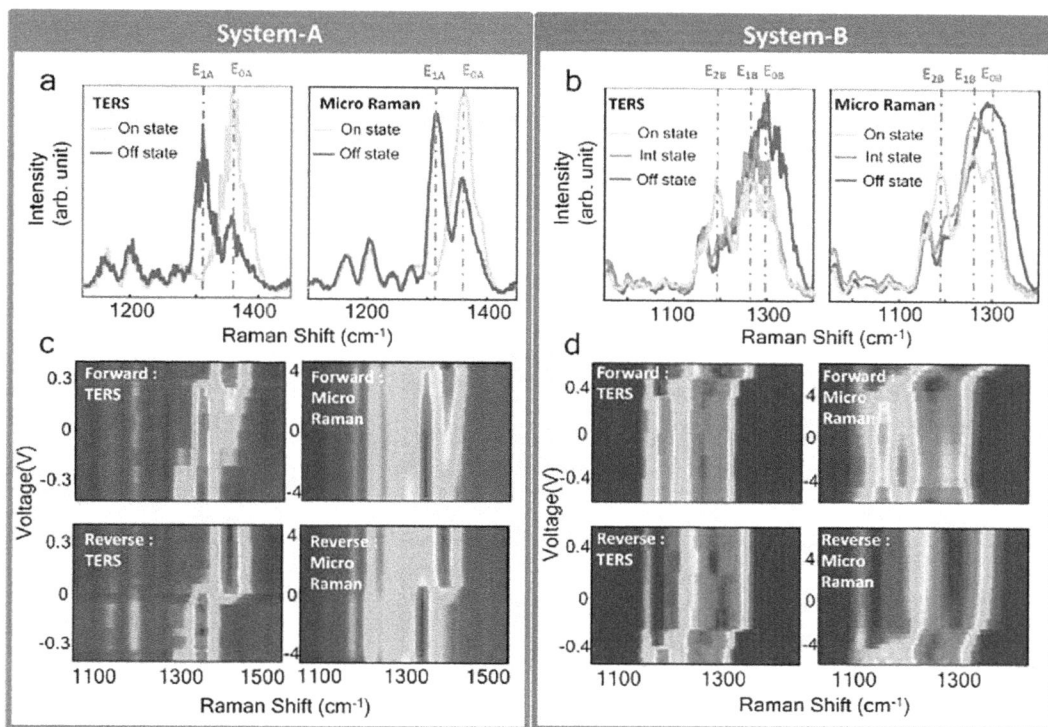

FIGURE 6.66 (a) System A: TER (left) and μRaman (right) spectra measured in-operando for two states. (b) System B: TER (left) and μRaman (right) spectra measured in-operando for three conductance states. (c) and (d) Intensity color plots of spectral voltage dependence in TERS (left) and μRaman (right) of Systems A and B for forward (top) and reverse (bottom) voltage sweeps. (Reprinted with permission from Ref. [1], Copyright © 2020 John Wiley & Sons, Ltd.)

behavior for both devices A&B at the micro and nanoscales. Another way to show the areal uniformity of the switching mechanism is to record TERS maps and monitor the intensity of ON/intermediate/OFF state characteristics Raman bands. Figures 6.67b and c show the variation of intensity in the range of 1,330–1,400 cm^{-1} for device A for scan area of $2 \times 2\,\mu m^2$ (pixel size = 50 nm) and $200 \times 200\,nm^2$ (pixel size = 7 nm), respectively in both ON and OFF conductance states of device A. The standard deviation $(A(2\sigma)/A_{mean})$ in the spectral distributions in all the states are <10%. In Figure 6.67d to f, we show the TERS uniformity for device B. The spectral range of detection is 1,250–1,380 cm^{-1} and the spectral weights in different states (i.e., OFF, intermediate, and ON).

6.4.2.3 Conclusions and Perspectives

This chapter shows how TERS mapping realized in-operando in a memristive organic device brings the key demonstration of the homogeneity of the molecular redox transition mechanism between conductance states with a sub-10 nm spatial resolution. With the complementary μRaman study a uniform 100% switching of an entire device area has been proved. Knowing that the molecular switching process scale from nanoscale to macroscale is a significant and long-awaited achievement in organic memristors and molecular resistance switches. The study paves the way for robust devices and implementation in ultralow energy digital electronics. This in-operando characterization method could be applied to a wide variety of molecule-enabled electronic devices and systems, such as molecular diodes, organic light emitters, and collectors.

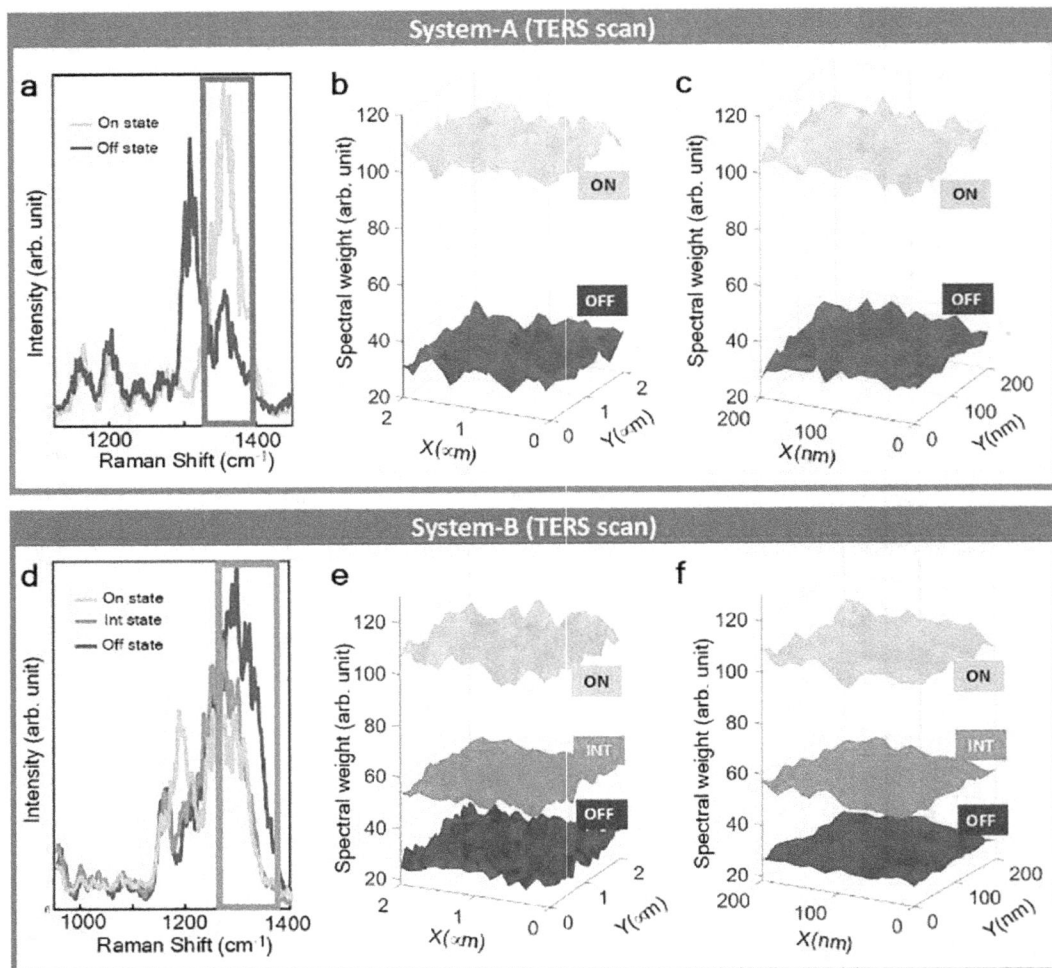

FIGURE 6.67 Areal uniformity in TERS mapping: (a) In situ TER spectra measured in ON and OFF states of system A. The purple area in spectra (a) represents the range (1,320–1,400 cm⁻¹) integrated in the TERS maps presented in panels (b) and (c). (b) A $2\times2\,\mu m^2$ TERS map in the on and off states. (c) A $200\times200\,nm^2$ TERS map of ON and OFF states with the same spectral range as in (b). (d–f) Same as (a)–(c) for the three states of system B. (Reprinted with permission from Ref. [1], Copyright © 2020 John Wiley & Sons, Ltd.)

REFERENCES

1. S. Goswami, D. Deb, A. Tempez, M. Chaigneau, S. Prasad Rath, M. Lal, A. R. Stanley Williams, S. Goswami, T. Venkatesan, *Adv. Mat. 32*, 2004370 (2020).
2. C.-F. Wang, B. T. O'Callahan, D. Kurouski, A. Krayev, and P. Z. El-Khoury *J. Phys. Chem. Lett. 11*, 3809–3814 (2020).
3. P. Z. El-Khoury and E. Aprà *J. Phys. Chem. C 124*, 17211–17217 (2020).
4. C. Toccafondi, G. Picardi, and R. Ossikovski, *J. Phys. Chem. C 120*, 18209–18219 (2016).
5. A. Merlen, M. Chaigneau, and S. Coussan, *Phys. Chem. Chem. Phys. 17*, 19134–19138 (2015).
6. G. Picardi, A. Krolikowska, R. Yasukuni, M. Chaigneau, M. Escude, V. Mourier, C. Licitra, and R. Ossikovski, *Chem. Phys. Chem. 15*, 276–282 (2014).
7. E. Sheremet, R. D. Rodriguez, D. R. T. Zahn, A. G. Milekhin, E. E. Rodyakina, and A. V. Latyshev, *J. Vac. Sci. Technol. B. 32*, 04E110 (2014).
8. S. Goswami, A. J. Matula, S. P. Rath, S. Hedström, S. Saha, M. Annamalai, D. Sengupta, A. Patra, S. Ghosh, and H. Jani, *Nat. Mater. 16*, 1216 (2017).

9. S. Goswami, S. P. Rath, D. Thompson, S. Hedström, M. Annamalai, R. Pramanick, B. R. Ilic, S. Sarkar, S. Hooda, C. A. Nijhuis, J. Martin, R. S. Williams, S. Goswami, and T. Venkatesan, *Nat. Nanotechnol.* *15*, 380 (2020).

10. S. Goswami, D. Thompson, R. S. Williams, S. Goswami, and T. Venkatesan, *Appl. Mater. Today 19*, 100626 (2020).

11. S. Goswami, S. Goswami, and T. Venkatesan, *Appl. Phys. Rev. 7*, 21303 (2020).

6.4.3 Bio-TERS

6.4.3.1 Introduction

Tip-enhanced Raman spectroscopy of biological systems (referred to as Bio-TERS) is an active and most challenging application of TERS due to the high complexity of biological "wet, warm and noisy" environment with weak Raman signals. Many biomolecules have common structural motifs that result in overlapping vibrational bands with spectral congestion. Autofluorescence background may overwhelm the Raman signals. Flexible structure and fluid environment may lead to temporal fluctuations of Raman spectra even at the same spatial location. Overcoming these challenges will be rewarding in realizing a big potential of TERS for rapid, sensitive, and noninvasive detection and identification of biological systems, including DNA and protein sequencing, probing local environment at the single biomolecule level in viruses, bacteria, cancer cells, and other systems. Several previous reviews of Bio-TERS highlighted various aspects of historical developments, challenges, and successes [1–3]. Here we discuss some key highlights using selected examples of published reports.

6.4.3.2 Spectral Complexity of Proteins

Raman spectroscopy is a useful tool for the chemical analysis of proteins that provides information about the composition and secondary structure. The high sensitivity and spatial resolution of TERS show promise for sequencing of proteins at the single molecule level and for determining their folding configurations. However, weak Raman signals and spectral complexity are the major challenges for developing practical applications. Several studies have demonstrated TERS of proteins by optimizing experimental configurations to address these challenges.

Functional cofactor groups such as heme in cytochrome c provide resonance enhancement of Raman signal. TERS of isolated [4] and membrane-embedded [5] cytochrome c proteins showed the ability to detect local signals of few molecules. The difference between the TERS and SERS experiments on heme proteins is illustrated in Figure 6.68a. SERS is typically performed on an ensemble of plasmonic nanoparticles and, therefore, provides enhanced but ensemble-averaged information, which has a high spectral congestion due to the mixture of signals from many molecules with different orientations. TERS, on the other hand, is performed using only one plasmonic particle at the tip apex, which enhances the signal from a single or a few molecules. This simplifies Raman signals and may be used to separate heme and amino acids [4].

Tip enhancement alone is not always sufficient for imaging biomolecules with weak Raman signals. Gap-mode plasmonic configuration is often used for single-molecule SERS and TERS, where a plasmonic cavity is formed between two nanoparticles or by image-dipole coupling of tip to a metallic substrate. The distance dependence of gap-mode coupling, however, shows the largest enhancement for tip-sample gap of less than 5–10 nm, which prohibits the use of large biomolecules, viruses, and cells. Another gap-plasmon-like approach was developed using functionalized gold nanoparticles (GMPs), which could be probed by TERS giving large signal enhancement without the metallic substrate [6–8]. Biotin-streptavidin interaction was studied first, followed by the detection of membrane-embedded integrin proteins (Figure 6.68b).

The ability to detect proteins at the nanoscale may be used in bioinspired materials science such as generation and characterization of hybrid protein nanofibers (hPNFs) for drug delivery,

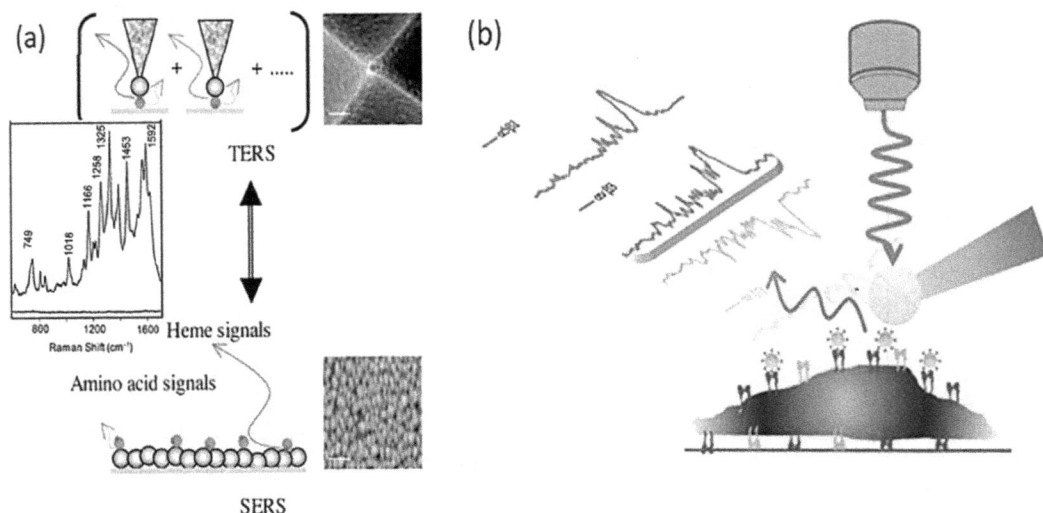

FIGURE 6.68 (a) TERS and SERS of a heme protein cytochrome c. (Reprinted with permission from Ref. [4], Copyright © 2008 *Am. Chem. Soc.*; Reprinted with permission from Ref. [5], Copyright © 2011 *Royal Chem. Soc.*) (b) TERS of integrin proteins bound to GNPs on intact cancer cell membranes. (Reprinted with permission from Ref. [7], Copyright © 2016 *Am. Chem. Soc.*)

biosensing, and tissue engineering applications [9]. For example, TERS was used for studying the self-assembly of human serum albumin (HSA) and human hemoglobin (HGB) proteins into hPNFs via ethanol-triggered denaturation (Figure 6.69). TERS signals of HGB were enhanced by resonance scattering at 532 nm excitation due to the presence of the heme group in HGB and its absence in HSA, which allowed for an improved distinction between the two proteins.

Tip-enhanced Raman spectroscopy was also used for nanoscale imaging of protein aggregates related to Alzheimer's and Parkinson's diseases such as amyloid fibrils [11]. TERS was used to investigate secondary structure and composition of insulin fibrils [12], nanoscale organization of hydrophilic and hydrophobic domains on individual insulin amyloid fibrils [13], toxic oligomers [14], and to study fibril polymorphism [15,16].

A combined SERS and TERS imaging approach was used to investigate the interactions between amyloid fibrils and gold nanoparticles (Au NPs) as a potential method of fibril degradation [10]. It was found that Au NPs interacted with fibrils by breaking the disulfide S-S bridges and reducing the cysteine residues on the surface of fibrils, which was confirmed by TERS.

6.4.3.3 TERS for DNA and RNA Sequencing

Imaging complex biomolecules such as proteins, lipids, and nucleic acids require nanoscale characterization techniques with subnanometer spatial resolution. Historically, TERS was first used for distinguishing nucleobases [17–19] (Figure 6.70a), DNA and RNA sensing [20–23], and, finally, sequencing [24–26] (Figure 6.70b). TERS-based DNA sequencing has been a major challenge due to the difficulty in the stable alignment of ssDNA and overlapping spectra of nucleotides. Recent developments in the commercial TERS instrumentation (for example, by Horiba Scientific) led to a breakthrough in DNA sequencing using under ambient conditions achieving the spatial resolution below 1 nm [25] (Figure 6.70b).

The next-envisioned steps may involve solving the challenge of gapless mode sequencing of double-stranded DNA and resolving the conformations of DNA aptamers, which are small segments of DNA used as drugs and sensors. First attempts of DNA aptamers imaging using TERS achieved resolution below 3 nm [27].

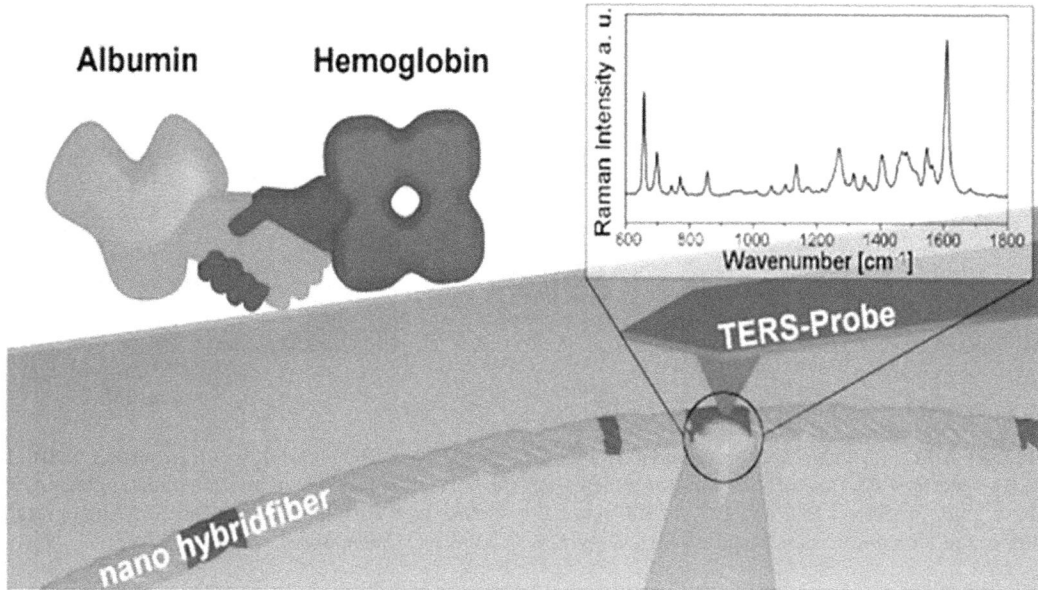

FIGURE 6.69 Protein "handshake" on the nanoscale detected by TERS imaging of albumin–hemoglobin hybrid protein nanofibers. (Reprinted with permission from Ref. [9], Copyright © 2018 *Am. Chem. Soc.*)

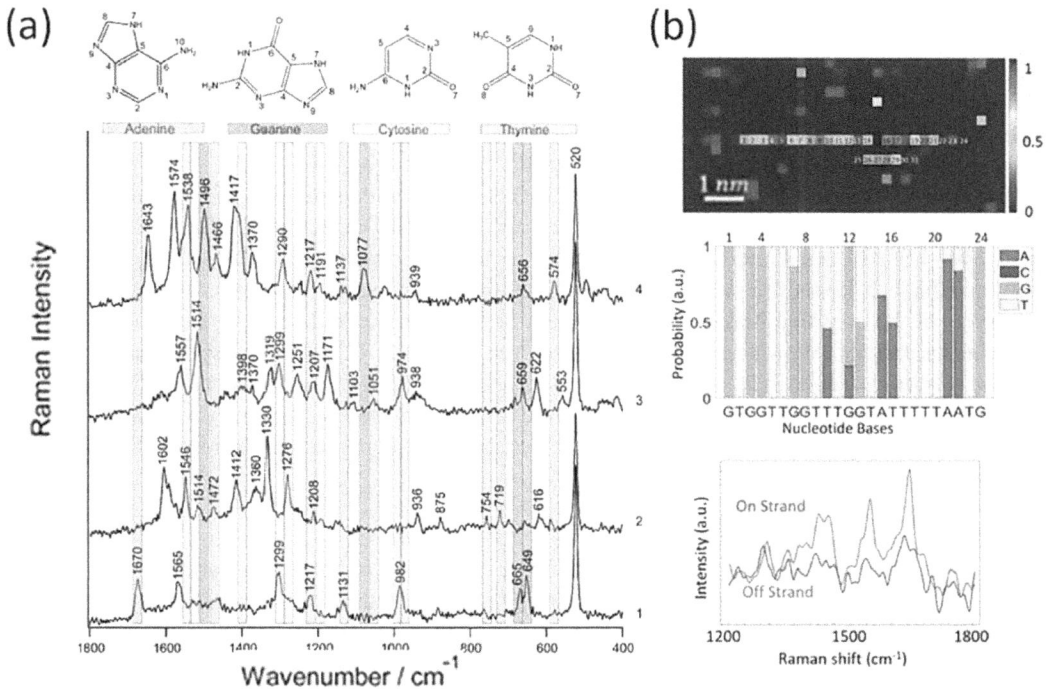

FIGURE 6.70 (a) TERS spectra of nucleobases. *Creative Commons* [18]. (b) TERS-based DNA sequencing: (top) TERS map of ssDNA; (middle) probabilities of bases from the spots in the TERS map; (bottom) TERS spectra on and off DNA strands. (Reprinted with permission from Ref. [25], Copyright © 2018 *Am. Chem. Soc.*)

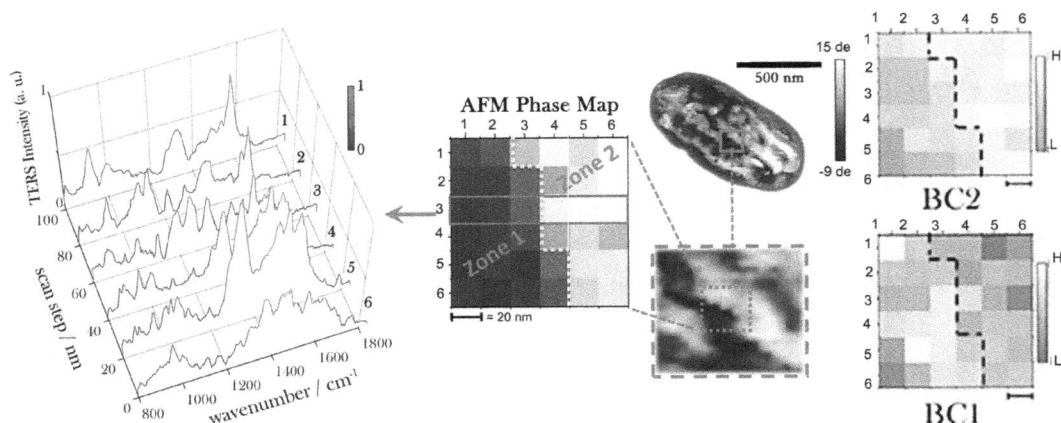

FIGURE 6.71 (a) TERS spectra recorded from a single location on *S. epidermidis* cell (from Ref. [28]). (b) TERS spectra of *Bacillus subtilis* spore recorded along the horizontal line shown by the red rectangle in AFM phase map. Processed BC_1 and BC_2 TERS map components show correlations with the AFM topography. (Reprinted with permission from Ref. [31], Copyright © 2014 *Am. Chem. Soc.*)

6.4.3.4 TERS of Bacteria

Bacteria play an important role in the functioning of organisms as well as whole ecosystems. While most conventional optical imaging techniques focused on studying the whole bacterial cell, TERS may provide a deeper understanding of the structure and chemical composition of the bacterial membrane at the nanoscale [28–32].

The first TERS studies of the gram-positive bacteria, *Staphylococcus epidermidis*, showed the nano-heterogeneous properties of the bacterial cell surface with high spatial resolution [28–30]. The measured TERS spectra showed dynamic variations even from the same location on the bacterial surface which was explained by the dynamics on the cell surface structure. This complicated the analysis and still presents a major challenge for the TERS imaging of bacteria.

The surface structure of the bacterial spores, on the other hand, is more rigid and has less dynamical fluctuations, leading to more stable and reproducible TERS spectra [31]. Bacterial spores can survive for 1,000 of years under extreme environments due to the protection by their external multi-shell coat made of lipids and proteins. TERS may provide new insights into the structure-functional properties of the coat that may be used to design novel biomimetic devices and develop new anti-bacterial treatments against spores such as *Bacillusanthracis*. TERS of the spore model system, *Bacillus subtilis*, showed rich spectral features across the spore coat that varied significantly based on location (Figure 6.71). TEM and AFM phase maps identified regions of variable stiffness as grooves and ridges, which correlated with the density of proteins measured by TERS, combined with the statistical analysis.

6.4.3.5 TERS of Viruses

Tip-enhanced Raman spectroscopy of viruses has been an active field of research for more than a decade and has become especially relevant during the recent pandemic outbreak. Fast and reliable identification of viruses has important security and healthcare applications. Nanoscale imaging may provide new insights into the virucidal treatment mechanisms. TERS can provide nanoscale chemical analysis of the virus surface, and, combined with the correlated morphological AFM analysis, it may become a powerful characterization tool for point-of-care applications.

The first proof-of-principle demonstrations were performed on a single tobacco mosaic virus (TMV) [33], on distinguishing different virus strains such as avipoxvirus (APV) and adeno-associated virus (AAV) [34], and discriminating between enveloped *Varicella-zoster virus* (VZV)

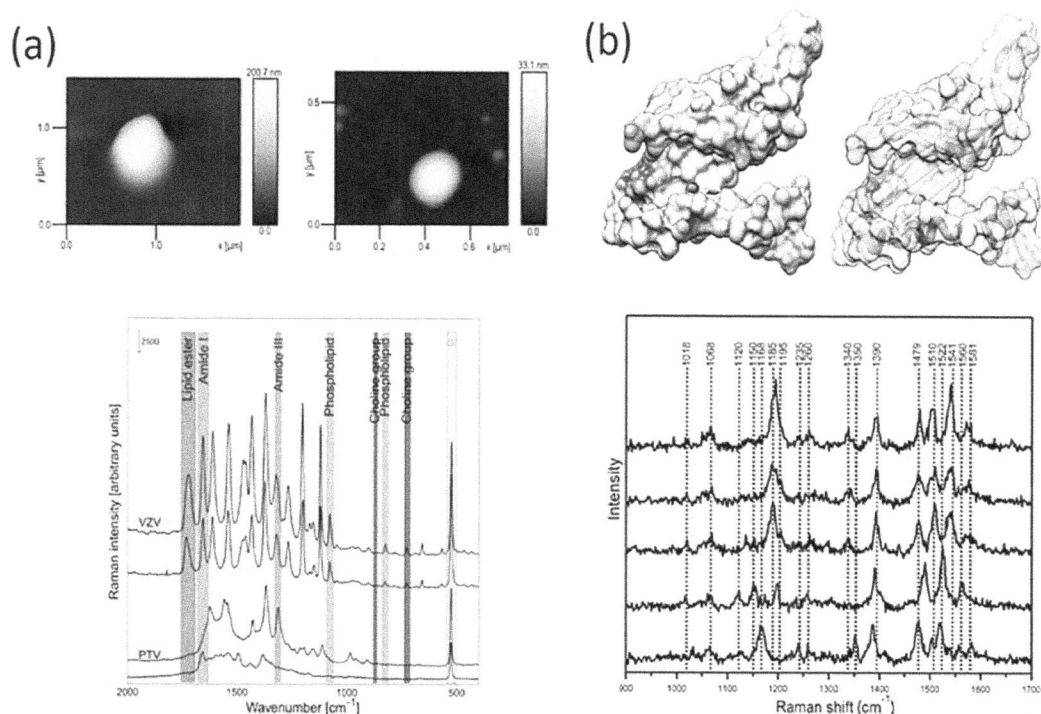

FIGURE 6.72 (a) AFM topographic images of (left) enveloped *Varicella-zoster virus* (VZV) and (right) non-enveloped *Porcine teschovirus* (PTV), and their TERS spectra. (Reprinted with permission from Ref. [35], Copyright © 2015 *Royal Soc. Chem.*) (b) Crystal structure of bacteriophage MS_2 surface protein charge map (left) and superimposed secondary structure and charge map (right), and TERS spectra from MS_2 particles. (Reprinted with permission from Ref. [36], Copyright © 2020 *Am. Chem. Soc.*)

and non-enveloped *Porcine teschovirus* (PTV) by detecting the lipid (highlighted in blue, present only in enveloped virus) and protein (highlighted in green, present in both viruses) peaks [35] (Figure 6.72a). Figure 6.72a also shows spectral variations for repeated measurements of each virus species. VZV showed larger spectral variations due to a more complex structure.

Bimodal imaging using atomic force microscopy infrared (AFM-IR) spectroscopy combined with TERS can provide an improved virus analysis based on the complementary information from these two techniques [36]. Such dual imaging was performed on MS_2 bacteriophage single particles revealing their protein secondary structure and amino acid composition, which may help to investigat the mode of interaction between virus and cell receptors by correlating the structure and amino acid charge distributions [36] (Figure 6.72b).

Future progress in TERS of virus development should be extended to the pandemic-related applications, where fast and user-friendly commercial TERS instruments (such as, for example, Horiba Scientific TERS systems) would be used in high-level biosafety facilities worldwide.

6.4.3.6 From Point Spectra Toward Nanoscale TERS Imaging of Whole Cells

One of the biggest remaining challenges of TERS is application to imaging of whole cells. The first results showed promising steps toward this goal [37–40]. One of the challenges is the limited signal enhancement due to the absence of the plasmonic gap mode because of the large tip-sample distance determined by the large cell height (hundreds of nanometers). This makes it not feasible to obtain chemical information of the intact cell organelles due to the large size of the cell. Since the average height of the cell membrane is ~5–10 nm, the main part of the TERS spectra of intact cells is from the membrane

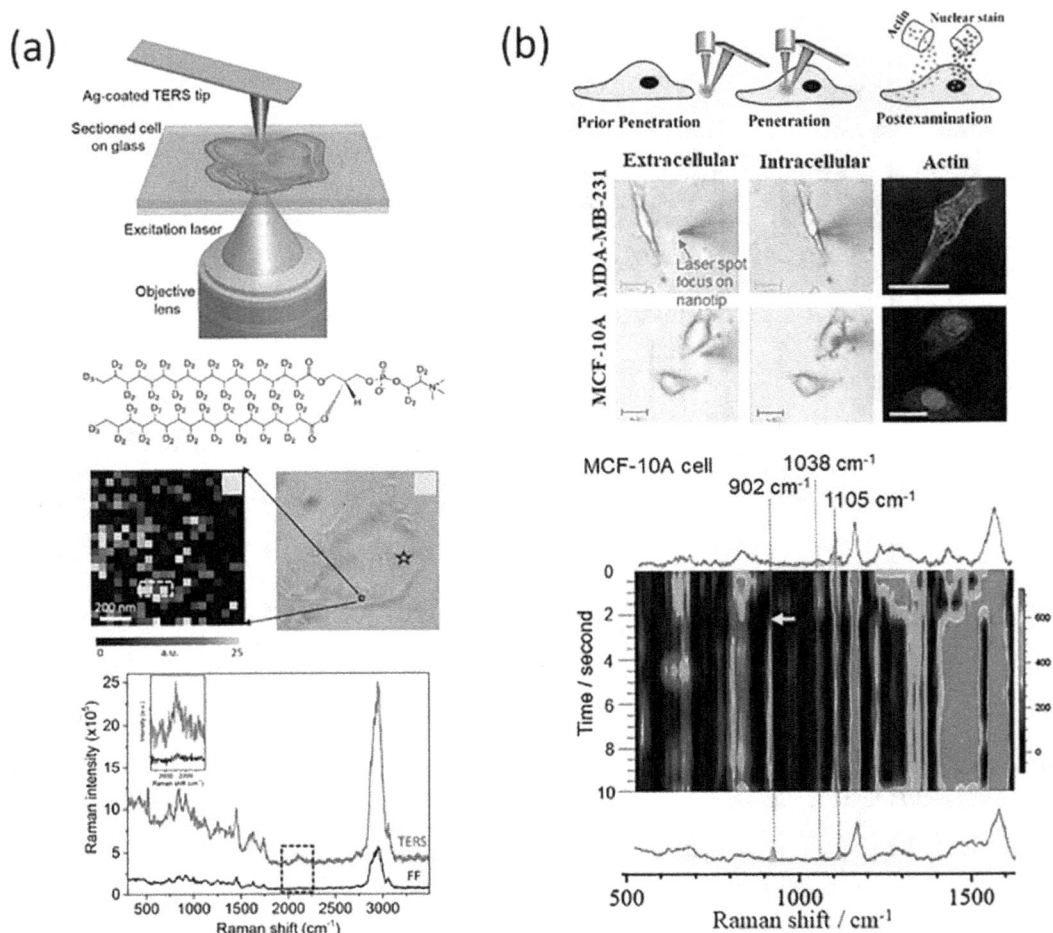

FIGURE 6.73 (a) Schematic of the TERS imaging of sectioned adipocyte cell. Molecular structure of a deuterated phospholipid. TERS map of the cell region within black square in optical image on a glass substrate. Comparison of TERS spectra (recorded from the region in the white dashed box in TERS map) and confocal far-field (FF) spectra. (Reprinted with permission from Ref. [38], Copyright © 2017 *Royal Soc. Chem.*) (b) Time series TERS measurements of penetrated drug-functionalized nanotip into normal MCF10a breast cell. White arrow shows the appearance of 902 cm^{-1} peak 2 s after tip penetration. (Reprinted with permission from Ref. [40], Copyright © 2021 *Am. Chem. Soc.*)

proteins and lipids. However, TERS mapping of sectioned adipocyte mouse cells was performed and showed nanoscale distribution of newly synthesized phospholipid molecules [38] (Figure 6.73a).

Tip-enhanced Raman spectroscopy of cancer drug generation in live cells was shown by functionalizing TERS tip with ferrocifen molecules and penetrating the normal (MCF-10A) and cancer (MDA-MB-231) breast cells [39,40] (Figure 6.73b). Dynamic TERS spectra were recorded during the penetration and revealed the kinetics of the in situ bio-induced quinone methide (QM) drug formation (via formation of 902 cm^{-1} peak show by arrow in Figure 6.73b). Due to high oxidative metabolism of cancer cells, conversion of ferrocifen to QM was higher in MDA-MB-231 cancer cells than in MCF-10A cells, which was indicated by increased intensity of the TERS peak at 902 cm^{-1} and simultaneous reduction at 1,038 cm^{-1}. TERS spectra provide rich spectral information with other peaks that may be used to confirm and improve biological interpretation. Nanoscale analysis of biomolecules in whole cells is crucial for understanding complex biomolecular processes within cell's extracellular and intracellular environment that may help to improve target-specific drug delivery and medical treatments. TERS provides this critical information as nondestructive point spectra or mapping with a high spatial resolution.

REFERENCES

1. Sharma, G.; Deckert-Gaudig, T.; Deckert, V. Tip-enhanced Raman scattering—targeting structure-specific surface characterization for biomedical samples. *Advanced Drug Delivery Reviews* 2015, *89*, 42–56.
2. Bonhommeau, S.; Lecomte, S. Tip-enhanced Raman spectroscopy: A tool for nanoscale chemical and structural characterization of biomolecules. *ChemPhysChem* 2018, *19*(1), 8–18.
3. Kurouski, D.; Dazzi, A.; Zenobi, R.; Centrone, A. Infrared and Raman chemical imaging and spectroscopy at the nanoscale. *Chemical Society Reviews* 2020, *49*(11), 3315–3347.
4. Yeo, B.-S.; Mädler, S.; Schmid, T.; Zhang, W.; Zenobi, R. Tip-enhanced Raman spectroscopy can see more: The case of cytochrome c. *The Journal of Physical Chemistry C* 2008, *112*(13), 4867–4873.
5. Böhme, R.; Mkandawire, M.; Krause-Buchholz, U.; Rösch, P.; Rödel, G.; Popp, J.; Deckert, V. Characterizing cytochrome c states–TERS studies of whole mitochondria. *Chemical Communications* 2011, *47*(41), 11453–11455.
6. Wang, H.; Schultz, Z. D. The chemical origin of enhanced signals from tip-enhanced Raman detection of functionalized nanoparticles. *Analyst* 2013, *138*(11), 3150–3157.
7. Xiao, L.; Wang, H.; Schultz, Z. D. Selective detection of RGD-integrin binding in cancer cells using tip enhanced Raman scattering microscopy. *Analytical Chemistry* 2016, *88*(12), 6547–6553.
8. Wang, H.; Schultz, Z. D. TERS detection of Avβ_3 integrins in intact cell membranes. *ChemPhysChem* 2014, *15*(18), 3944–3949.
9. Helbing, C.; Deckert-Gaudig, T.; Firkowska-Boden, I.; Wei, G.; Deckert, V.; Jandt, K. D. Protein handshake on the nanoscale: How albumin and hemoglobin self-assemble into nanohybrid fibers. *ACS Nano* 2018, *12*(2), 1211–1219.
10. Capocefalo, A.; Deckert-Gaudig, T.; Brasili, F.; Postorino, P.; Deckert, V. Unveiling the interaction of protein fibrils with gold nanoparticles by plasmon enhanced nano-spectroscopy. *Nanoscale* 2021, *13* (34), 14469–14479.
11. Kurouski, D.; Van Duyne, R. P.; Lednev, I. K. Exploring the structure and formation mechanism of amyloid fibrils by Raman spectroscopy: A review. *Analyst*, 2015, *140*(15), 4967–4980.
12. Kurouski, D.; Deckert-Gaudig, T.; Deckert, V.; Lednev, I. K. Structure and composition of insulin fibril surfaces probed by TERS. *Journal of the American Chemical Society* 2012, *134*(32), 13323–13329.
13. Deckert-Gaudig, T.; Kurouski, D.; Hedegaard, M. A.; Singh, P.; Lednev, I. K.; Deckert, V. Spatially resolved spectroscopic differentiation of hydrophilic and hydrophobic domains on individual insulin amyloid fibrils. *Scientific Reports* 2016, *6*(1), 1–9.
14. Bonhommeau, S.; Talaga, D.; Hunel, J.; Cullin, C.; Lecomte, S. Tip-enhanced Raman spectroscopy to distinguish toxic oligomers from Aβ_1–42 fibrils at the nanometer scale. *Angewandte Chemie International Edition* 2017, *129*(7), 1797–1800.
15. Kurouski, D.; Deckert-Gaudig, T.; Deckert, V.; Lednev, I. K. Surface characterization of insulin protofilaments and fibril polymorphs using tip-enhanced Raman spectroscopy (TERS). *Biophysical Journal* 2014, *106*(1), 263–271.
16. Krasnoslobodtsev, A. V.; Deckert-Gaudig, T.; Zhang, Y.; Deckert, V.; Lyubchenko, Y. L. Polymorphism of amyloid fibrils formed by a peptide from the yeast prion protein sup 35: AFM and tip-enhanced Raman scattering studies. *Ultramicroscopy* 2016, *165*, 26–33.
17. Rasmussen, A.; Deckert, V. Surface-and tip-enhanced Raman scattering of DNA components. *Journal of Raman Spectroscopy: An International Journal for Original Work in all Aspects of Raman Spectroscopy, Including Higher Order Processes, and also Brillouin and Rayleigh Scattering* 2006, *37*(1–3), 311–317.
18. Treffer, R.; Lin, X.; Bailo, E.; Deckert-Gaudig, T.; Deckert, V. Distinction of nucleobases – A tip-enhanced Raman approach. *Beilstein Journal of Nanotechnology* 2011, *2*(1), 628–637.
19. Zhang, R.; Zhang, X.; Wang, H.; Zhang, Y.; Jiang, S.; Hu, C.; Zhang, Y.; Luo, Y.; Dong, Z. Distinguishing individual DNA bases in a network by non-resonant tip-enhanced Raman scattering. *Angewandte Chemie International Edition* 2017, *56*(20), 5561–5564.
20. Domke, K. F.; Zhang, D.; Pettinger, B. Tip-enhanced Raman spectra of picomole quantities of DNA nucleobases at Au(111). *Journal of the American Chemical Society* 2007, *129*(21), 6708–6709.

21. Bailo, E.; Deckert, V. Tip-enhanced Raman spectroscopy of single RNA strands: Towards a novel direct-sequencing method. *Angewandte Chemie International Edition* 2008, *47*(9), 1658–1661.

22. Najjar, S.; Talaga, D.; Schué, L.; Coffinier, Y.; Szunerits, S.; Boukherroub, R.; Servant, L.; Rodriguez, V.; Bonhommeau, S. Tip-enhanced Raman spectroscopy of combed double-stranded DNA bundles. *The Journal of Physical Chemistry C* 2014, *118*(2), 1174–1181.

23. Lipiec, E.; Sekine, R.; Bielecki, J.; Kwiatek, W. M.; Wood, B. R. Molecular characterization of DNA double strand breaks with tip-enhanced Raman scattering. *Angewandte Chemie International Edition* 2014, *126*(1), 173–176.

24. Lin, X.-M.; Deckert-Gaudig, T.; Singh, P.; Siegmann, M.; Kupfer, S.; Zhang, Z.; Gräfe, S.; Deckert, V. Direct base-to-base transitions in SsDNA revealed by tip-enhanced Raman scattering. *arXiv preprint arXiv:1604.06598* 2016.

25. He, Z.; Han, Z.; Kizer, M.; Linhardt, R. J.; Wang, X.; Sinyukov, A. M.; Wang, J.; Deckert, V.; Sokolov, A. V.; Hu, J. Tip-enhanced Raman imaging of single-stranded DNA with single base resolution. *Journal of the American Chemical Society* 2018, *141*(2), 753–757.

26. He, Z.; Qiu, W.; Kizer, M. E.; Wang, J.; Chen, W.; Sokolov, A. V.; Wang, X.; Hu, J.; Scully, M. O. Resolving the sequence of RNA strands by tip-enhanced Raman spectroscopy. *ACS Photonics* 2020, *8*(2), 424–430.

27. He, S.; Li, H.; Gomes, C. L.; Voronine, D. V. Tip-enhanced Raman scattering of DNA aptamers for listeria monocytogenes. *Biointerphases* 2018, *13*(3), 03C402.

28. Neugebauer, U.; Rösch, P.; Schmitt, M.; Popp, J.; Julien, C.; Rasmussen, A.; Budich, C.; Deckert, V. On the way to nanometer-sized information of the bacterial surface by tip-enhanced Raman spectroscopy. *ChemPhysChem: A European Journal of Chemical Physics and Physical Chemistry* 2006, *7*(7), 1428–1430.

29. Neugebauer, U.; Schmid, U.; Baumann, K.; Ziebuhr, W.; Kozitskaya, S.; Deckert, V.; Schmitt, M.; Popp, J. Towards a detailed understanding of bacterial metabolism—spectroscopic characterization of staphylococcus epidermidis. *ChemPhysChem* 2007, *8*(1), 124–137.

30. Budich, C.; Neugebauer, U.; Popp, J.; Deckert, V. Cell wall investigations utilizing tip-enhanced Raman scattering. *Journal of Microscopy* 2008, *229*(3), 533–539.

31. Rusciano, G.; Zito, G.; Isticato, R.; Sirec, T.; Ricca, E.; Bailo, E.; Sasso, A. Nanoscale chemical imaging of bacillus subtilis spores by combining tip-enhanced Raman scattering and advanced statistical tools. *ACS Nano* 2014, *8*(12), 12300–12309.

32. Deckert-Gaudig, T.; Böhme, R.; Freier, E.; Sebesta, A.; Merkendorf, T.; Popp, J.; Gerwert, K.; Deckert, V. Nanoscale distinction of membrane patches – A TERS study of halobacterium salinarum. *Journal of Biophotonics* 2012, *5*(7), 582–591.

33. Cialla, D.; Deckert-Gaudig, T.; Budich, C.; Laue, M.; Möller, R.; Naumann, D.; Deckert, V.; Popp, J. Raman to the limit: Tip-enhanced Raman spectroscopic investigations of a single tobacco mosaic virus. *Journal of Raman Spectroscopy: An International Journal for Original Work in all Aspects of Raman Spectroscopy, Including Higher Order Processes, and also Brillouin and Rayleigh Scattering* 2009, *40*(3), 240–243.

34. Hermann, P.; Hermelink, A.; Lausch, V.; Holland, G.; Möller, L.; Bannert, N.; Naumann, D. Evaluation of tip-enhanced Raman spectroscopy for characterizing different virus strains. *Analyst* 2011, *136*(6), 1148–1152.

35. Olschewski, K.; Kämmer, E.; Stöckel, S.; Bocklitz, T.; Deckert-Gaudig, T.; Zell, R.; Cialla-May, D.; Weber, K.; Deckert, V.; Popp, J. A manual and an automatic TERS based virus discrimination. *Nanoscale* 2015, *7*(10), 4545–4552.

36. Dou, T.; Li, Z.; Zhang, J.; Evilevitch, A.; Kurouski, D. Nanoscale structural characterization of individual viral particles using atomic force microscopy infrared spectroscopy (AFM-IR) and tip-enhanced Raman spectroscopy (TERS). *Analytical Chemistry* 2020, *92*(16), 11297–11304.

37. Wood, B. R.; Bailo, E.; Khiavi, M. A.; Tilley, L.; Deed, S.; Deckert-Gaudig, T.; McNaughton, D.; Deckert, V. Tip-enhanced Raman scattering (TERS) from hemozoin crystals within a sectioned erythrocyte. *Nano Letters*, 2011, *11*(5), 1868–1873.

38. Kumar, N.; Drozdz, M. M.; Jiang, H.; Santos, D. M.; Vaux, D. J. Nanoscale mapping of newly-synthesised phospholipid molecules in a biological cell using tip-enhanced Raman spectroscopy. *Chemical Communications* 2017, *53*(16), 2451–2454.

39. Lin, Y.-C.; Ke, Z.-Y.; Liao, P.-H.; Tseng, C.-Y.; Kong, K. V. Reversible detection of phosphorylation and dephosphorylation by tip-enhanced Raman spectroscopy using a cyclopentadienyl ruthenium nanotag functionalized tip. *Chemical Communications* 2020, *56*(6), 936–939.

40. Hsu, K.-J.; Hsieh, C.-L.; Tsai, C.-J.; Kong, K. V. Probing molecular-scale oxidative generation of quinone methides and their transformation using tip-enhanced Raman spectroscopy. *The Journal of Physical Chemistry Letters* 2021, *12*(3), 1110–1115.

6.4.4   APPLICATION OF TERS TO HETEROGENOUS CATALYTIC SYSTEMS

Rational development of novel catalytic materials requires highly sensitive analytical tools capable of delivering molecular information at the nanoscale. In this regard, the ultrahigh sensitivity and spatial resolution of TERS have a unique advantage over conventional methods of catalyst characterization that only provides ensemble information at the bulk scale or the vibrational spectroscopic methods such as UV-vis, Raman, and IR, whose lateral resolution is diffraction-limited to the microscale. In the last two decades, TERS has established itself as a promising nanoanalytical tool for the hyperspectral investigation of heterogeneous catalysts at nanometer length scales [1,2]. Herein, we review the application of TERS to both model and real-life catalytic systems.

6.4.4.1   Plasmon-driven Photocatalysis

Plasmon-driven photocatalytic conversion of 4-nitrothiophenol (4-NTP) and 4-amiothiophenol (4-ATP) top, p′-dimercaptoazobisbenzene (DMAB) is the most widely studied model catalytic system using TERS [3]. This is primarily due to the high Raman cross section of the reactant and product molecules, which makes it easier to visualize the reaction over time and space. The capability of TERS to monitor a photocatalytic reaction over time was first demonstrated by van Schrojenstein Lantman et al. [4]. Catalytic transformation of 4NTP → DMAB was successfully triggered and monitored at the apex of a Ag coated TERS probe placed at an ultrathin Au nanoplate using 532 nm laser irradiation, while no reaction was observed under 633 nm laser irradiation. Kumar et al. demonstrated the ability of TERS to spatially resolve catalytic active sites of 4ATP → DMAB photocatalytic coupling on a rough Ag film [5]. Since Ag coated TERS probe itself was catalytically active, it was first passivated using a 3–5 nm alumina coating. The dielectric protection of the TERS probe prevented catalytic interference while maintaining sufficient plasmonic enhancement for imaging of 4ATP → DMAB reaction hotspots with a spatial resolution of ca. 20 nm. Going a step further, in another study Kumar et al. showed that it is possible to resolve photocatalytic reaction hotspots in an aqueous environment using TERS [6]. This time, multilayer metal-coated TERS probes [7] were protected using an ultrathin zirconia layer, which not only preserved the plasmonic activity of the probe and eliminated catalytic interference with the reaction, but also made the rather fragile Ag coating of the TERS probes ultrastable in liquid environments. Using these plasmonically active, chemical inert, durable, and robust TERS probes 4-ATP → DMAB hotspots were successfully mapped with a spatial resolution of ca. 14 nm on a rough Ag film.

More recently, oxidative and reductive photocatalytic coupling of 4ATP → DMAB and 4NTP → DMAB on well-defined Au(111) and Ag(111) surfaces was investigated by Sun et al. [8]. Interestingly, the authors observed that none of these catalytic coupling reactions occurred on Ag(111) surface. Furthermore, reductive coupling of 4-NTP on well-defined Au(111) surface was found to be excitation laser-dependent proceeding only under 532 nm light irradiation. Authors rationalized their experimental observations with different energies of the 532 nm and 633 nm lasers and the variable orientation of the reactant molecules adsorbed on the Ag and Au substrates calculated using DFT. Bhattarai and El-Khoury discovered the presence of cis-DMAB together with trans-DMAB during nanoscale chemical imaging of 4-NTP functionalized Au nanoplates in an aqueous environment [9]. Moreover, 4-NTP → DMAB reaction was found to be confined to distinct hotspots on the substrate suggesting possible involvement of molecular crowding and stearic effects in the photocatalytic dimerization process. In another study, Wang et al. observed the formation of 4-NTP thiolate (anionic form of 4-NTP) in their TERS measurements [10]. Using DFT, finite-difference time-domain, and finite element method calculations, the authors demonstrated that maximum temperature in the TERS near-field shouldn't exceed 305K and therefore formation of 4-NTP thiolate occurred more likely via hot carriers rather than thermal desorption. In a separate study, the same group reported observing 4-NTP thiolate species on 4-NTP functionalized faceted Ag nanoparticles more frequently than the usual DMAB product [11].

6.4.4.2 Bimetallic Catalysts

Yet another category of heterogenous catalysts investigated intensely using TERS are bimetallic catalytic systems. Yin et al. examined active sites of hydrogenation reaction on a Pd(submonolayer)/ Au(111) bimetallic catalyst using selective hydrogenation of chloronitrobenzenethiol (CNBT) to chloroaminobenzenethiol as a model system [12]. Using high-resolution TERS imaging and supporting DFT modeling, the authors discovered hydrogen spillover effect by identifying CABT molecules on Au(111) surface, which is not supposed to catalyze the reaction. The spatial range of hydrogen spillover was estimated to be 15–30 nm. Su et al. studied active oxygen species (AOS) generated on a Pd/Au(111) surface using 4′-(pyridin-4-yl)biphenyl-4-yl)-methanethiol (PBT) as TERS probe [13]. AOS including OH radicals were generated by immersing PBT functionalized bimetallic surface in H_2O_2 solution, which oxidized the thiolate and desorbed it from the surface. Using TERS imaging, authors confirmed that Pd was the active site of AOS generation and Pd step edge had an enhanced catalytic activity compared to the Pd terrace. Zhong et al. probed electronic and catalytic properties of a well-defined Pd(submonolayer)/Au(111) bimetallic model catalyst using phenyl isocyanide (PIC) as a probe molecule [14]. Using TERS line scan across a Pd step, the authors observed a weakened N≡C bond and enhanced reactivity of PIC at the Pd step edge compared to the Pd terrace with an impressive spatial resolution of 3 nm under ambient conditions. Furthermore, the same group used TERS to distinguish electronic properties of the terrace, step, edge, kink, and corner sites on a Pt nanoisland/Au(111) bimetallic surface with a spatial resolution of ca. 2.5 nm [15].

6.4.4.3 Organometallic Phthalocyanine Catalysts

Van Duyne group has produced notable research using TERS to study organometallic phthalocyanine catalysts. Using UHV-TERS (UHV-TERS), Nguyen et al. probed adsorption of O_2 on cobalt(II) phthalocyanine (CoPc) supported on Ag(111), which is the first step in oxygen reduction reaction (ORR) [16]. Using a combination of $^{16}O_2$ and $^{18}O_2$ isotopologue substitution, STM imaging, TERS, and DFT modeling, the authors identified two distinct adsorption configurations: O_2/CoPc/Ag(111) and O/CoPc/Ag(111). Additionally, distinct TERS signals were observed for the $^{16}O_2$, $^{18}O_2$, ^{16}O, and ^{18}O species adsorbed on the Co center demonstrating high chemically sensitivity of TERS for probing catalytic systems at the molecular scale. Jiang et al. studied electrochemical (EC) behavior of catalytically active CoPc at solid/liquid interface under realistic ORR conditions using EC-TERS [17]. At increasingly negative substrate potential, CoPc molecules were found to organize into highly ordered monolayers on Au(111) surface until an ordered → diffused state transition was triggered accompanied with the disappearance of TERS signals. The authors concluded that partially reduced CoPc molecules are the dominant species formed under steady-state ORR. *Operando* characterization of iron (II) phthalocyanine (FePc) deactivation during ORR was investigated by Chen et al. using EC-TERS [18]. The authors affirmed that FePc demetallation during ORR occurred by a direct loss of Fe^{2+} and not by carbon corrosion. Furthermore, FePc demetallation was correlated with a loss of ORR activity demonstrating EC-TERS as a promising tool for the *operando* monitoring of electrocatalytic reactions.

6.4.4.4 Fluid Cracking Catalyst (FCC) Particles

Kumar et al. applied hyperspectral tip-enhanced fluorescence (TEFL) microscopy to study industrially spent FCC particles. To probe internal structure, three-dimensional FCC particles were cut into ultrathin slices using microtome sectioning [19]. Catalytically active Brønsted acid sites present in the zeolite domains were visualized via thiophene staining, which produced fluorescent conjugated oligomeric species. High-resolution TEFL imaging revealed significant inter- and intra-particle variation in catalytic activity and size of zeolite domains. Although the size of zeolite domains varied from 0.002 to 0.28 μm², 75% of the domains had a size of < 0.04 μm². Furthermore, inside individual FCC particles, the average size of active zeolite domains was found to decrease from center to edge. This study demonstrated the capability of tip-enhanced optical microscopy to probe degradation of real-life multicomponent complex catalytic systems at the nanometer length scales.

REFERENCES

1. Kumar, N.; Mignuzzi, S.; Su, W.; Roy, D., Tip-enhanced Raman spectroscopy: Principles and applications. *EPJ Tech. Instrum.* 2015, *2*(1), 9.
2. Kumar, N.; Weckhuysen, B. M.; Wain, A. J.; Pollard, A. J., Nanoscale chemical imaging using tip-enhanced Raman spectroscopy. *Nature Protocols* 2019, *14*(4), 1169–1193.
3. Hartman, T.; Wondergem, C. S.; Kumar, N.; van den Berg, A.; Weckhuysen, B. M., Surface- and tip-enhanced Raman spectroscopy in catalysis. *Journal of Physical Chemistry Letters* 2016, *7*(8), 1570–1584.
4. van Schrojenstein Lantman, E. M.; Deckert-Gaudig, T.; Mank, A. J.; Deckert, V.; Weckhuysen, B. M., Catalytic processes monitored at the nanoscale with tip-enhanced Raman spectroscopy. *Nature Nanotechnology* 2012, *7*(9), 583–586.
5. Kumar, N.; Stephanidis, B.; Zenobi, R.; Wain, A. J.; Roy, D., Nanoscale mapping of catalytic activity using tip-enhanced Raman spectroscopy. *Nanoscale* 2015, *7*(16), 7133–7137.
6. Kumar, N.; Wondergem, C. S.; Wain, A. J.; Weckhuysen, B. M., In situ nanoscale investigation of catalytic reactions in the liquid phase using zirconia-protected tip-enhanced Raman spectroscopy probes. *Journal of Physical Chemistry Letters* 2019, *10*(8), 1669–1675.
7. Kumar, N.; Su, W. T.; Vesely, M.; Weckhuysen, B. M.; Pollard, A. J.; Wain, A. J., Nanoscale chemical imaging of solid-liquid interfaces using tip-enhanced Raman spectroscopy. *Nanoscale* 2018, *10*(4), 1815–1824.
8. Sun, J. J.; Su, H. S.; Yue, H. L.; Huang, S. C.; Huang, T. X.; Hu, S.; Sartin, M. M.; Cheng, J.; Ren, B., Role of adsorption orientation in surface plasmon-driven coupling reactions studied by tip-enhanced Raman spectroscopy. *Journal of Physical Chemistry Letters* 2019, *10*(10), 2306–2312.
9. Bhattarai, A.; El-Khoury, P. Z., Nanoscale chemical reaction imaging at the solid-liquid interface via TERS. *Journal of Physical Chemistry Letters* 2019, *10*(11), 2817–2822.
10. Wang, R.; Li, J.; Rigor, J.; Large, N.; El-Khoury, P. Z.; Rogachev, A. Y.; Kurouski, D., Direct experimental evidence of hot carrier-driven chemical processes in tip-enhanced Raman spectroscopy (TERS). *The Journal of Physical Chemistry C* 2020, *124*(3), 2238–2244.
11. Wang, C.-F.; O'Callahan, B. T.; Kurouski, D.; Krayev, A.; Schultz, Z. D.; El-Khoury, P. Z., Suppressing molecular charging, nanochemistry, and optical rectification in the tip-enhanced Raman geometry. *The Journal of Physical Chemistry Letters* 2020, *11*(15), 5890–5895.
12. Yin, H.; Zheng, L.-Q.; Fang, W.; Lai, Y.-H.; Porenta, N.; Goubert, G.; Zhang, H.; Su, H.-S.; Ren, B.; Richardson, J. O.; Li, J.-F.; Zenobi, R., Nanometre-scale spectroscopic visualization of catalytic sites during a hydrogenation reaction on a Pd/Au bimetallic catalyst. *Nature Catalysis* 2020, *3*(10), 834–842.
13. Su, H. S.; Feng, H. S.; Zhao, Q. Q.; Zhang, X. G.; Sun, J. J.; He, Y. H.; Huang, S. C.; Huang, T. X.; Zhong, J. H.; Wu, D. Y.; Ren, B., Probing the local generation and diffusion of active oxygen species on a Pd/Au bimetallic surface by tip-enhanced Raman spectroscopy. *Journal of the American Chemical Society* 2020, *142*(3), 1341–1347.
14. Zhong, J. H.; Jin, X.; Meng, L. Y.; Wang, X.; Su, H. S.; Yang, Z. L.; Williams, C. T.; Ren, B., Probing the electronic and catalytic properties of a bimetallic surface with 3 nm resolution. *Nature Nanotechnology* 2017, *12*(2), 132–136.
15. Su, H. S.; Zhang, X. G.; Sun, J. J.; Jin, X.; Wu, D. Y.; Lian, X. B.; Zhong, J. H.; Ren, B., Real-space observation of atomic site-specific electronic properties of a Pt nanoisland/Au(111) bimetallic surface by tip-enhanced Raman spectroscopy. *Angewandte Chemie-International Edition* 2018, *57*(40), 13177–13181.
16. Nguyen, D.; Kang, G.; Chiang, N.; Chen, X.; Seideman, T.; Hersam, M. C.; Schatz, G. C.; Van Duyne, R. P., Probing molecular-scale catalytic interactions between oxygen and cobalt phthalocyanine using tip-enhanced Raman spectroscopy. *Journal of the American Chemical Society* 2018, *140*(18), 5948–5954.
17. Jiang, S.; Chen, Z.; Chen, X.; Nguyen, D.; Mattei, M.; Goubert, G.; Van Duyne, R. P., Investigation of cobalt phthalocyanine at the solid/liquid interface by electrochemical tip-enhanced Raman spectroscopy. *Journal of Physical Chemistry C* 2019, *123*(15), 9852–9859.
18. Chen, Z.; Jiang, S.; Kang, G.; Nguyen, D.; Schatz, G. C.; Van Duyne, R. P., Operando characterization of iron phthalocyanine deactivation during oxygen reduction reaction using electrochemical tip-enhanced Raman spectroscopy. *Journal of the American Chemical Society* 2019, *141*(39), 15684–15692.
19. Kumar, N.; Kalirai, S.; Wain, A. J.; Weckhuysen, B. M., Nanoscale chemical imaging of a single catalyst particle with tip-enhanced fluorescence microscopy. *Chemcatchem* 2019, *11*(1), 417–423.

6.4.5   **EXCITON-ENHANCED NANOSCALE SECOND-HARMONIC GENERATION AND IMAGING**
           OF TWO-DIMENSIONAL TRANSITION METAL DICHALCOGENIDE STACKING

Second-harmonic generation (SHG) is a nonlinear optical response arising exclusively from broken inversion symmetry in the electric-dipole limit. Recently, SHG has attracted widespread interest as a versatile and noninvasive tool for characterization of crystal symmetry and emerging ferroic or topological orders in quantum materials. However, conventional far-field optics is unable to probe local symmetry at the deep subwavelength scale. Here, we demonstrate near-field SHG imaging of 2D semiconductors and heterostructures with the spatial resolution down to 20 nm using a scattering-type nano-optical apparatus. We show that near-field SHG efficiency is greatly enhanced by excitons in atomically thin TMDs. Furthermore, by correlating nonlinear and linear scattering-type nano-imaging, we resolve nanoscale variations of interlayer stacking order in bilayer WSe_2, and reveal the stacking-tuned excitonic light-matter-interactions. Our work demonstrates nonlinear optical interrogation of crystal symmetry and structure-property relationships at the nanometer length scales relevant to emerging properties in quantum materials.

6.4.5.1   Introduction

A wide range of emerging electronic, magnetic, and topological properties in quantum materials are intimately linked to spatial inversion symmetry. In Van der Waals' heterostructures, the crystal symmetry can be precisely controlled by interlayer stacking and twisting techniques, providing a knob for programming the electronic bandstructure [1–3], magnetic orders [4], optical [5–11], thermal [12], phonon properties [13], as well as mechanical relaxation and atomic reconstruction [14]. SHG is the lowest-order nonlinear optical response, where the incident laser beam interacts with the material, emitting a frequency-doubled signal. The leading-order SHG arises from electric-dipole polarization, and is exclusively allowed in systems with a broken spatial inversion symmetry [15]. In addition to the crystallographic lattice structure, hidden orders of charge and spin configurations can also break inversion symmetry, such as ferroelectricity [16], antiferromagnetism [17–19], and even electron nematicity [20]. Therefore, SHG microscopy has grown in importance as a tool for directly accessing and visualizing crystal orientation and ferroic domain patterns. Since the spatial resolution attainable by far-field optics is limited by optical diffraction, developing scanning near-field SHG (i.e., "nano-SHG") imaging is highly desirable for investigation of symmetry properties at the nano- and meso-scale. This is the critical length scale relevant to many emerging quantum material systems, such as the lateral size of moiré superlattice [21,22], Van der Waals' multiferroic domain patterns [23], and strain-localized 2D QD [24].

Development of scanning near-field SHG imaging has proven to be a formidable task. Specifically, nano-SHG imaging of atomically thin material is challenging because only an extraordinarily small area of sample material is available under the tip for nanoscale frequency conversion. In previous work, the low efficiency of SHG, exacerbated by weak nano-optical interactions, has generally required a fundamental pump laser with high pulse fluence of about 10 mJ/cm² [25] exceeding the reported damage threshold of plasmonic nanostructures [26] and most 2D semiconductors including TMDs [27]. Hampered by tip and sample degradation, the achieved spatial resolution for nano-SHG has typically been ~100 nm [25,28–31], with few exceptions on nonlinear ferroelectric oxide films [32] and plasmonic nanocubes [33]. In fact, sub-100-nm-level spatial resolution has not been achieved in SHG studies of 2D materials to date. Additionally, previous experiments also suffered from large background signals, including both the far-field SHG response from the sample and the SHG generated by the apex of the scanning probe (i.e., tip-SHG) [25,32]. As the background SHG is at the same wavelength with the desired near-field signal, disentangling the contributions is difficult, often requiring complex post-processing of image data based on first-principle models [34].

In this work, we realize nano-SHG imaging of atomically thin TMD semiconductors and heterostructures with a spatial resolution of 20 nm. The aforementioned efficiency bottleneck is overcome by two-photon pumping of the band-nesting excitons, providing SHG enhancement over a relatively broad range of frequency. We also generate near-field SHG signals with significantly lower background as evidenced by approach curve data. As a proof-of-principle demonstration, we image bilayer WSe_2 grown by chemical vapor deposition (CVD), resolving crystal domains with different interlayer stacking orders, and reveal the effects of stacking symmetry on nano-optical nonlinear light scattering.

6.4.5.2 Exciton-Enhanced Nano-SHG Results

Figure 6.74a shows a schematic of exciton-enhanced nano-SHG. The plasmonic tip is an AFM probe coated with Ti/Au (10/150 nm) layers for plasmonic enhancement. For our fabrication of plasmonic scanning probes for nano-SHG, we start with commercial silicon AFM probes with a resonance frequency of 300 (or 60) kHz and spring constant of 100 (or 2.7) N/m. The stiffer cantilevers are more stable in scanning probe operation with the pulsed laser, but in principle, both kinds of AFM probes can work. Layers of Ti/Au are coated by electron beam evaporation, with a rate of 2 Angstrom/s in a chamber with base pressure of 10^{-8} to 10^{-7} Torr. The coating thickness of Ti/Au is 2/30 nm. Note that since the AFM tip is mounted with the apex facing towards the crucible, we estimate that the actual coating thickness on the sidewall of the pyramid-shaped shaft is about 1/5

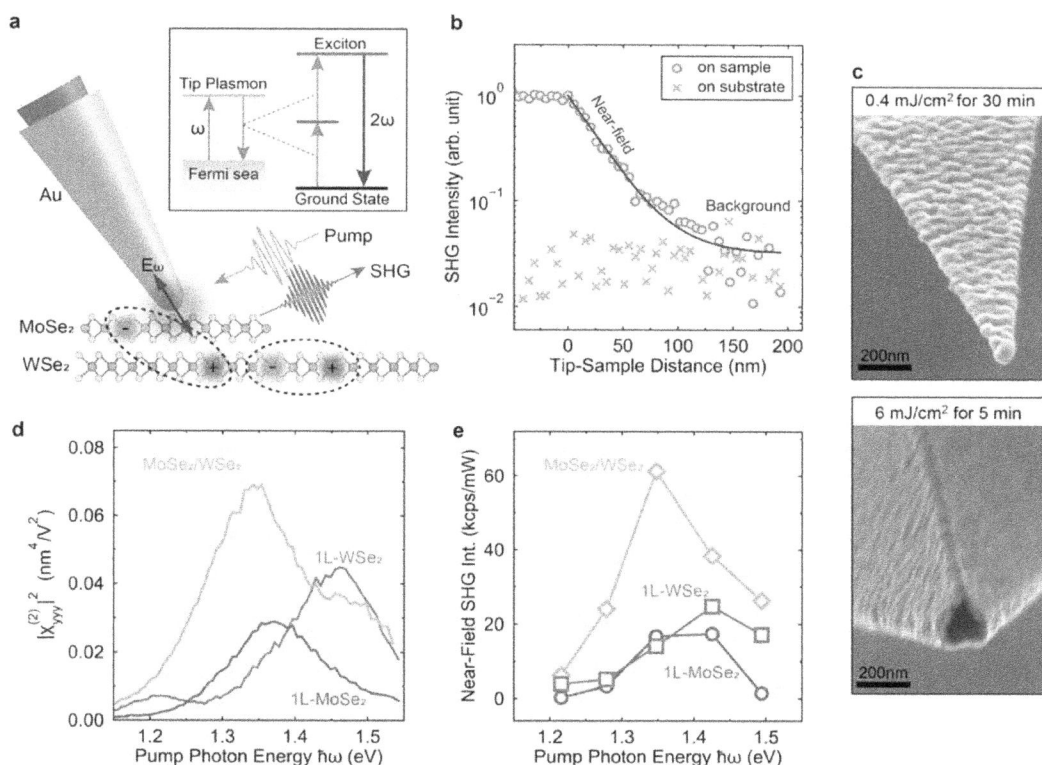

FIGURE 6.74 Exciton-enhanced near-field second-harmonic generation (SHG) from two-dimensional semiconductors and heterostructures. (a) Schematic of near-field SHG (i.e., nano-SHG) from 2D semiconductors. The tip plasmon is excited by a femtosecond pump pulse at ω, and SHG is enhanced by the exciton resonance at 2ω. (b) Approaching curves showing SHG intensity as a function of tip-sample distance. Solid blue line (1) is a fit using exponential decay plus a flat background. (c) Scanning electron microscopy (SEM) of Au-coated plasmonic tips after scanning with low (upper) and high (lower) pump fluence. (d) Dependence of nonlinear susceptibility $|\chi^{(2)}|$ on pump photon energy as measured in the far-field, showing enhancement by band-nesting excitons. (e) Dependence of near-field SHG efficiency on pump photon energy. kcps: kilo counts per second; mW: miliWatt. (Reprinted with permission from Ref. [44], Copyright © 2022 John Wiley & Sons, Ltd.)

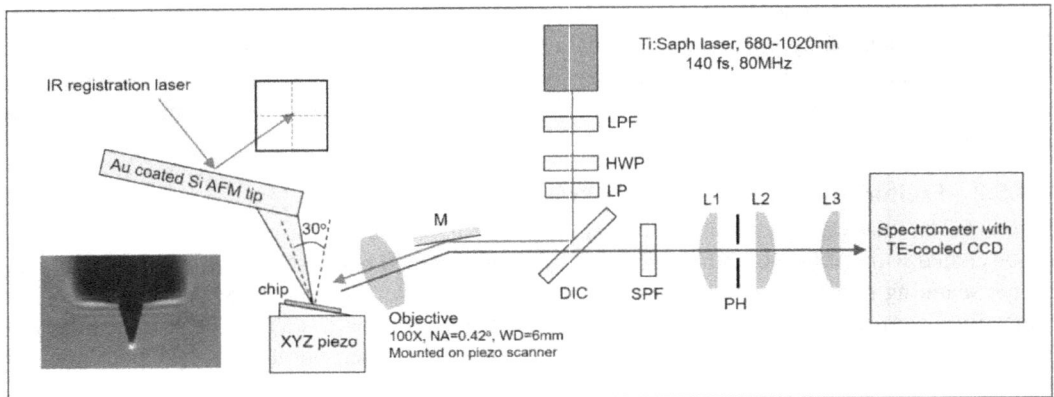

ᵃEffective NA considering the partial blocking of objective view angle.
L1, L2, L3: lenses
PH: pinhole, 100 μm diameter
SPF: 700 nm short pass filter
DIC: 803 nm short pass dichroic beam splitter
LPF: 800 nm long pass filter
HWP: half wave plate
LP: linear polarizer
M: mirror

FIGURE 6.75 Schematics of nano-SHG measurement setup. Fundamental pump laser is focused onto the apex of the Au-coated AFM tip, producing plasmonic enhancement over a nanoscale area on the sample. Inset shows the camera view of pump laser (900 nm) focused onto the tip apex. The back-reflected SHG is collected by a scanned objective and a series of free-space optics. Signals are detected by a cooled-CCD spectrometer. (Reprinted with permission from Ref. [44], Copyright © 2022 John Wiley & Sons, Ltd.)

of the nominal values. Thus, the evaporation thickness of Ti/Au as registered by the crystal monitor is set to be 10/150 nm. Tips coated with different thicknesses of Au have been tested, and 150 nm is found to provide low laser ablation damage and negligible SHG from the tip itself at our typical power levels for nano-SHG imaging.

Our nano-SHG setup is home-built based on an atomic force microscope (Omegascope, Horiba Inc.). Schematics of the full setup are shown in Figure 6.75. The fundamental pump laser is a Ti:Saph laser (Coherent) with 140 fs pulse width and 80 MHz repetition rate. The objective lens is 100X with effective (beam is partially blocked by substrate) NA of 0.42. Sample is placed on a tilted wedge for controlling tip angle. The collected signal is passed through appropriate spectral filters, a 100 μm pinhole, and dispersed by a 150 g/mm holographic grating onto an electron magnifying charge-coupled device (EMCCD, Horiba Synapse FIVS). The system collection efficiencies are determined to be 2.8, 4.4, 7.9, 13.4, and 18.1% for wavelengths at 415, 435, 460, 485, and 510 nm, respectively, which are used for calculating nano-SHG efficiency reaching the first lens (in Figure 6.74e). In nano-SHG imaging, the AFM operates under tapping mode when going from one pixel to the next. At each pixel, the tip goes into contact with the sample for a set integration time during which the SHG is measured.

The tip is tilted with respect to the sample plane's normal direction at an angle of 30°. The fundamental pump laser has a pulse width of 140 fs and a repetition rate of 80 MHz. The incidence direction (i.e., k-vector direction) of the pump is perpendicular to the axial orientation of the tip, and the electric field is set as p-polarized (i.e., parallel to the tip axis) to effectively excite the tip plasmon (Figure 6.76). With this configuration, the tip generates a localized plasmonic hot spot with both in-plane and out-of-plane electric field components at the fundamental frequency of ω. The back-reflected SHG is collected by the same objective (effective NA, NA=0.42) used for focusing the pump (Figure 6.75). The second-harmonic nature of the collected signal is verified by examining spectra position and the quadratic dependence of its intensity on pump power (Figure 6.77).

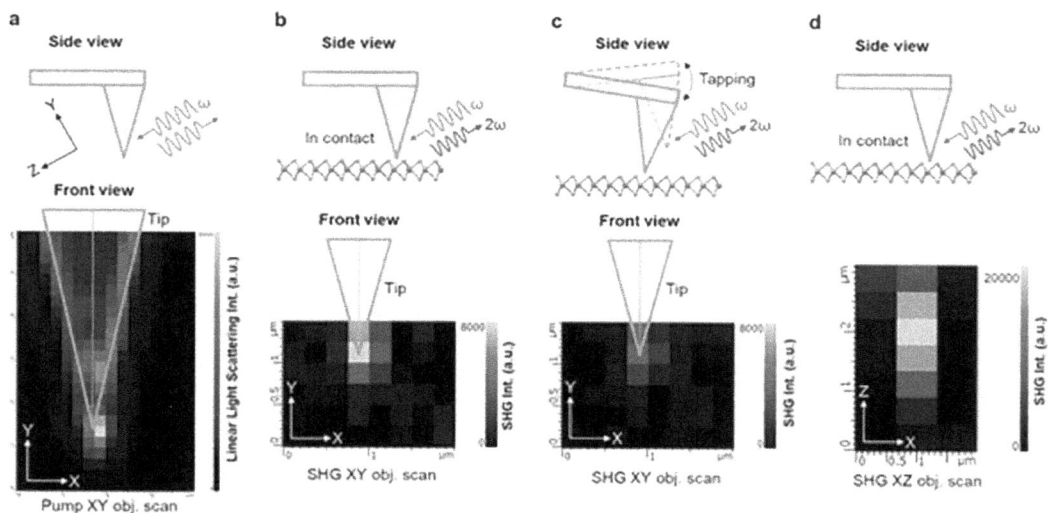

FIGURE 6.76 Alignment procedure for nano-SHG scanning. (a) The fundamental pump beam is raster-scanned with a piezo across the tip in the XY plane (coordinates are drawn in the figures), and the backscattered pump beam is collected. The tip apex can be located by identifying the brightest spot near the bottom of the image, aided by following the brightly scattering line along the edge of the pyramid shaft. In this step, we locate the rough position of tip apex prior to landing on sample. (b) The tip is landed onto the sample and stays in contact with normal force feedback. The pump beam is then raster-scanned while the backscattered SHG signal is collected. The brightest spot corresponds to the tip apex location with strongest nano-SHG signal is identified. (c) Same as in (b), but with the tip in tapping mode. Here the tip apex is, on average, a distance of ~20–30 nm away from the sample. Since the nano-SHG intensity decays exponentially with tip-sample distance, here we observe significantly reduced nano-SHG intensity from the tip apex. Along with approach curves, this serves as evidence of near-field enhancement. (d) In the last alignment step, the objective is XZ scanned, so that we can find the best focal position along the Z axis to collect maximum SHG. Because of chromatic aberration existent in our optics, the optimal focal lengths for the pump and the SHG wavelengths are slightly different, so this step is necessary. (Reprinted with permission from Ref. [44], Copyright © 2022 John Wiley & Sons, Ltd.)

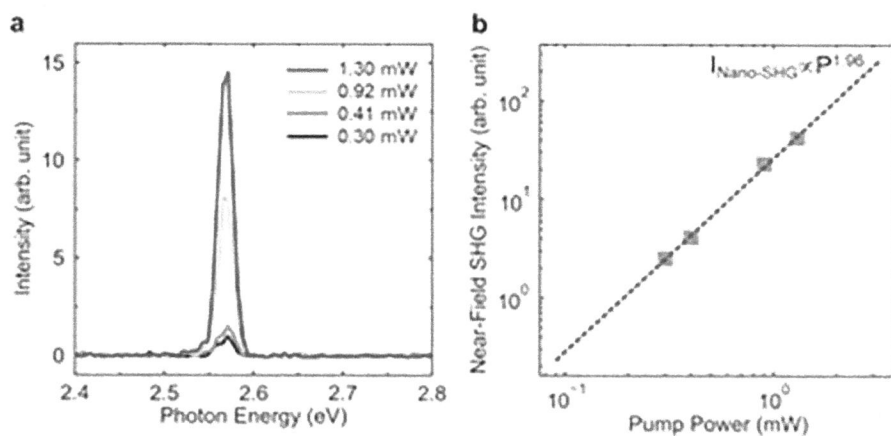

FIGURE 6.77 Nano-SHG as a function of pump power. (a) Nano-SHG spectra acquired at different pump powers at 1.29 eV (960 nm). (b) Dependence of integrated SHG intensity on pump power. The dashed line represents a power-law fit with power coefficient of 1.96. (Reprinted with permission from Ref. [44], Copyright © 2022 John Wiley & Sons, Ltd.)

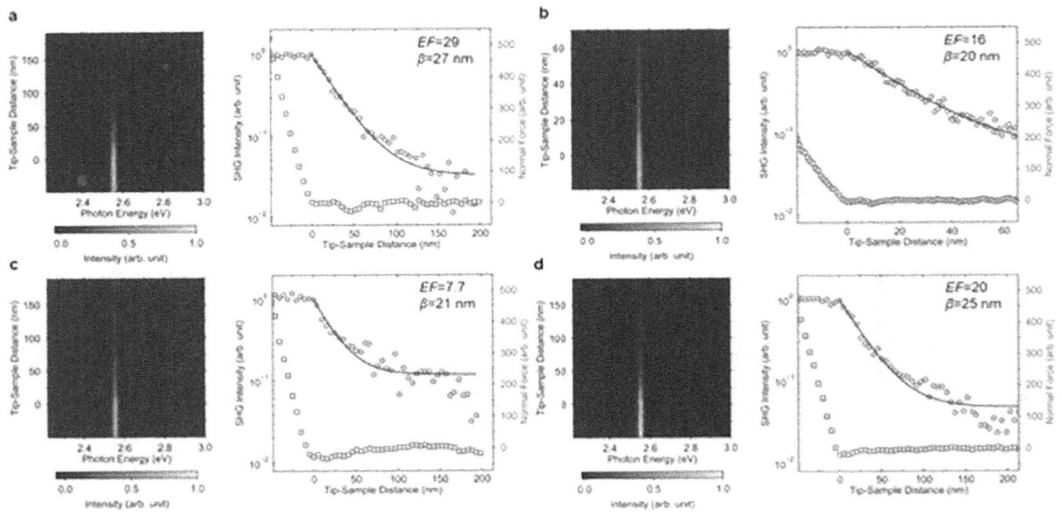

FIGURE 6.78 Near-field enhancement factors of nano-SHG from multiple approach curves. The panels (a–d) show data from multiple measurements. The left side of each panel shows a heat map of emission spectra measured as a function of tip-sample distance. A sharp emission line at the SHG energy is observed, without significant spurious spectral background. The right of each panel shows the dependence of SHG intensity on tip-sample distance (blue dots), together with fitting (blue solid line, 1). The fit function of intensity I over tip-sample distance z is: $I(z) = I_{NF}*\exp(-z/\beta) + I_{BG}$. Here I_{NF} is the near-field SHG intensity, I_{BG} is the background, β the evanescent decay length. The nano-SHG near-field enhancement factor is evaluated as $EF = I_{NF}/I_{BG}$. The right of each panel also shows the force-distance curves recorded simultaneously with the nano-SHG approaching curves, from which the contact point (i.e., tip-sample distance is zero) can be identified. (Reprinted with permission from Ref. [44], Copyright © 2022 John Wiley & Sons, Ltd.)

To establish the near-field nature of SHG, we measure the dependence of SHG intensity as a function of tip-sample distance on a monolayer WSe_2 sample (Figure 6.74b). SHG intensity follows an exponential decay as the tip is retracted away from the sample, providing compelling evidence for the evanescent-wave nature of the tip-localized plasmonic hot spot. At further distance, a slight deviation from the exponential trend is observed, which is attributed to a weak residual far-field background. We fit the approach curve data with contributions from the near-field (a single exponential decay) and the background (a flat baseline), as shown by the solid line in Figure 6.74b. The 1/e exponential decay length is found to be 23±3 nm. Analogous to tip-enhanced Raman and photoluminescence (PL) [35,36], the near-field enhancement factor for nano-SHG can be defined as the ratio of near-field signal over the background. The average enhancement factor is deduced to be 18±7 (see Figure 6.78 for more details), indicating that the collected signal is dominated by the local response within the evanescent field of the tip, with minimal interference from far-field SHG or tip-SHG background. Therefore, no post-processing is required for data interpretation. With the large pump incident angle (~60°), the reflected far-field SHG will primarily propagate in the forward direction (towards the left in Figure 6.74a) due to lateral phase matching rather than back into the collection objective.

To achieve the highest possible spatial resolution, we find it critical to avoid ablation damage of the tip apex. Therefore, the pump fluence should be well controlled. Previous work reported that with the femtosecond laser, damage threshold of Au plasmonic nanostructure is on the order of 1 mJ/cm² [26]. In our experiments, when the pulse fluence is set at 0.4 mJ/cm², the coated metal layer remains continuous and conformally covers the entire apex with Au grains, as shown by scanning electron microscopy (SEM) images in Figure 6.74c, upper panel. In comparison, when the pulse fluence is increased to 6 mJ/cm², the plasmonic apex is severely damaged after only 5 min of exposure (Figure 6.74c, lower panel).

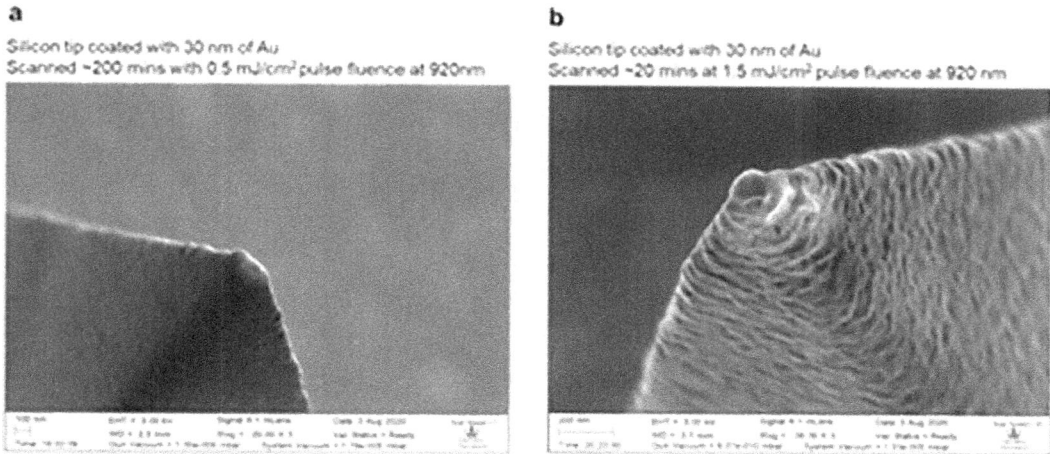

FIGURE 6.79 Additional scanning electron microscopy (SEM) images of near-field tips after scanning under different levels of pump fluence. Silicon tip coated with 30 nm of Au after (a) 200 min of exposure to 0.5 mJ/cm²pulse fluence, and (b) 20 min of exposure to 1.5 mJ/cm²pulse fluence. Laser wavelength is at 920 nm. (Reprinted with permission from Ref. [44], Copyright © 2022 John Wiley & Sons, Ltd.)

The second-order nonlinear susceptibilities of atomically thin TMDs show large enhancement on exciton resonances in the visible and near-infrared range [37]. Figure 6.74d shows the far-field SHG efficiency measured for representative monolayer (1L-) WSe_2, $1L$-$MoSe_2$, and 3R-stacked $MoSe_2$/WSe_2 heterobilayer samples used in this work. Here we quantify the SHG efficiency as the square of sheet nonlinear susceptibility, $|\chi^{(2)}|^2$, which is proportional to SHG intensity under normalized pump power. The spectra are highly dispersive with peaks corresponding to two-photon resonances of band-nesting excitons in the respective materials [37]. Figure 6.74e shows the near-field SHG efficiency at different pump photon energies, quantified as kilocounts per second (kcps) per miliwatt (mW) reaching the first collection lens. Exciton enhancement is also pronounced in the near-field data, with nano-SHG efficiency improving by an order of magnitude when the pump photon energy is in two-photon resonance with band-nesting excitons.

Nano-SHG scanning probe imaging of a $MoSe_2$/WSe_2 heterobilayer sample is shown in Figure 6.80. The sample is created by transferring a CVD-grown triangular monolayer $MoSe_2$ flake onto a larger triangular WSe_2 flake, with 3R-type interlayer stacking order. The TMD layers are on a hexagonal boron nitride (hBN) substrate ~20 nm thick. The entire Van der Waals stack is on a SiO_2/Si substrate. Figure 6.80a shows the far-field SHG microscopy. The overlapped $MoSe_2$/WSe_2 heterobilayer area is bounded by the white dashed triangle. The brightest small triangle on the top-left side of the heterobilayer is a nucleation region with 3R-stacked bilayer WSe_2, and the stacking effects on SHG are discussed in latter part of this work. The $MoSe_2$/WSe_2 heterobilayer shows strong SHG intensity compared to $1L$-WSe_2, consistent with measurements of the nonlinear susceptibility shown in Figures 6.74d and e. As SHG intensity from the hBN substrate is highly thickness and pump wavelength dependent [5], the hBN substrate used here is properly chosen so that the substrate SHG intensity is more than one order of magnitude lower than that from $1L$-WSe_2. The right panel of Figure 6.80a shows a zoom-in view of the confocal SHG image in the boxed region where, from top to bottom, the material changes from $MoSe_2$/WSe_2 heterobilayer to $1L$-WSe_2 and finally to the hBN substrate. The intensity transitions within a sub-micrometer length scale and thus appear blurred in the far-field image due to diffraction-limited resolution.

Figure 6.80b shows the nano-SHG image of the same boxed region. The image has 30 by 60 pixels with an integration time of 1 s/pixel, and photon energy of the fundamental pump is set at 1.27 eV. In this case, the sharp SHG contrast transition across these materials is clearly resolved,

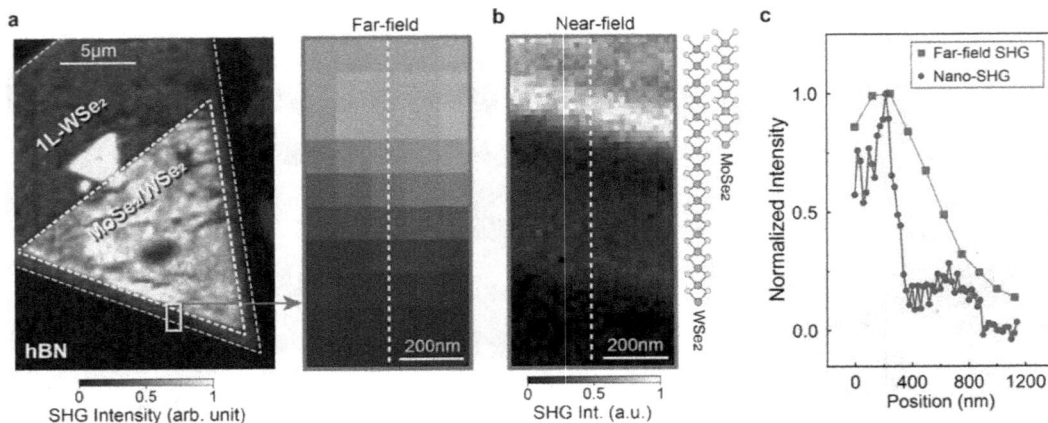

FIGURE 6.80 Nano-SHG imaging of a 2D semiconductor heterostructure with sub-diffraction-limited spatial resolution. (a) Far-field confocal SHG imaging of a 3R-stacked $MoSe_2$/WSe_2 heterobilayer on hBN substrate. The heterobilayer region is enclosed by the white dashed triangle. The right panel shows a zoom-in view of the boxed region. (b) Nano-SHG imaging of the same boxed region as in (a), with schematic illustration of the $MoSe_2$ and WSe_2 layers on the right. (c) Comparing far-field and nano-SHG intensities along the dashed white lines in (a, b). (Reprinted with permission from Ref. [44], Copyright © 2022 John Wiley & Sons, Ltd.)

demonstrating the nanoscale spatial resolution of SHG. Figure 6.80c compares the SHG intensity profiles along the line cuts in far- and near-field images. A 20 nm sharp step transition of SHG intensity is observed from 1L-WSe_2 to the hBN substrate, which is estimated as the spatial resolution. The resolution agrees well with the radius of curvature of tip apex seen in SEM images, and matches the exponential decay length of evanescent wave extracted from the approach curve data in Figure 6.74b. In addition, the near-field SHG intensity is enhanced along the edge of the $MoSe_2$ layer. Edge-enhanced SHG has previously been observed by far-field SHG microscopy, showing bright edges with full-width half magnitude (FWHM) on the order of several hundreds of nm, limited by diffraction [38,39]. It was proposed that atomic termination and reconstruction along the crystal edge may introduce one-dimensional midgap states, causing resonant enhancement of a nonlinear optical transition. In the nano-SHG images shown here, we find the bright edge region has an FWHM of ~70 and ~110 nm for the samples presented in Figures 6.80b and 6.81c, respectively. Both are significantly wider than the near-field spatial resolution (~20 nm). Therefore, the edge enhancement in this work occurs within a finite-width ribbon-like region, rather than an atomic-scale 1D edge state. Our results suggest that edge-related mesoscale disorders, such as increased defect density [40], could also play a role in edge-enhanced nonlinear response.

Next we demonstrate that local interlayer stacking order can be identified and visualized by nano-SHG. Figure 6.81a shows an optical microscope image of a CVD-grown WSe_2 sample transferred onto SiO_2 substrate. The flake of interest includes both monolayer and bilayer regions. The dark-colored bilayer region has a jigsaw-like pattern, which is formed by multiple triangular domains with 3R and 2H stacking order [41], as confirmed by following nonlinear nano-SHG and linear s-SNOM scattering. AFM topography of the boxed area is shown in Figure 6.81b, where the bilayer and monolayer regions can be clearly distinguished by height. No appreciable topographic contrast is observed within the bilayer region. To investigate the stacking orders of these bilayer domains, the nano-SHG image is acquired from the same region, as shown by Figure 6.81c. This image is 40 by 40 pixels in size with an integration time of 0.4 s/pixel, and photon energy of the fundamental pump is 1.41 eV. In stark contrast to the topographic image showing a homogeneous bilayer, a prominent SHG contrast is observed. The upper bright region corresponds to a 3R-stacked bilayer with broken inversion symmetry, where the nonlinear dipole polarization from top and bottom

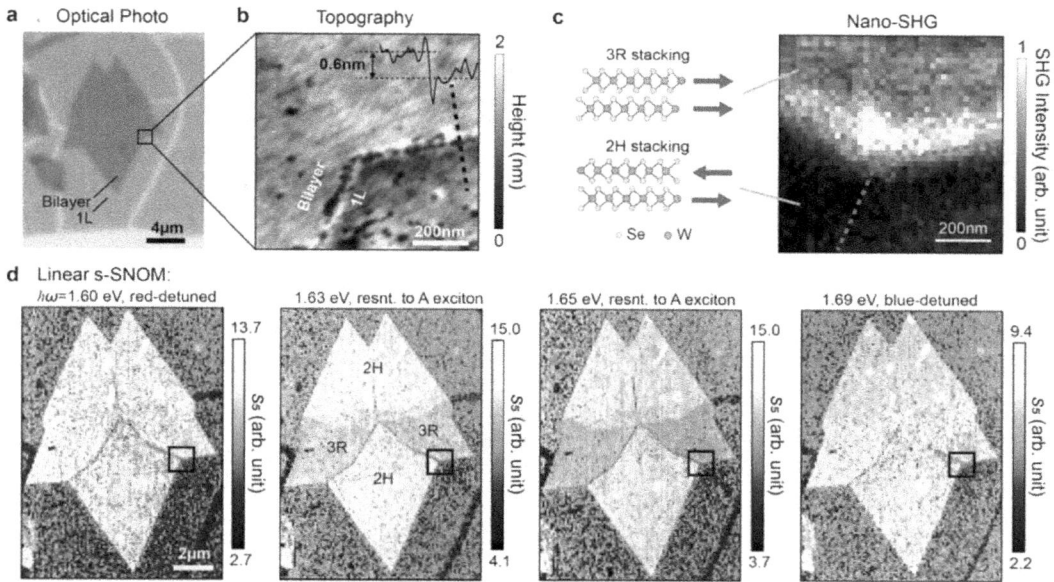

FIGURE 6.81 Revealing local interlayer stacking order and stacking effects on excitonic light-matter-interaction. (a) Optical microscope photo of the WSe_2 sample with bilayer and monolayer (1L) regions. (b) AFM topography and (c) Nano-SHG image of the boxed area in (a). Height profile along the black dashed line in (b) reveals the monolayer-to-bilayer step. The parallel (anti-parallel) interlayer lattice alignment for 3R(2H)-bilayer domains are shown by (c), leading to constructive (destructive) interference of local SHG response. (d) Images of linear scattering SNOM amplitude acquired with the incident photon energies on- and off-resonance with the A excitons. Clear contrast between 3R and 2H stacking is observed on exciton resonance. Boxed regions in (d) correspond to the nano-SHG scanning area in (c). S_5: fifth harmonic of scattering amplitude. (Reprinted with permission from Ref. [44], Copyright © 2022 John Wiley & Sons, Ltd.)

layers are in-phase, thus resulting in the constructive SHG response. The 3R-stacking is further confirmed by polarization-resolved far-field SHG measurements. In contrast, no SHG is detected from the bottom-left region, consistent with the 2H-stacked bilayer structure that preserves inversion symmetry.

Quantitatively, the collected nano-SHG intensity from the 3R-bilayer and the monolayer have average values of 620 cps and 160 cps, respectively, with the intensity ratio being close to four. The intensity ratio indicates that nano-SHG response is dominated by in-plane dipole moments as sketched by Figure 6. 6.81c, where the 3R-bilayer has aligned dipole moments with constructive interference, giving four times the intensity of a monolayer. Out-of-plane $\chi^{(2)}$ components ($\chi^{(2)}_{xyz}$, $\chi^{(2)}_{xzy}$, $\chi^{(2)}_{zxy}$) are allowed in principle based on the point group symmetry (D_3) of 3R-homobilayer TMD. However, if the out-of-plane contribution was significant, we would expect the intensity ratio to be significantly higher than four, since the additional out-of-plane $\chi^{(2)}$ components are absent in a monolayer. We also find that nano-SHG intensity from the centrosymmetric 2H-bilayer is negligible (well below our detection noise floor of ~20 cps). This finding confirms that the presence of the scanning probe does not significantly alter or break local symmetry, or at least is insignificant compared to the intrinsic crystal stacking order.

With the 3R and 2H crystal domains being visualized, we further investigate the effect of different stacking symmetry on light-matter interactions by performing linear optical scattering SNOM measurements. The scattered signal is governed by the dielectric function of a material, thus allowing one to investigate the exciton resonance energy and oscillator strength [42]. Figures 6.81d and 6.82 show the scattering-SNOM images of the scattering amplitude modulated at the fifth harmonic of tip tapping frequency, S_5, which is proportional to local reflectance

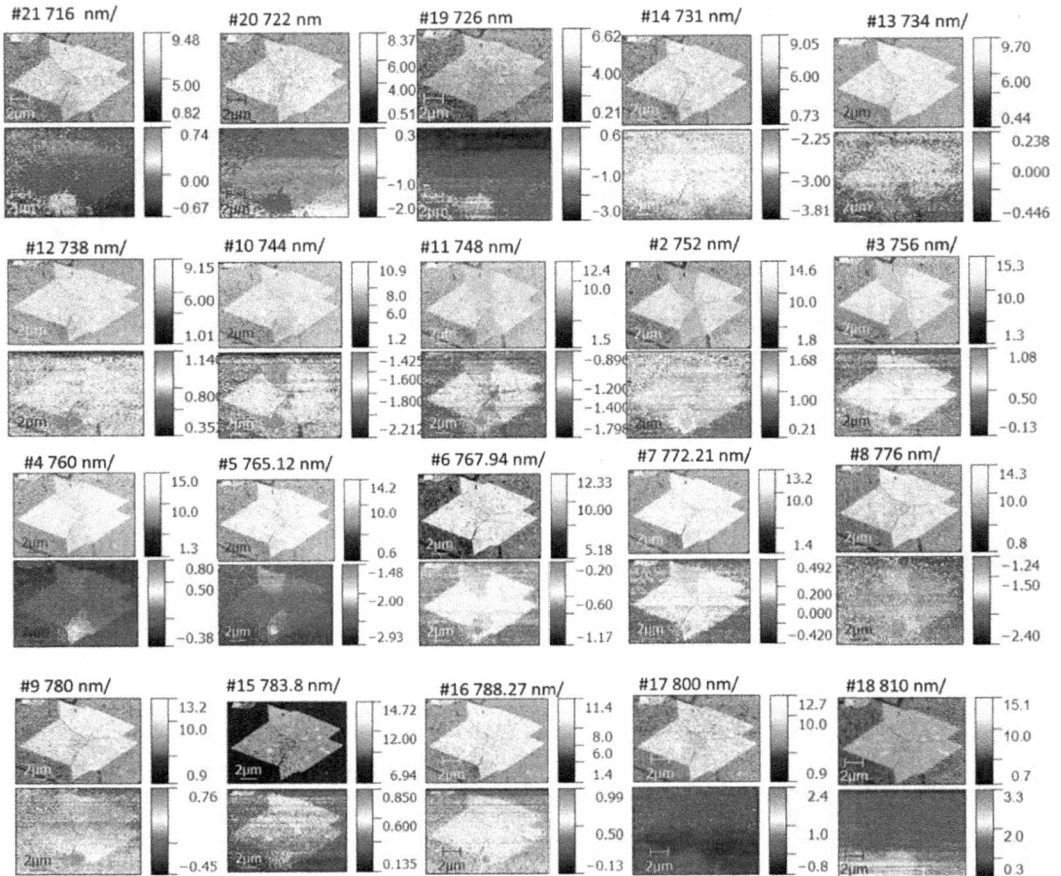

FIGURE 6.82 Linear scattering-SNOM images at additional scanning wavelengths. For each wavelength, the top shows scattering amplitude S_5 and the bottom shows scattering phase φ_5. (Reprinted with permission from Ref. [44], Copyright © 2022 John Wiley & Sons, Ltd.)

contrast within the weak resonance limit [42]. A range of incident photon energies is applied around the A exciton resonance of bilayer WSe_2 which is expected to be at around 1.65 eV. Recent far-field spectroscopy showed variation of optical reflectance between 2H and 3R bilayers [41]. Here in the near-field, we observed wavelength-dependent contrast in nano-optical light scattering across-the-crystal domains. When the photon energy is either red-detuned (1.60 eV) or blue-detuned (1.69 eV) from the A excitons, the entire bilayer flake shows uniform scattering amplitude, suggesting that the stacking effect on the optical dielectric function is rather weak in the non-resonant frequency range. However, when the incident laser resonantly excited the A exciton (1.63 eV and 1.65 eV data), contrast between the domains clearly emerges. In accordance to previous theoretical calculations [43], the breaking of inversion symmetry in the 3R-bilayer leads to additional splitting of the valance band maximum around K points, which further results in split exciton resonances. Thus, correlating these complementary nonlinear optical imaging modalities allows us to unambiguously establish the crystal-symmetry-based origins of the contrast in local optical response.

6.4.5.3 Future and Outlook

This chapter reports on nano-SHG imaging of atomically thin TMDs and their heterostructures, and demonstrates its application for nanoscale visualization of interlayer crystal stacking orders.

Our strategy of exploiting the strong exciton enhancement of nonlinear optical response to boost nano-SHG efficiency and perform scanning probe imaging enables an order of magnitude lower pump fluence compared to non-resonant cases, avoiding laser ablation damage to the plasmonic tip. Further, we demonstrate the power of combining nonlinear and linear nanooptics for correlating local variations of atomic registration and interlayer stacking to changes in excitonic light-matter-interaction, revealing intimate structure-property relationships at the length scales relevant to quantum materials. Since SHG is sensitive to a variety of symmetry-breaking orders, we expect that the nano-SHG approach highlighted by these results can be widely applied to visualize subwavelength domains of crystal structure, ferroelectric and magnetic orders. More generally, it should be applicable to the expansive collection of novel heterostructures exhibiting inherent and programmably broken symmetries [5], whose exceptional properties have yet to be fully unlocked.

REFERENCES

1. Du, L., Hasan, T., Castellanos-Gomez, A., Liu, G. B., Yao, Y., Lau, C. N., and Sun, Z., "Engineering symmetry breaking in 2D layered materials," *Nature Reviews Physics*, vol. 3, no. 3, pp. 193–206, 2021.
2. Finney, N. R., Yankowitz, M., Muraleetharan, L., Watanabe, K., Taniguchi, T., Dean, C. R., and Hone, J., "Tunable crystal symmetry in graphene–boron nitride heterostructures with coexisting moiré superlattices," *Nature Nanotechnology*, vol. 14, no. 1, pp. 1029–1034, 2019.
3. Suzuki, R., Sakano, M., Zhang, Y. J., Akashi, R., Morikawa, D., Harasawa, A., Yaji, K., Kuroda, K., Miyamoto, K., Okuda, T., and Ishizaka, K., "Valley-dependent spin polarization in bulk MoS_2 with broken inversion symmetry," *Nature Nanotechnology*, vol. 9, no. 8, pp. 611–617, 2014.
4. Sivadas, N., Okamoto, S., Xu, X., Fennie, C. J., and Xiao, D., "Stacking-dependent magnetism in bilayer CrI_3," *Nano Letters*, vol. 18, no. 12, pp. 7658–7664, 2018.
5. Yao, K., Finney, N. R., Zhang, J., Moore, S. L., Xian, L., Tancogne-Dejean, N., Liu, F., Ardelean, J., Xu, X., Halbertal, D., Watanabe, K., Taniguchi, T., Ochoa, H., Asenjo-Garcia, A., Zhu, X. Y., Basov, D. N., Rubio, A., Dean, C. R., Hone, J., and Schuck, P. J., "Enhanced tunable second harmonic generation from twistable interfaces and vertical superlattices in boron nitride homostructures," *Science Advances*, vol. 7, no. 10, p. eabe8691, 2021.
6. Yang, F., Song, W., Meng, F., Luo, F., Lou, S., Lin, S., Gong, Z., Cao, J., Barnard, E. S., Chan, E., Yang, L., and Yao, J., "Tunable second harmonic generation in twisted bilayer graphene," *Matter*, vol. 3, no. 4, pp. 1361–1376, 2020.
7. Sunku, S. S., Ni, G., Jiang, B. Y., Yoo, H., Sternbach, A., McLeod, A. S., Stauber, T., Xiong, L., Taniguchi, T., Watanabe, K., Kim, P., Fogler, M. M., and Basov, D. N., "Photonic crystals for nano-light in moiré graphene superlattices," *Science*, vol. 362, no. 6419, pp. 1153–1156, 2018.
8. Jin, C., Regan, E. C., Yan, A., Utama, M. I. B., Wang, D., Zhao, S., Qin, Y., Yang, S., Zheng, Z., Shi, S. Watanabe, K., Taniguchi T., Tongay S., Zettl A., and Wang F., Observation of moiré excitons in WSe_2/WS_2 heterostructure superlattices, *Nature*, vol. 567, pp. 76–80, 2019.
9. Seyler, K. L., Rivera, P., Yu, H., Wilson, N. P., Ray, E. L., Mandrus, D. G., Yan, J., Yao, W., and Xu, X., "Signatures of moiré-trapped valley excitons in $MoSe_2$/WSe_2 heterobilayers," *Nature*, vol. 7746, no. 567, p. 66, 2019.
10. Tran, K., Moody, G., Wu, F., Lu, X., Choi, J., Kim, K., Rai, A., Sanchez, D. A., Quan, J., Singh, A. Embley, J., et al., Evidence for moiré excitons in van der Waals heterostructures, *Nature*, vol. 567, p. 71, 2019.
11. Lin, K. Q., Junior, P. E. F., Bauer, J. M., Peng, B., Monserrat, B., Gmitra, M., Fabian, J., Bange, S., and Lupton, J. M., "Twist-angle engineering of excitonic quantum interference and optical nonlinearities in stacked 2D semiconductors," *Nature Communications*, vol. 12, pp. 1–7, 2021.
12. Sunku, S. S., Halbertal, D., Stauber, T., Chen, S., McLeod, A. S., Rikhter, A., Berkowitz, M. E., Lo, C. F. B., Gonzalez-Acevedo, D. E., Hone, J. C., Dean, C. R., Fogler, M. M., and Basov, D. N., "Hyperbolic enhancement of photocurrent patterns in minimally twisted bilayer graphene," *Nature Communications*, vol. 12, no. 1, pp. 1–7, 2021.
13. Moore, S. L., Ciccarino, C. J., Halbertal, D., McGilly, L. J., Finney, N. R., Yao, K., Shao, Y., Ni, G., Sternbach, A., Telford, E. J., Kim, B. S., Rossi, S. E., Watanabe, K., Taniguchi, T., Pasupathy, A., Dean, C. R., Hone, J., Schuck, P. J., Narang, P., and Basov, D, "Nanoscale lattice dynamics in hexagonal boron nitride moiré superlattices," *Nature Communications*, vol. 12, no. 1, p. 5741, 2021.

14. Rosenberger, M. R., Chuang, H. J., Phillips, M., Oleshko, V. P., McCreary, K. M., Sivaram, S. V., Hellberg, C. S., and Jonker, B. T., "Twist angle-dependent atomic reconstruction and moiré patterns in transition metal dichalcogenide heterostructures," *ACS Nano*, vol. 14, no. 4, pp. 4550–4558, 2020.

15. Y. Shen, *The Principles of Nonlinear Optics.* New York: Wiley-Interscience, 1984.

16. Trassin, M., Luca, G. D., Manz, S., and Fiebig, M., "Probing ferroelectric domain engineering in $BiFeO_3$ thin films by second harmonic generation," *Advanced Materials*, vol. 27, no. 33, pp. 4871–4876, 2015.

17. Sun, Z., Yi, Y., Song, T., Clark, G., Huang, B., Shan, Y., Wu, S., Huang, D., Gao, C., Chen, Z., McGuire, M., Cao T., Xiao D., Liu W-T., Yao W., Xu X., and Wu S., "Giant nonreciprocal second-garmonic generation from antiferromagnetic bilayer CrI_3," *Nature*, vol. 572, pp. 497–501, 2019.

18. Cheong, S. W., Fiebig, M., Wu, W., Chapon, L., and Kiryukhin, V., "Seeing is believing: Visualization of antiferromagnetic domains," *NPJ: Quantum Materials*, vol. 5, no. 1, pp. 1–10, 2020.

19. Lee, K., Dismukes, A. H., Telford, E. J., Wiscons, R. A., Wang, J., Xu, X., Nuckolls, C., Dean, C. R., Roy, X., and Zhu, X., "Magnetic order and symmetry in the 2D semiconductor CrSBr," *Nano Letters*, vol. 21, no. 8, pp. 3511–3517, 2021.

20. Harter, J. W., Zhao, Z. Y., Yan, J. Q., Mandrus, D. G., and Hsieh, D., "A parity-breaking electronic nematic phase transition in the spin-orbit coupled metal $Cd_2Re_2O_7$," *Science*, vol. 356, no. 6335, pp. 295–299, 2017.

21. McGilly, L. J., Kerelsky, A., Finney, N. R., Shapovalov, K., Shih, E. M., Ghiotto, A., Zeng, Y., Moore, S. L., Wu, W., Bai, Y., Watanabe, K., Taniguchi, T., Stengel, M., Zhou, L., Hone, J., Zhu, X. Y., Basov, D. B., Dean, C., Dreyer, C., and Pasupathy, A., "Visualization of moiré superlattices," *Nature Nanotechnology*, vol. 15, no. 7, pp. 580–584, 2020.

22. Li, H., Li, S., Naik, M. H., Xie, J., Li, X., Wang, J., Regan, E., Wang, D., Zhao, W., Zhao, S., Kahn, S., Yumigeta, K., Blei, M., Taniguchi, T., Watanabe, K., Tongay, S., Zettl, A., Louie, S. G., Wang, F., and Crommie, M. F., "Imaging moiré flat bands in three-dimensional reconstructed WSe_2/WS_2 superlattices," *Nature Materials*, vol. 20, no. 7, pp. 1–6, 2021.

23. Sun, Q. C., Song, T., Anderson, E., Brunner, A., Förster, J., Shalomayeva, T., Taniguchi, T., Watanabe, K., Gräfe, J., Stöhr, R., Xu, X., and Wrachtrup, J., "Magnetic domains and domain wall pinning in atomically thin CrB_3 revealed by nanoscale imaging," *Nature Communications*, vol. 12, no. 1, pp. 1–7, 2021.

24. Darlington, T. P., Carmesin, C., Florian, M., Yanev, E., Ajayi, O., Ardelean, J., Rhodes, D. A., Ghiotto, A., Krayev, A., Watanabe, K., Taniguchi, T., Kysar, J., Pasupathy, A., Hone, J., Jahnke, F., Borys, N. J., and Schuck, P. J., "Imaging strain-localized excitons in nanoscale bubbles of monolayer WSe_2 at room temperature," *Nature Nanotechnology*, vol. 15, no. 10, pp. 854–860, 2020.

25. Park, K. D., and Raschke, M. B., "Polarization control with plasmonic antenna tips: A universal approach to optical nanocrystallography and vector-field imaging," *Nano Letters*, vol. 18, no. 5, pp. 2912–2917, 2018.

26. Albrecht, G., Kaiser, S., Giessen, H., and Hentschel, M., "Refractory plasmonics without refractory materials," *Nano Letters*, vol. 17, no. 10, pp. 6402–6408, 2017.

27. Chen, Y. L., Tseng, Y. H., Chen, Y. C., Chang, W. H., Her, T. H., and Luo, C. W., "Femtosecond-laser ablation of monolayer tungsten diselenide (WSe_2) on sapphire," *CLEO: QELS_Fundamental Science*, pp. JTu2A-4, 2018.

28. Smolyaninov, I. I., Zayats, A. V., and Davis, C. C., "Near-field second harmonic generation from a rough metal surface," *Physical Review B*, vol. 56, no. 15, p. 9290, 1997.

29. Bozhevolnyi, S. I., Pedersen, K., Skettrup, T., Zhang, X., and Belmonte, M., "Far-and near-field second-harmonic imaging of ferroelectric domain walls," *Optics Communications*, vol. 152, no. 4, pp. 221–224, 1998.

30. Zavelani-Rossi, M., Celebrano, M., Biagioni, P., Polli, D., Finazzi, M., Duo, L., Cerullo, G., Labardi, M., Allegrini, M., Grand, J., and Adam, P. M., "Near-field second-harmonic generation in single gold nanoparticles," *Applied Physics Letters*, vol. 92, no. 9, p. 093119, 2008.

31. Johnson, J. C., Yan, H., Schaller, R. D., Petersen, P. B., Yang, P., and Saykally, R. J., "Near-field imaging of nonlinear optical mixing in single zinc oxide nanowires," *Nano Letters*, vol. 2, no. 4, pp. 279–283, 2002.

32. Wang, C.-F., and El-Khoury, P. Z., "Multimodal tip-enhanced nonlinear optical nanoimaging of plasmonic silver nanocubes," *The Journal of Physical Chemistry Letters*, vol. 12, no. 44, p. 10761, 2021.

33. Neacsu, C. C., van Aken, B. B., Fiebig, M., and Raschke, M. B., "Second-harmonic near-field imaging of ferroelectric domain structure of $YMnO_3$," *Physical Review B*, vol. 79, no. 10, p. 100107, 2009.

34. Mahieu-Williame, L., Gresillon, S., Cuniot-Ponsard, M., and Boccara, C., "Second harmonic generation in the near field and far field: A sensitive tool to probe crystalline homogeneity," *Journal of Applied Physics*, vol. 101, no. 8, p. 083111, 2007.

35. Hartschuh, A., "Tip-enhanced near-field optical microscopy," *Angewandte Chemie International Edition*, vol. 47, no. 43, pp. 8178–8191, 2008.

36. Umakoshi, T., Yano, T. A., Saito, Y., and Verma, P., "Fabrication of near-field plasmonic tip by photo reduction for strong enhancement in tip-enhanced Raman spectroscopy," *Applied Physics Express*, vol. 5, no. 5, p. 052001, 2012.

37. Yao, K., Yanev, E., Chuang, H. J., Rosenberger, M. R., Xu, X., Darlington, T., McCreary, K. M., Hanbicki, A. T., Watanabe, K., Taniguchi, T., Jonker, B. T., Zhu, X., Basov, D. N., Hone, J., and Schuck, P. J., "Continuous wave sum frequency generation and imaging of monolayer and heterobilayer two-dimensional semiconductors," *ACS Nano*, vol. 14, no. 1, pp. 708–714, 2019.

38. Yin, X., Ye, Z., Chenet, D. A., Ye, Y., O'Brien, K., Hone, J. C., and Zhang, X., "Edge nonlinear optics on a MoS_2 atomic monolayer," *Science*, vol. 344, no. 6183, p. 488, 2014.

39. Lin, K. I., Ho, Y. H., Liu, S. B., Ciou, J. J., Huang, B. T., Chen, C., Chang, H. C., Tu, C. L., and Chen, C. H., "Atom-dependent edge-enhanced second-harmonic generation on MoS_2 monolayers," *Nano Letters*, vol. 18, no. 2, pp. 793–797, 2018.

40. Bao, W., Borys, N. J., Ko, C., Suh, J., Fan, W., Thron, A., Zhang, Y., Buyanin, A., Zhang, J., Cabrini, S., Ashby, P. D., Weber-Bargioni, A., Tongay, S., Aloni, S., Ogletree, D. F., Wu, J., Salmeron. M., and Schuck, P. J., "Visualizing nanoscale excitonic relaxation properties of disordered edges and grain boundaries in monolayer molybdenum disulfide," *Nature Communications*, vol. 6, no. 1, pp. 1–7, 2015.

41. McCreary, K., Phillips, M., Chuang, H.-J., Wickramaratne, D., Rosenberger, M., Hellberg, C., and Jonker, B., "Stacking-dependent optical properties in bilayer WSe_2," *arXiv:2111.05704*, 2021.

42. Zhang, S., Li, B., Chen, X., Ruta, F., Shao, Y., Sternbach, A., McLeod, A., Sun, Z., Xiong, L., Moore, S., Xu, X. Wu, W., Shabani, S., Zhou, L., Wang, Z., Mooshammer, F., Ray, E., Wilson, N., Basov, D., et al., "Nano-spectroscopy of excitons in atomically thin transition metal dichalcogenides," *Research Square*, vol. 13, no. 1, p. 542, 2021.

43. He, J., Hummer, K., and Franchini, C., "Stacking effects on the electronic and optical properties of bilayer transition metal dichalcogenides MoS_2, $MoSe_2$, WS_2, and WSe_2," *Physical Review B*, vol. 89, no. 7, p. 075409, 2014.

44. Yao, K., Zhang, S., Yanev, E., McCreary, K., Chuang, H.-J., Rosenberger, M., Darlington, T. P., Krayev, A., Jonker, B., Hone, J. C., Basov, D. N., and Schuck, P. J., "Nanoscale optical imaging of 2D semiconductor stacking orders by exciton-enhanced second harmonic generation," *Advanced Optical Materials*, accepted, doi:10.1002/adom.202200085, 2022.

6.5 TIP-ENHANCED PHOTOLUMINESCENCE (TEPL) SPECTROSCOPY

6.5.1 LIGHT-MATTER INTERACTIONS AND EXCITONS IN SEMICONDUCTORS

Photoluminescence is one of the prominent radiative processes resulting from the light-matter interactions through which the generation and recombination of electron-hole pairs take place in a matter upon its interaction with light. Suppose the matter is a single molecule and is excited upon interacting with light. In that case, the radiative transition takes place between the energy levels of the highest occupied molecular orbital (HOMO) and the lowest unoccupied molecular orbital (LUMO) separated by a gap [1,2]. If the matter is a solid crystal, then electron-hole pairs generate and recombine radiatively between conduction and valence bands separated by a band gap while interacting with light.

The excited electron-hole pairs are bound by Coulomb interactions and behave as quasi-particles, called excitons. The excitons in organic and inorganic semiconductors are strongly influenced by their dielectric constant, carrier generation and recombination rate, carrier mobility, and defects or impurities. In the organic molecular crystals having low dielectric constant, excitons are localized to a single-molecule (Frenkel type) or an adjacent molecule (charge transfer type) and have high binding energies (100s of meV) due to the strong Coulomb interactions and poor screening of charges. Whereas in the inorganic semiconductor crystals having a large dielectric constant, exciton binding energies are comparatively low (10s of meV) (Wannier–Mott type). The intensity of excitonic PL spectra in the stationary limit under low exciton densities is given by [3],

$$I_{PL}(\omega) \propto \text{Im}\left[\sum \frac{\left|\varnothing_\lambda\left(r=0\right)\right|^2 N_\lambda}{E_\lambda - \hbar\omega - i\delta_\lambda}\right] \tag{6.5.1}$$

where φ_λ; $N_\lambda = N_{\lambda\,\text{carrier}} + N_{\lambda\,\text{excitation}}$ is bound-state solutions of the Wannier equation, and $N_\lambda = N_{\lambda\,\text{carrier}} + N_{\lambda\,\text{excitation}}$ is the population densities.

Excitons in the nanoscale materials; 0D QD, 1D nanowires and nanotubes, and 2D materials, have additional confinement and delocalization effects, which modify the exciton binding energy, lifetimes, exchange interactions, diffusion length, and emission energy with respect to their bulk counterpart [2,3]. Different excitonic models describe the excitons in bulk materials having different dielectric constants and their characteristics in nanoscale materials [2]. PL spectroscopy and imaging provide a deep understanding of the above-mentioned excitonic characteristics with respect to spatial and temporal variations. The conventional PL spectroscopic imaging technique uses excitation of sample volumes down to 1 μm or sub-μm due to the diffraction-limited far-field optics and, thus, averages the excitonic luminescence characteristics over this volume. However, luminescence characteristics are susceptible to local (confined to a few nm or sub-nm region) compositional variation, defects or impurities, and strain in the individual semiconductor nanostructures.

The development of near-field optics advances diffraction-limited far-field micro-PL imaging to probe locally in the individual nanostructures and organic molecules [4–7]. There are three types of experimental schemes namely, (i) scanning near-field optical microscope (SNOM or NSOM) using aperture probe, (ii) apertureless SNOM (ASNOM) or tip-enhanced probe, (iii) STM-based probe, which has their advantages and disadvantages in enhancing the luminescence strength locally and to achieve the submolecular resolution imaging [1,6–9]. In the following sections, we will discuss the recent development in near-field PL spectroscopic imaging of different nanostructures from QD to 2D materials, including the subnanometer resolution in the single-molecule.

6.5.2 Quantum Emitters (Single Molecules and Semiconductor Quantum Dots)

Organic molecules have been explored for several electronic and optoelectronic applications, including organic LED (OLED), photovoltaic devices, simultaneous energy harvesting, and storage devices [2,10,11]. Many of these devices are based on the light-matter interactions in which the absorption and emission of light upon interacting with molecules such as several dye molecules, conjugated polymers, etc. Ever since the development of SNOM to image the individual carbocyanine dye molecules in a submonolayer with a resolution of ~λ/50 (where λ being is the wavelength of light), the near-field nanospectroscopy technique evolved in many directions to study the single molecules [12]. The vibrational properties of several organic molecules (porphyrin molecules, R6G, phthalocyanine molecules, etc.) have been mapped on a single and even submolecular level using STM-based TERS probes [13–17]. The development of UHV and low temperature-based near-field spectroscopy techniques to stabilize the single molecules and the development of atomistic probes to confine the light in nano- and picocavities have been made it possible to achieve submolecular imaging [17–19].

Unlike the near-field Raman scattering where the Raman intensity can be enhanced by decreasing the tip-sample distances (up to few Å), luminescence enhancement under near-field excitation is quite difficult due to many-body interactions of excited charges in plasmonic tip and sample when the tip-sample separation is <5–10 nm. Particularly, in organic molecules, these interactions are more prominent due to energy transfer between molecule and plasmonic surface. Fluorescence quenching was observed in lissamine dye molecules that were chemically attached to the smallest Au nanoparticles (radius 1 nm) due to the increased nonradiative decay rate and decreased radiative decay rates of dye molecules [20].

Later, experimental and theoretical studies on the fluorescence rate of a single molecule (nile blue molecule) as a function of its distance to a laser-irradiated Au nanoparticle (Figure 6.83a–d),

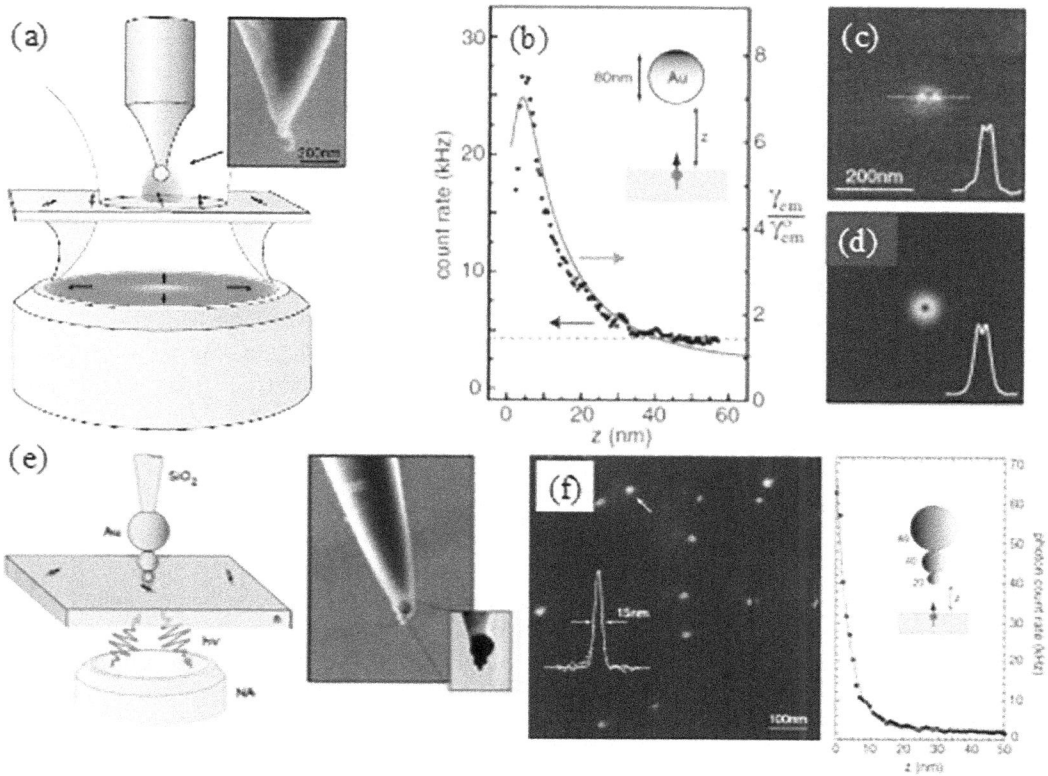

FIGURE 6.83 (a) Experimental scheme of tip-enhanced fluorescence from single molecule. Inset shows the SEM image of a gold particle attached to the end of a pointed optical fiber. (b) Fluorescence rate as a function of particle-surface distance z for a vertically oriented molecule. (c) and (d) Experimental and theoretical fluorescence rate image of a single molecule acquired for $z=2$ nm. The dip in the center indicates fluorescence quenching. (Reprinted with permission from Ref. [21], Copyright © 2006 *Am. Phys. Soc.*) (e) Experimental scheme (left) of antenna made of discrete colloidal gold nanoparticles of decreasing size is placed next to a sample with randomly distributed nile blue molecules and SEM image of a colloidal trimer antenna (right). (f) Tip-enhanced fluorescence image (left) of the single-molecule and fluorescence rate as a function of the distance between vertically oriented molecule and trimer antenna. (Reprinted with permission from Ref. [22], Copyright © 2012 *Am. Phys. Soc.*)

verified the continuous transition from fluorescence enhancement to fluorescence quenching [21]. The study concluded that the local field enhancement led to an increased excitation rate (γ_{exc}); whereas nonradiative energy transfer to the particle led to a decrease of the quantum yield causing fluorescence quenching. These two competing processes control the final emission rate (γ_{emis}),

$$\gamma_{emis} = \gamma_{exc}\left[\gamma_r / \gamma\right] = \gamma_{exc}\left[1-\left(\gamma_{nr} / \gamma_0\right)/\left(\gamma_r / \gamma_0\right)\right] \tag{6.5.2}$$

where $\gamma_{nr}=\gamma-\gamma_{ris}$ nanoradiative decay rate, and γ and γ_0 are total and free-space decay rates, respectively.

Using a self-similar trimer antenna (Figure 6.83e and f) consisting of 80, 40, and 20 nm Au nanoparticles, a fluorescence enhancement of 40 and spatial confinement of 15 nm is achieved for molecules with high intrinsic quantum yield efficiency (>80%) [22]. Compared with a single Au nanoparticle, the self-similar trimer antenna improved the enhancement-confinement ratio by more than an order of magnitude.

Fluorescence quenching from a single molecule or a quantum emitter is largely suppressed by designing the nanocavities using two plasmonic nano-antennas separated by thin insulating spacer material [23]. Such nanocavities enhance the emitter excitation and allow the single emitter to strongly couple with hybrid-plasmonic modes, resulting in a significant reduction of the nonradiative decay and fluorescence quenching [24,25]. A similar technique is implemented for simultaneous tip-enhanced Raman and fluorescence spectroscopy measurements on Cy5 fluorescent molecules to enhance Raman and fluorescence signals [26]. An ultrathin 1 nm silica shell is coated onto Ag tip to separate it from the bottom plasmonic substrates on which the Cy5 molecules are spread. This shell-isolated tip-enhanced nanospectroscopy not only suppresses the nonradiative energy loss but also avoids tip contamination by molecules and the surrounding environment. The Raman signals can be reduced while enhancing only the fluorescence of the molecules by increasing the SiO_2 shell thickness to 2 nm. The resultant fluorescence enhancement factor (EF_{flu}) of 1.4×10^3 is calculated using the following equation [26],

$$EF_{flu} = |E|^2 \left(\frac{q}{q_0} \right) \qquad (6.5.3)$$

where E, q, and q^0 are the EM field enhancement, the quantum yield with electromagnetic enhancement, and the quantum yield without electromagnetic enhancement, respectively.

The excitation and luminescence emission from single molecules can also be studied using STM-based nanospectroscopy. The molecules are confined precisely in nano- or picocavities formed between atomistic STM probe plasmonic surface. The gap plasmons in such nanocavities can be controlled by a voltage applied to the tip [27–32]. In such configuration, an ultrathin insulating spacer layer (such as Al_2O_3, NaCl, etc.) must be used between the molecules and supporting plasmonic substrate to avoid the inelastic electron tunneling from the tip to substrate; thus, the interaction between molecule and plasmonic substrate can be reduced [28–32]. By controlled inelastic electron/hole tunneling from the tip, single molecules of tetraphenyl porphyrin (TPP), Zn(-II)-etioporphyrinI, PTCDA (3,4,9,10-perylenetetracarboxylic dianhydride), phthalocyanine (ZnPc, MgPc and H_2Pc) were excited at submolecular spatial resolution and the luminescence maps of individual molecules were collected to study the electronic and vibrational states of the molecule [27–30]. Such STM-induced electroluminescence (STM-EL) is used to probe the electronic and vibronic changes that result from tautomerization process (interconversion between two constitutional molecular isomers) in free-base phthalocyanine (H_2Pc) molecules deposited on NaCl-coated Ag(111) surface in a submolecular spatial resolution [33]. Despite its atomic resolution and local electronic excitation capability, STM-EL has a limitation in selective excitation of a single quantum state due to the energy spread in tunneling electrons from the tip, resulting in the broader emission line width [34].

The single molecules can be excited by illuminating the STM junction with external excitation laser light, which generates a strong local electromagnetic field of the nanocavity plasmons, resulting in enhanced luminesce (STM-PL) from a single molecule. By designing an STM-PL setup combing all the above-discussed developments (nanocavity, thin spacer layer, and atomic protrusion), PL imaging of a single phthalocyanine (ZnPc) molecule with a spatial resolution of ~8 Å can be achieved [35]. The latest development in STM-based luminescence of single molecules is the PL (fluorescence) excitation spectroscopy and imaging using STM (STM-PLE or TE-FE) [34,36]. By combining a narrow-line tunable laser with STM, the PL emission from an individual quantum state of a single phthalocyanine (H_2Pc) and single porphyrin (H_2P) molecules are probed and imaged, respectively, with submolecular spatial and micro-electron volt (0.5 meV peak width) energy resolution [34,36]. These studies have demonstrated the capability of TEPL excitation spectroscopy to resolve the Franck-Condon (FC) and Herzberg-Teller (HT) active modes, submolecular transitions between vibrational states in the first singlet electronic excited state (S_1) and the vibrational states in the electronic ground state (S_0) such as $S_0 \rightarrow S_1(0–0)$, $(0–1)$, $(0–1')$, etc.

The other classes of quantum emitters are 0D semiconductor nanocrystal or QDs, which are strong light-absorbing and bright narrowband emitting inorganic nanocrystals. The variation in optical absorption and emission properties of individual QDs due to their distribution of size, shape, and surface chemistry can be probed and studied by the TEPL nanospectroscopy [37,38]. Another topic of interest in semiconductor QDs is the origin and control of PL intermittency, also known as "blinking", which can be studied using near-field PL spectroscopy. The PL quenching occurs due to the nonradiative Auger recombination and interception of photoexcited carriers at the surface of QD [39].

The fluorescence emission from single CdSe(ZnS) core-shell QD coupled to rough Au surface is enhanced fivefold compared to the emission from QDs on smooth Au surface [40]. The blinking is very effective for the QDs on smooth Au surface due to the charging of QDs, just like the organic molecules on metal surface. However, the surface-enhanced fluorescence on rough Au surface effectively competes with Auger recombination and quenches the blinking, allowing the observation of both neutral and charged exciton emission from the QDs [40]. A similar fivefold intensity in fluorescence from single CdSe(ZnS) core-shell QD embedded in the evanescent optical field above a prism surface is observed in the presence of a Si ANSOM probe [41]. In such a configuration, the spatial resolution is limited by the sharpness of the tip (~10 nm). In addition, the external field used to induce enhancement led to a substantial background signal due to the large tip-sample gap. To overcome this problem, the tip can be operated in the intermittent contact mode with the sample for tightly confined field enhancement. As a result, the strong enhancement in near-field fluorescence of CdSe(ZnS) QD is observed with a lateral resolution <10 nm in the tip-enhanced fluorescence imaging [42].

The near-field PL from a single quantum dot can be further enhanced by establishing a strong coupling between QD and the near-field generated by the plasmonic nanoresonator (PNR) [43]. The strong coupling occurs when the interaction rate ($\sqrt{1/V}$) is larger than the cavity loss rate ($\propto$ $1/Q$) and emitter decay rate (Γ). Here, V and Q are mode volume and quality factor of the cavity, respectively. Such strong coupling is achieved at RT between CdSeTe(ZnS) QD, and PNR fabricated at the apex of the scanning probe and observed the neutral and charged exciton emissions [43]. The strongly coupled QDs and nanocavities can also generate the hybrid state of an exciton and plasmon at RT, known as plexcitonic state [44]. However, the fixed PNR slot configuration at tip apex has a limitation in the cavity mode volume and spatial resolution. This problem can be overcome by a simple design of antenna-type nanotip and a metal mirror substrate separated by a thin vinsulating Al_2O_3 spacer layer to form and tune the confined nanocavity field between tip and substrate [44]. In this configuration, it is also possible to dynamically control the coupling between nanocavity and QD, which allows to measure and control the light-matter entanglement and single-photon quantum gates. The tip-enhanced plexciton PL spectra recorded at RT using this strong coupling configuration on a single CdSe/ZnS core-shell QD covered with protecting Al_2O_3 (0.5 nm) layer showed maximum Rabi splitting up to 160 meV [44]. By dynamically controlling the coupling strength using tip-QD distance variation in both lateral and vertical directions, the variations in TEPL intensity and the mode splitting of the plexciton state from a single QD in the nanocavity are also demonstrated. The spatial resolution in TEPL mapping is further improved by Au-coated Si scanning probe instead of a smooth Au antenna tip [45]. The plasmonic nanocorrugated regions at the apex of Au tip enhanced the spatial resolution up to 5 nm in TEPL imaging of single CdSe(ZnS) QD in ambient conditions. It should be also noted that in this work TEPL effect was observed when the probe was brought in direct contact with the QDs [45].

6.5.3 1D Semiconductor Nanowires and Nanotubes

The local compositional, phase, and strain variations coupled to electroninc and vibrational properties along the individual 1D nanowires and nanotubes can be explored by near-field nanospectroscopy

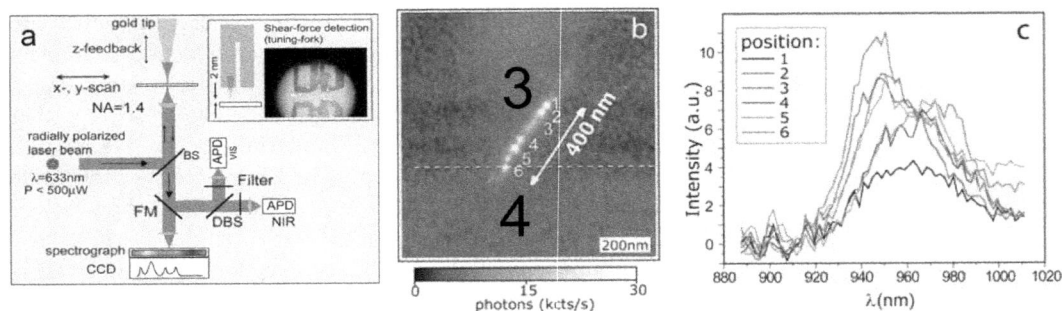

FIGURE 6.84 Experimental scheme of the TEPL setup (a) The corresponding TEPL map (b) and position-dependent TEPL spectra (c) on single SWCNT. (Reprinted with permission from Ref. [52], Copyright © 2006 Wiley-VCH.)

[46–50]. However, the TEPL studies on 1D semiconductor nanostructures are limited due to the challenges associated with far-field background, low quantum yield, and poor coupling between far-field excitation wavelengths that are used to excite the semiconductor nanostructure and the plasmonic antenna probes.

Semiconducting single-walled carbon nanotubes (SWCNT) show PL from excitons with Bohr-radii of 2–5 nm, however, with low quantum yield (~10^{-4}) and fast decay (~ps time-scales) due to the competing nonradiative transitions associated with exciton trapping at impurity sites and surface defects. The bandgap also changes with the nanotube's radii. These local defects are also found to limit the delocalized excitonic states along the CNT. The simultaneous TERS and TEPLnanospectroscopy is used to study such local variation within a single SWCNT with a spatial resolution < 15 nm [51,52] (Figure 6.84). The enhancement in PL due to the near-field is then estimated as $\approx (\eta_{NF}/\eta_{inc})(E_{NF}/E_{inc})^2$, where η_{NF}, and η_{inc} are PL quantum yield in the presence of near-field excitation and incident light. E_{inc}, and E_{NF} are incident and near-field generated electric fields. In addition, the study also showed the local variations in PL emission at different locations along the SWCNT. Another TEPL study on a single SWCNT revealed the exciton energy transfer and propagation in SWCNTs [53,54]. The exciton energy transfer is found to occur between a pair of SWCNTs, from a larger bandgap nanotube to a smaller bandgap nanotube limit, with a limitation to a few nm due to the nonradiative transitions. The PL decay is also observed at the end of the nanotube due to the exciton propagation and nonradiative relaxation at the end of the tube. The high-resolution TEPL imaging on SWCNT has revealed the localized excitons in DNA-wrapped SWCNT on mica substrate at RT [55]. The localization is identified due to the spatially confined exciton energy minima with depths of 15 meV and lateral energy gradients exceeding 2 meV/nm [55].

The size-dependent quantum confinement in semiconductor nanostructures is known to modify the electronic bandgap. In addition, the crystal phase can also control the bandgap in semiconductor nanostructures. The quasi-1D CdSe NWs are known to have alternate wurtzite (bandgap ≈ 1.797 eV) and zinc blende (bandgap ≈1.712 eV) crystal phases in single NW, resulting in PL broadening. Nanoscale phase variations in single and branched CdSe NWs having diameters < 10 nm are studied using TEPL nanospectroscopy and imaging [56,57] (Figure 6.85). The study revealed strong spatial fluctuations in both PL intensities and energies. The PL energies of 1.772 and 1.880 eV are found in thicker and thinner NWs, respectively. In addition, 22 meV shift in PL energies of single NW having a diameter of 8 nm and a length of 400 nm is observed [56]. The local variations in excitonic emission due to trap states within a single InP NWis studied using campanile probe hyperspectral PL imaging operated in near-field excitation and near-field collection configuration [58]. The local variations due to trap states along the NW are observed to be below the exciton diffusion lengths which are around 100s of nm in InP NWs.

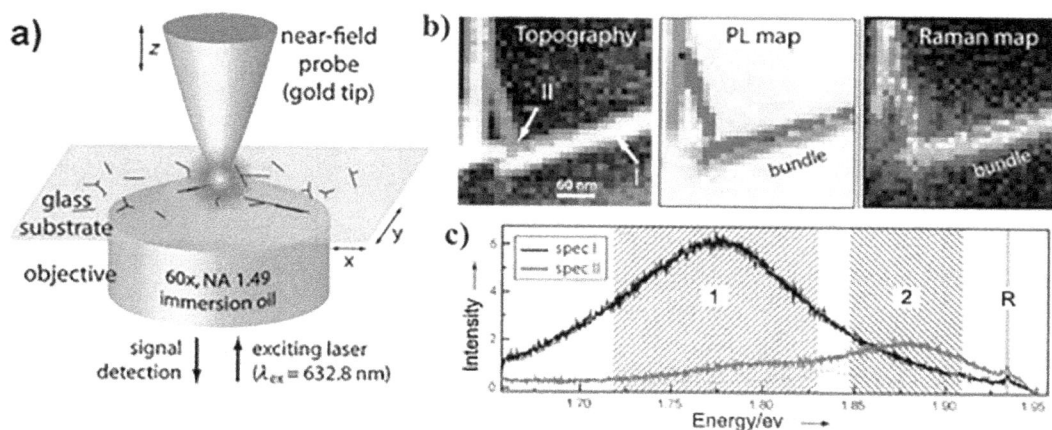

FIGURE 6.85 (a) Experimental scheme of the TEPL setup. (b) The topography (left), TEPL map (middle), and TERS map of branched CdSe NWs. (c) The position-dependent TEPL spectra at two locations showed in topography map in (b). (Reprinted with permission from Ref. [56], Copyright © 2011 John Wiley & Sons, Ltd.)

6.5.4   2D Semiconductors

Transition metal dichalcogenides such as MX_2 (M: Mo, W and X: S, Se, Te) are widely explored 2D semiconductor crystals among the family of 2D materials beyond graphene due to their direct bandgap nature at monolayer limit and their strong interaction with light [59]. The excitonic emission properties of these atomically thin semiconductors are highly sensitive towards layer thickness, layer order, including rotation and slide angle, dielectric environment, impurities or dopants, grain boundaries, local strain generated by wrinkles and bubbles, and heterostructure interfaces. Near-field naospectroscopy has been shown to be a very promising tool to study these local variations at nanometer resolution [38,60].

The CVD-grown monolayer TMD are shown to have several edge disorders and grain boundaries which modify the exciton emissions locally [61–63]. These nanoscale PL variations were studied using Campanile probe-based sub-diffraction hyperspectral imaging on CVD-grown monolayer MoS_2 [63]. The study revealed the nanoscale heterogeneity in exciton emissions was observed due to the pristine interior, disordered edges, and intra- and inter-flake grain boundaries in the MoS_2 layer. Such variations are difficult to resolve by diffraction-limited micro-PL imaging. The nanospectroscopic PL imaging revealed the edge regions and grain boundaries in CVD grown MoS_2 layers were optically defective, and quenching of the exciton emission at these locations is due to the presence of sulfur vacancies [63]. The nanoscale heterogeneities in exciton emissions due to several sources can be decoupled by combining TERS, TEPL, and local strain control by hybrid nano-optomechanical tip-enhanced spectroscopy and imaging with a spatial resolution of ~15 nm [64]. Both PL quenching and blueshift in emission energies are observed at edges and nucleation sites (NS) on the WSe_2 monolayer due to energy funneling. Whereas only PL quenching is observed without any change in emission energy at the twin boundaries (TBs) over ~30 nm length scales [64]. The strain can also change the bandgap resulting in modification of PL emission energies. Local nanoscale strain engineering is also demonstrated by controlling the tip-sample interaction forces allowing to tune of the reversible bandgap up to 24 meV and irreversible bandgap up to 48 meV [64]. By separating the effect of strain on PL from the other sources, it is suggested that the primary sources of local variation in PL emission are defects and stoichiometry variations.

The local strain induces nanoscale confinement potential, and the localized excitonic states in 2D TMD crystals that appear as a result, may demonstrate single-photon emission at cryogenic temperatures. The local strain induced by nanobubbles in monolayers of WSe_2 was studied by room-temperature TEPL nanospectroscopy and imaging [65]. The study revealed that the strain-localized

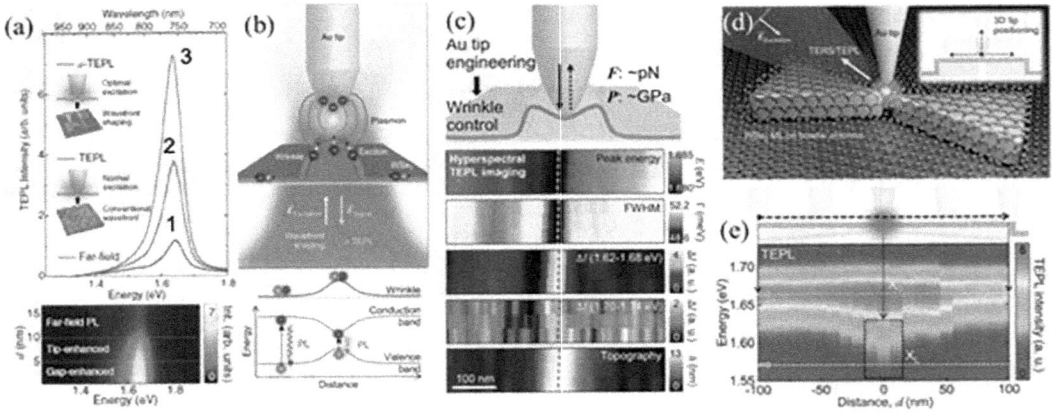

FIGURE 6.86 (a) Comparison of far-field PL (black, 1) and TEPL spectra of a WSe₂ monolayer with conventional excitation (blue, 2) and with wavefront shaping by the optimized spatial phase mask (top), and the spectral evolution of *a*-TEPL response with respect to the distance between Au tip and the crystal on Au film (bottom). *Creative Commons* [66]. Adaptive tip-enhanced PL (*a*-TEPL) spectroscopy based on dynamic wavefront shaping. (b) Illustration of *a*-TEPL measurements on a wrinkled WSe₂ monolayer showing the plasmon–exciton coupling at the apex of Au tip and WSe₂ (top), and the energy band diagram of a wrinkled WSe₂ describing bandgap reduction (bottom). (c) Illustration of the tip-induced nano-engineering of wrinkles to control strain, bandgap, and exciton dynamics (top), and hyperspectral TEPL images and corresponding AFM topography image of a wrinkle (bottom). (Reprinted with permission from Ref. [67], Copyright © 2021 John Wiley & Sons, Ltd.) (d) Experimental scheme of tip-enhanced nanospectroscopy and imaging of localized excitons in WSe₂ placed on triple-sharp tips structure cavity and (e) the corresponding TEPL map of radiative emission of localized excitons in WSe₂. (Reprinted with permission from Ref. [68], Copyright © 2021 John Wiley & Sons, Ltd.)

excitons (~10 nm length scales) around the periphery of a nanobubble in WSe₂ featured emission (X_L) energy (~1.56 eV) which was red-shifted from the neutral exciton (X_0) emission (~1.65 eV) [66]. The effective coupling between tip-excitation light, adaptive control of enhanced sensitivity, and polarization in TEPL are achieved by maximizing the in-plane polarization component of the LSP of the tip through dynamically shaping the wavefront of the excitation field using a sequence feedback algorithm. The *a*-TEPL enhanced the PL signal of the WSe₂ monolayer over ~10⁴ fold as compared to normal TEPL measurements without wavefront shaping [66] (Figure 6.86a). Naturally formed wrinkles in monolayer TMD causes local strain, which modify excitonic states, resulting in variation of bandgap and quantum yield of the exciton emission. The nanoscale wrinkles in WSe₂ are probed by hyperspectral *a*-TEPL spectroscopy and imaging with <15 nm spatial resolution using the near-field wavefront shaping technique [67] (Figure 6.86b and c). The study reveals that the uniaxial tensile strain at wrinkle apex in WSe₂ monolayer caused the enhancement in PL quantum yield and redshift (10 meV) in energy. The combination of enhanced excitation rate ($|E_{NF}/E_{FF}|^2$) and Purcell enhancement ($\gamma_{PF} = \gamma/\gamma_0$) results in total TEPL enhancement factor, $EF = |E_{NF}/E_{FF}|^2 \times (\gamma_{PF})$. The dynamic wavefront shaping using *a*-TEPL measurements higher enhancement factor (6.1×10^3) of about 30% is compared to normal TEPL EF (4.4×10^3). The local strain at the wrinkle apex is dynamically controlled by pressing the plasmonic tip allowing to tune the exciton dynamics and emission properties through nano-mechanical strain engineering [67] (Figure 6.86c). In general, the localized excitons (X_L) in 2D TMD have low quantum yield and subwavelength emission regions at RT due to their small binding energies. However, it is possible to create robust X_L states at RT using local strain engineering and near-field enhancement. The evolution of X_L emission with respect to dynamically induced strain on WSe₂ monolayer is studied by placing the WSe₂ layer on top of Au tips in a bowtie antenna having a plasmonic cavity size of 5–20 nm [68] (Figure 6.86d and e). The

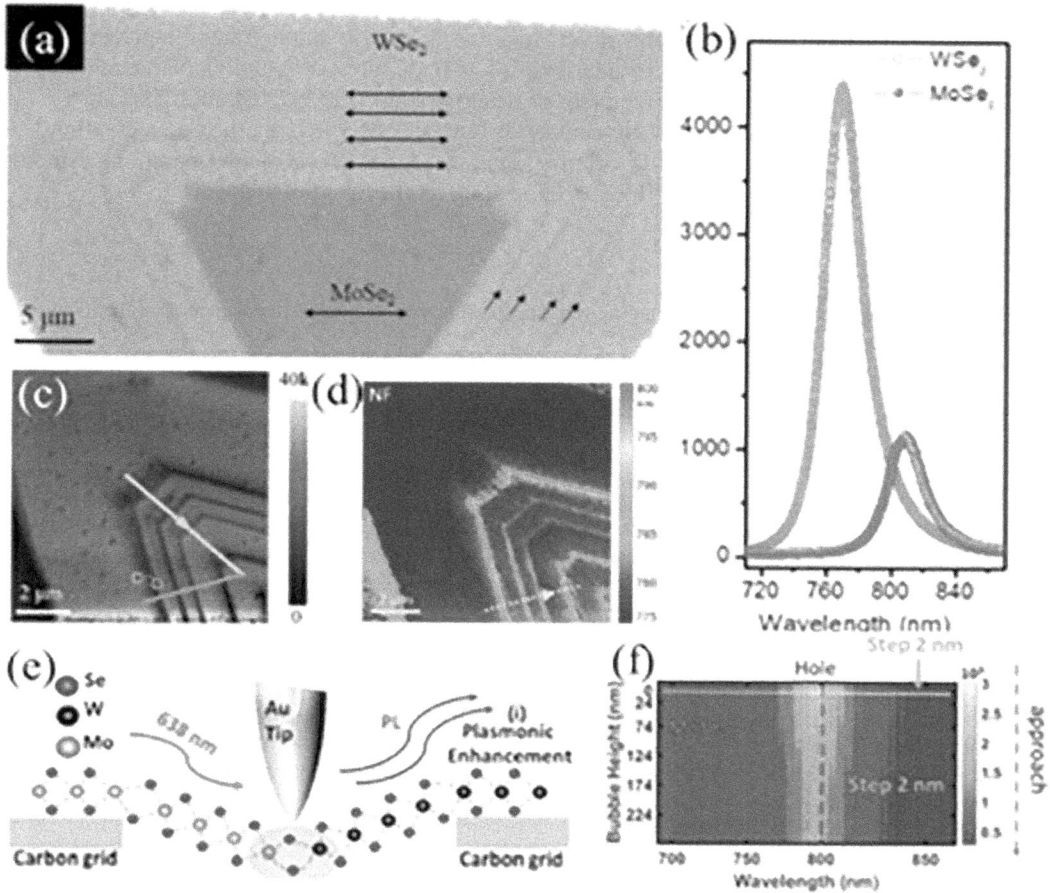

FIGURE 6.87 Lateral heterostructures of 2D TMD. (a) Optical image of a nine-junction monolayer lateral heterostructure formed by alternate MoSe$_2$ and WSe$_2$. (b) Far-field PL spectra of individual domains. (c) and (d) TEPL maps of PL integral intensity and PL position, respectively. (Reprinted with permission from Ref. [70], Copyright © 2019 Optica.) (e) Schematic of the TEPL measurement on freestanding WSe$_2$–MoSe$_2$ heterostructure scanned by a plasmonic Au tip on a carbon grid. (f) Tip-sample distance-dependent PL of freestanding MoSe$_2$ on hole. (Reprinted with permission from Ref. [71], Copyright © 2022 American Chemical Society.)

bowtie antenna itself induces a tensile strain of 0.3% in <30 nm region leading to strong X_L emission from WSe$_2$ monolayer. The X_L emission is further enhanced and probed by approaching the scanning tip closer to the strained WSe$_2$ layer forming a triple-sharp-tips cavity which generates the additional Purcell enhancement. The resultant total TEPL EF is calculated as ≈4.0×10^4, using EF=$|E_Z/E_0|^2 \times Q_z/Q_0$, where Q_z and Q_0 are quantum yields for the z-polarized dipole emitters located within the triple-sharp-tips cavity and without a cavity.

Lateral and vertical heterostructures of 2D TMD fabricated by the edge-epitaxy method are very promising for tuning the exciton emission with atomically sharp interfaces [69]. However, these multijunction later heterostructures are found to show aging effects over time. The TEPL spectroscopy and imaging on air-aged MoSe$_2$–WSe$_2$ multijunction later heterostructures showed 40 nm sharp transition (WSe$_2$ → MoSe$_2$), and 230 nm smooth interface (MoSe$_2$ → WSe$_2$) [70] (Figure 6.87a–d). The nanoscale TEPL imaging also revealed the alloying signatures at the multi-junctions, and aging-induced nanoparticle-suppressed exciton emissions. Due to localized strain around the bubble, naturally formed nanoscale bubbles in monolayer TMD generate exciton funnels. The intrinsic excitonic nature of the bubbles in freestanding 2D WSe_2–$MoSe_2$ lateral heterostructures are studied by TEPL nanospectroscopy and imaging [71] (Figure 6.87e and f). Nanoscale bubble generation and nanoindentation of the bubbles in heterostructure are carried out by plasmonic tip. The PL enhancement as a function of tip-sample distance is observed and the corresponding EFs are described in the limit of classical (tip-sample distance > 1 nm, EF ≈10^2) and quantum plasmonic regimes (tip-sample distance < 1 nm, EF ≈10^4) depending on the tip-sample distance.

REFERENCES

1. K. Kuhnke, et al., Atomic-scale imaging and spectroscopy of electroluminescence at molecular interfaces. *Chem. Rev. 117*, 5174–5222 (2017).
2. G. D. Scholes, et al., Excitons in nanoscale systems. *Nat. Mater. 5*, 683–696 (2006).
3. S. W. Koch, et al., Semiconductor excitons in new light. *Nat. Mater. 5*, 523–531 (2006).
4. A. Lewis, et al., Developmentof a 500 Å spatial resolution light microscope. *Ultramicroscopy 13*, 227–231 (1984).
5. A. Gustafsson, et al., Local probe techniques for luminescence studiesof low-dimensional semiconductor structures. *J. Appl. Phys. 84*, 1715 (1998).
6. D Courjon, et al., Near field microscopy and near field optics. *Rep. Prog. Phys. 57*, 989–1028 (1994).
7. E. Betzig, Nobel lecture: Single molecules, cells, and super-resolution optics. *Rev. Mod. Phys. 87*, 1153–1168 (2015).
8. L. Novotny, B. Hecht, Principles of Nano-Optics. London: Cambridge University Press (2012).
9. A. Hartschuh, et al., Tip-enhanced near-field optical microscopy. *Angew. Chem. Int. Ed. 47*, 8178–8191 (2008).
10. G. Malliaras, et al., An organic electronics primer. *Phys. Today 58*, 53–58 (2005).
11. K. Kato, et al., Light-assisted rechargeable lithium batteries: Organic moleculesfor simultaneous energy harvesting and storage. *Nano Lett. 21*, 907–913 (2021).
12. E. Betzig, et al., Single molecules observed by near-field scanning optical microscopy. *Science 262*, 1422–1425 (1993).
13. J. Steidtner, et al., Tip-enhanced Raman spectroscopy and microscopy on single dye moleculeswith 15 nm resolution. *Phys. Rev. Lett. 100*, 236101 (2008).
14. R. Zhang, et al., Chemical mapping of a single molecule byplasmon-enhanced Raman scattering. *Nature 498*, 82–86 (2013).
15. S. Jiang, et al., Distinguishing adjacent molecules on a surfaceusing plasmon-enhanced Raman scattering. *Nat. Nanotechnol. 10*, 865–869 (2015).
16. P. Liu, et al., Single-molecule imaging using atomistic near-field tip-enhanced Raman spectroscopy. *ACS Nano 11*, 5094–5102 (2017).
17. J. Lee, et al., Visualizing vibrational normal modes of a singlemolecule with atomically confined light. *Nature 568*, 78–82 (2019).
18. F. Benz, et al., Single-molecule optomechanicsin "picocavities". *Science 354*, 726–729 (2016).
19. E. A. Pozzi, et al., Ultrahigh-vacuum tip-enhanced Raman spectroscopy. *Chem. Rev. 117*, 4961–4982 (2017).
20. E. Dulkeith, et al., Fluorescence quenching of dye molecules near gold nano particles: Radiative and nonradiative effects. *Phys. Rev. Lett. 89*, 203002 (2002).
21. P. Anger, et al., Enhancement and quenching of single-molecule fluorescence. *Phys. Rev. Lett., 96*, 113002 (2006).
22. C. Hoppener, et al., Self-similar gold-nanoparticle antennas for a cascaded enhancement of the optical field. *Phys. Rev. Lett. 109*, 017402 (2012).
23. R. Faggiani, et al., Quenching, plasmonic, and radiative decays in nanogap emitting devices. *ACS Photonics 2*, 1739–1744 (2015).
24. R. Chikkaraddy, et al., Single-molecule strong coupling at roomtemperature in plasmonic nanocavities. *Nature 535*, 127–130 (2016).
25. N. Kongsuwan, et al., Suppressed quenching and strong-coupling of purcell-enhanced single-molecule emission in plasmonic nanocavities. *ACS Photonics 5*, 186–191 (2018).
26. Y.-P. Huang, et al., Shell-isolated tip-enhanced Raman and fluorescence spectroscopy. *Angew. Chem. Int. Ed. 57*, 7523–7527 (2018).

27. Z. C. Dong, et al., Generation of molecular hot electro luminescence by resonant nanocavity plasmons. *Nat. Photon. 4*, 50–54 (2009).

28. X. H. Qiu, et al., Vibrationally resolved fluorescence excited with sub-molecular precision. *Science 299*, 542–546 (2003).

29. H. Imada, Real-space investigation of energy transfer inheterogeneous molecular dimers. *Nature 538*, 364–367 (2016).

30. H. Imada, et al., Single-molecule investigation of energy dynamics in a coupled plasmon–exciton system. *Phys. Rev. Lett. 119*, 013901 (2017).

31. B. Doppagne, et al., Vibronic spectroscopy with sub-molecular resolution from STM-induced electroluminescence. *Phys. Rev. Lett.118*, 127401 (2017).

32. K. Kimura, et al., Selective triplet exciton formation in a singlemolecule. *Nature 570*, 210–213 (2019).

33. B. Doppagne, et al., Single-molecule tautomerization trackingthrough space- and time-resolved fluorescence spectroscopy. *Nat. Nanotechnol. 15*, 207–211 (2020).

34. H. Imada, et al., Single-molecule laser nanospectroscopywith micro-electron volt energy resolution. *Science 373*, 95–98 (2021).

35. B. Yang, et al., Sub-nanometre resolution in single-molecule photo luminescence imaging. *Nat. Photon. 14*, 693–699 (2020).

36. F. Qiu, et al., Optical images of molecular vibronic couplings from tip-enhanced fluorescence excitation spectroscopy. *JACS Au 2*, 150–158 (2022).

37. M. Nirmal, et al., Luminescence photophysics in semiconductor nanocrystals. *Acc. Chem. Res. 32*, 407–414 (1999).

38. H. Lee, et al., Tip-enhanced photoluminescence nano-spectroscopy and nano-imaging. *Nanophotonics 9*, 3089–3110 (2020).

39. A. L. Efros, D. J. Nesbitt, Origin and control of blinking in quantum dots. *Nat. Nanotechnol. 11*, 661–671 (2016).

40. K. T. Shimizu, W. K. Woo, B. R. Fisher, H. J. Eisler, M. G. Bawendi, Surface-enhanced emission from single semiconductor nanocrystals. *Phys. Rev. Lett. 89*, 117401 (2002).

41. V. V. Protasenko, et al., Fluorescence of single ZnS overcoated CdSe quantum dots studied by apertureless near-field scanning optical microscopy. *Opt. Commun. 210*, 11–23 (2002).

42. J. M. Gerton, L. A. Wade, G. A. Lessard, Z. Ma, S. R. Quake, Tip-enhanced fluorescence microscopy at 10 nanometer resolution. *Phys. Rev. Lett. 93*, 180801 (2004).

43. H. Groß, et al., Near-field strong coupling of single quantum dots. *Sci. Adv. 4*, eaar4906 (2018).

44. K.-D. Park, et al., Tip-enhanced strong coupling spectroscopy, imaging, and control of a single quantum emitter. *Sci. Adv. 5*, eaav5931 (2019).

45. C.-F. Wang, et al., Ambient tip-enhanced photoluminescence with 5 nm spatial resolution. *J. Phys. Chem. C 125*, 12251–12255 (2021).

46. A. Patsha, S. Dhara, S. Chattopadhyay, K.-H. Chen, L.-C. Chen, Optoelectronic properties of single and array of 1-D III-nitride nanostructures: An approach to light-driven device and energy resourcing. *J. Mat. Nanosci. 5*, 1–22 (2018).

47. L. G. Cancado, et al., Mechanism of near-field Raman enhancement in one-dimensional systems. *Phys. Rev. Lett. 103*, 186101 (2009).

48. A. Patsha, S. Dhara, A. K. Tyagi, Localized tip enhanced Raman spectroscopic study of impurity incorporated single GaN nanowire in the sub-diffraction limit. *Appl. Phys. Lett. 107*, 123108 (2015).

49. A. Patsha, S. Dhara, Size-dependent localized phonon population in semiconducting Si nanowires. *Nano Lett. 18*, 7181–7187 (2018).

50. S. Parida, A. Patsha, K. K. Madapu, S. Dhara, Nano-spectroscopic and nanoscopic imaging of single GaN nanowires in the sub-diffraction limit. *J. Appl. Phys. 127*, 173103 (2020).

51. A. Hartschuh, et al., Nanoscale optical imaging of excitonsin single-walled carbon nanotubes. *Nano Lett. 5*, 2310–2313 (2005).

52. H. Qian, et al., Near-field imaging and spectroscopy of electronic statesin single-walled carbon nanotubes. *Phys. Stat. Sol. (B) 243*, 3146–3150 (2006).

53. H. Qian, et al., Exciton transfer and propagation in carbon nanotubes studied by near-field optical microscopy. *Phys. Stat. Sol. (B) 245*, 2243–2246 (2008).

54. A. Hartschuh, et al., Tip-enhanced near-field optical microscopyof carbon nanotubes. *Anal. Bioanal. Chem. 394*, 1787–1795 (2009).

55. C. Georgi, et al., Probing exciton localization in single-walled carbon nanotubes using high-resolution near-field microscopy. *ACS Nano 4*, 5914–5920 (2010).

56. M. Bohmler, et al., Optical imaging of CdSe nanowires with nanoscale resolution. *Angew. Chem. Int. Ed. 50*, 11536–11538 (2011).

57. M. Bohmler, et al., Tip-enhanced near-field optical microscopy of quasi-1D nanostructures. *Chem. Phys. Chem. 13*, 927–929 (2012).

58. W. Bao, et al., Mapping local charge recombination heterogeneity bymultidimensional nanospectroscopic imaging. *Science 338*, 1317–1321 (2012).

59. S. Z. Butler, et al., Progress, challenges, and opportunities in two-dimensional materials beyond graphene. *ACS Nano 7*, 2898–2926 (2013).

60. Y. Lee, et al., Characterization of the structural defects in CVD-grown monolayered MoS_2 using near-field photoluminescence imaging. *Nanoscale 7*, 11909–11914 (2015).

61. A. Cohen, et al., Growth-etch metal-organic chemical vapor deposition approach of WS_2 atomic layers. *ACS Nano 15*, 526–538 (2021).

62. A. Patsha, K. Ranganathan, M. Kazes, D. Oron, A. Ismach, Halide chemical vapor deposition of 2D semiconducting atomically-thin crystals: From self-seeded to epitaxial growth. *Appl. Mater. Today 26*, 101379 (2022).

63. W. Bao, et al., Visualizing nanoscale excitonic relaxation properties of disordered edges and grain boundaries in monolayer molybdenum disulfide. *Nat. Commun. 6*, 7993 (2015).

64. K. D. Park, et al., Hybrid tip-enhanced nanospectroscopy and nanoimaging of monolayer WSe_2 with local strain control. *Nano Lett. 16*, 2621–2627 (2016).

65. T. P. Darlington, et al., Imaging strain-localized excitons in nanoscale bubbles of monolayer WSe_2 at room temperature. *Nat. Nanotechnol. 15*, 854–860 (2020).

66. D. Y. Lee, et al., Adaptive tip-enhanced nano-spectroscopy. *Nat. Commun. 12*, 3465 (2021).

67. Y. Koo, et al., Tip-induced nano-engineering of strain, bandgap, and exciton funneling in 2D semiconductors. *Adv. Mater. 33*, e2008234 (2021).

68. H. Lee, et al., Inducing and probing localized excitons in atomically thin semiconductors via tip-enhanced cavity-spectroscopy. *Adv. Funct. Mater. 31*, 2102893 (2021).

69. P. K. Sahoo, S. Memaran, Y. Xin, L. Balicas, H. R. Gutierrez, One-pot growth of two-dimensional lateral heterostructures via sequential edge-epitaxy. *Nature 553*, 63–67 (2018).

70. P. K. Sahoo, et al., Probing nano-heterogeneity and aging effects in lateral 2D heterostructures using tip-enhanced photoluminescence. *Opt. Mater. Exp. 9*, 1620 (2019).

71. A. Albagami, et al., Tip-enhanced photoluminescence of freestanding lateral heterobubbles. *ACS Appl. Mater. Interfaces 14*, 11006 (2022).

7 Conclusions and Future Directions

The book starts with a solid foundation of the inelastic scattering process of Raman scattering in Chapter 1. The in-depth knowledge of optical phonon (vibrational modes) observed in Raman spectroscopy helps understand the near-field nano-spectroscopic phenomena discussed in this book. At the end of the chapter, acoustic phonon confinement is also briefly touched upon to comprehend various critical physical phenomena like anomalous heat capacity, radial breathing mode, and collective spin wave in the low-dimensional system.

Imaging at the nanoscale using optical microscopy ushers a new dimension in the characterization tool to monitor many important physical phenomena in a very fast mode without much sample preparation, at low cost, and most importantly, without affecting its major physical and chemical properties. Nanoscopy, near-field optical microscopy, is enabled by exploiting noble metal-induced plasmonic, as described in the book in Chapters 2 and 4. However, noble metal-induced electric field enhancement for the plasmonic effect suffers a setback as the electron screening effect of noble metal, obstructing penetration of incident light, restricts the depth of study in the sample. Thus, alternative material of heavily doped wide-bandgap semiconductor is surfacing as new plasmonic material with a resonance frequency in the ultraviolet to the infrared region for nanoscopy. Wide-bandgap semiconductors allow light to penetrate without any electron screening effect of noble metal and hold huge promise to the exciting research area of plasmonics and its numerous applications. Optical confinement using dielectric and metallo-dielectric nanophotonics is also an alternative low-loss technique being used in nanoscopy. In the context of near-filed measurements, superlens using surface plasmon polariton (SPP) makes nanoscopy an exciting area of research, as SPP is slow light that is confined to the electron (metal or heavily doped wideband semiconductor) excess surface and possesses a large wave vector compared to the free space electromagnetic wave. Thus, a high resolution (~sub-nm) is possible if the imaging is carried out with SPP. Superlensing using both Ag and heavily doped semiconductors is already demonstrated. Light confined in the photonic bandgap, metallo-dielectric nanostructures, and Anderson localization can demonstrate wave-guiding and hyperlensing as low-light all-optical devices for ultra-fast data processing. These materials fall in the category of meta-materials with colossal potential in developing the next-generation materials world.

In nano-spectroscopy, an enhanced localized electric field using plasmonic or optical confinement of the incident electromagnetic wave is utilized for observing surface-enhanced Raman spectroscopy (SERS) which carries important chemical and phase information of material. In Chapter 5, the subject is dealt with the latest developments in the field of research. Fabrication of hotspots in the nanogaps is one of the prime objectives for the low-cost and reproducible application of SERS with a significant enhancement factor of the Raman intensity. However, many experimental results involving hotspots need computational understanding to rationalize the characteristics of the near-field completely. SERS study requires many improvements for realizing its full potential in biomedical applications, namely, bright nanotag, fiber-optics-based methods with SERS to avoid opacity of biological tissues and portable sensitive device with the ability to reliable and fast data collection so that it can be used in extreme conditions.

Tip-enhanced Raman spectroscopy (TERS) using a noble metal-coated surface probe microscopic nanotip is analogous to SERS with control over hotspot formation and the capability of spatial imaging at the atomic scale. As described in Chapter 6, the TERS enables one to study various aspects of molecular and submolecular information depending on the design of probe, mode

DOI: 10.1201/9781003248323-7

of its operation (force feedback or current feed), and environment (vacuum and temperature). An increase in the enhancement factor by at least one to three orders of presently achieved values using better probe design, comprehensive use of pulsed lasers, and performing the TERS imaging in an electrochemical environment can envisage large-scale applications of the TERS technique. Low-temperature ultrahigh-vacuum scanning tunneling microscopy-based TERS (LT-UHV-STM-TERS) studies, achieving submolecular resolution, have tested some of the current theoretical understandings necessitating innovative approaches and explanations, which lead to the exploration of new physical understanding of specific on-going processes. Various other tip-enhanced techniques of fluorescence (TEFL) and photoluminescence (TEPL) can be used to understand localized variation in composition, catalytic behavior, defect distribution, and strain in the sub-nm scale. The spatial resolution and spectral information can be further improved by adopting the LT-UHV technique. TERS study of inorganic samples, including advanced (quantum dots and layered material system) and strategic (strained Si) materials and organic molecules, shows the possibility of implementing the technique in the in situ device characterization. The methodology can be applied to a wide variety of molecule-enabled electronic devices and systems. In the study of biological molecules, TERS provides information as localized spectra or mapping with a high spatial resolution in the non-destructive study conveniently. As second-harmonic generation (SHG) is sensitive to a variety of symmetry-breaking orders, one can expect that the nano-SHG approach can be widely functional to envisage sub-wavelength domains of crystal structure, ferroelectric and magnetic orders, including novel heterostructures exhibiting inherent and tailored broken symmetries with new physical properties.

Index

Index 259

For Product Safety Concerns and Information please contact our EU
representative GPSR@taylorandfrancis.com
Taylor & Francis Verlag GmbH, Kaufingerstraße 24, 80331 München, Germany